Lecture Notes in Networks and Systems

Volume 136

Series Editor

Janusz Kacprzyk, Systems Research Institute, Polish Academy of Sciences, Warsaw, Poland

Advisory Editors

Fernando Gomide, Department of Computer Engineering and Automation—DCA, School of Electrical and Computer Engineering—FEEC, University of Campinas—UNICAMP, São Paulo, Brazil

Okyay Kaynak, Department of Electrical and Electronic Engineering, Bogazici University, Istanbul, Turkey

Derong Liu, Department of Electrical and Computer Engineering, University of Illinois at Chicago, Chicago, USA; Institute of Automation, Chinese Academy of Sciences, Beijing, China

Witold Pedrycz, Department of Electrical and Computer Engineering, University of Alberta, Alberta, Canada; Systems Research Institute, Polish Academy of Sciences, Warsaw, Poland

Marios M. Polycarpou, Department of Electrical and Computer Engineering, KIOS Research Center for Intelligent Systems and Networks, University of Cyprus, Nicosia, Cyprus

Imre J. Rudas, Óbuda University, Budapest, Hungary

Jun Wang, Department of Computer Science, City University of Hong Kong, Kowloon, Hong Kong

The series "Lecture Notes in Networks and Systems" publishes the latest developments in Networks and Systems—quickly, informally and with high quality. Original research reported in proceedings and post-proceedings represents the core of LNNS.

Volumes published in LNNS embrace all aspects and subfields of, as well as new challenges in, Networks and Systems.

The series contains proceedings and edited volumes in systems and networks, spanning the areas of Cyber-Physical Systems, Autonomous Systems, Sensor Networks, Control Systems, Energy Systems, Automotive Systems, Biological Systems, Vehicular Networking and Connected Vehicles, Aerospace Systems, Automation, Manufacturing, Smart Grids, Nonlinear Systems, Power Systems, Robotics, Social Systems, Economic Systems and other. Of particular value to both the contributors and the readership are the short publication timeframe and the world-wide distribution and exposure which enable both a wide and rapid dissemination of research output.

The series covers the theory, applications, and perspectives on the state of the art and future developments relevant to systems and networks, decision making, control, complex processes and related areas, as embedded in the fields of interdisciplinary and applied sciences, engineering, computer science, physics, economics, social, and life sciences, as well as the paradigms and methodologies behind them.

** Indexing: The books of this series are submitted to ISI Proceedings, SCOPUS, Google Scholar and Springerlink **

More information about this series at http://www.springer.com/series/15179

Tatiana Antipova
Editor

Integrated Science in Digital Age 2020

 Springer

Editor
Tatiana Antipova
Institute of Certified Specialists (ICS)
Perm, Russia

ISSN 2367-3370   ISSN 2367-3389 (electronic)
Lecture Notes in Networks and Systems
ISBN 978-3-030-49263-2   ISBN 978-3-030-49264-9   (eBook)
https://doi.org/10.1007/978-3-030-49264-9

This Springer imprint is published by the registered company Springer Nature Switzerland AG
The registered company address is: Gewerbestrasse 11, 6330 Cham, Switzerland

Preface

This book contains a selection of papers accepted for presentation and discussion at the 2020 International Conference on Integrated Science in Digital Age. This conference had the support of the Institute of Certified Specialists, Russia, and Springer. It was held on May 01–03, 2020.

The main idea of this conference is that the world of science is unified and united allowing all scientists/practitioners to be able to think, analyze and generalize their thoughts. An important characteristic feature of conference should be the short publication time and world-wide distribution. This conference enables fast dissemination so conference participants can publish their papers in print and electronic format, which is then made available worldwide and accessible by numerous researchers.

The Scientific Committee of 2020 International Conference on Integrated Science in Digital Age was composed of a multidisciplinary group of 107 experts who are intimately concerned with Integrated Science in Digital Age have had the responsibility for evaluating, in a 'double-blind review' process, the papers received for each of the main themes proposed for the conference: Blockchain & Cryptocurrency; Computer Law & Security; Digital Accounting & Auditing; Digital Business & Finance; Digital Economics; Digital Education; Digital Engineering; Machine Learning; Smart Cities in Digital Age; Health Policy & Management; Information Management.

The papers accepted for presentation and discussion at the conference are published by Springer (this book) and will be submitted for indexing by ISI, SCOPUS, among others. We acknowledge all of those that contributed to the staging of this conference (authors, committees, reviewers, organizers and sponsors). We deeply appreciate their involvement and support that was crucial for the success of the 2020 International Conference on Integrated Science in Digital Age.

May 2020

Tatiana Antipova

Contents

Blockchain and Cryptocurrency

Trade Route: An Anonymous P2P Online Exchange Market Using Blockchain Technology

Ahmed Ben Ayed[1]([envelope]) and Mohamed Amin Belhajji[2]

[1] University of the Cumberlands, Williamsburg, KY 40769, USA
Ahmed.BenAyed@ucumberlands.edu
[2] University of Quebec at Rimouski, Rimouski, QC, Canada
MohamedAmin.Belhajji@uqar.ca

Abstract. Due to the Internet rapid growth, e-commerce is becoming a new way for businesses to market their listings online and not be limited to local customers. Many buyers like to shop online because of the convenience. However, when clients shop online, many concerns emerge, such as worrying about how personal data will be handled, how secure/private the online site is, and whether or not the website should be trusted in the first place.

In this paper, we propose a novel approach leveraging the blockchain technology to create a platform that helps peers exchange goods without the need to go through an intermediary or the use of any traditional currency. All peers in this platform will have to sell their goods in exchange for "Trade Coins" that are only usable in the "Trade Route" market.

Keywords: Trade Route · Anonymous e-commerce · Trade Coin · Blockchain · Anonymous online market · Cryptocurrency

1 Introduction

Simply speaking, e-commerce is the purchasing and selling of products or services using the internet [1]. In the United States and since the mid-nineties e-commerce has grown exponentially [1] and became the newer and more sophisticated way of conducting business [2]. The e-commerce growth was not limited to the United States only, but other parts of the world are showing a similar growth as well. E-commerce sales in Taiwan were $6 billion in 2007 [3], and jumped to $117 billion in 2016 [4], which reflects a compounded annual growth rate of about 40%.

The data indicates that e-commerce is growing and will continue to grow; however, this growth has been overshadowed by privacy and security concerns.

2 Privacy and Security

Online privacy could be defined as "the rights of individuals and organizations to determine for themselves when, how, and to what extent information about them is to be transmitted to others" [5]. Privacy has an impact on the intent of customers to buy

© Springer Nature Switzerland AG 2021
T. Antipova (Ed.): ICIS 2020, LNNS 136, pp. 3–9, 2021.
https://doi.org/10.1007/978-3-030-49264-9_1

online. If there is a real or perceived lack of privacy, consumers are reluctant to reveal financial and private information to online retailers [6]. In some cases, online shoppers make up personal data as a protection mechanism [7]. Privacy affected online consumers' trust, which was a determinant of their willingness to purchase online. Online security is becoming the main concern of clients willing to purchase online. Online security could be defined as "the protection of data against accidental or intentional disclosure to unauthorized persons, or unauthorized modifications or destruction" [5]. Online security usually includes transactional encryption to prevent fraud or theft [8]. Even if systems are technically secure, many online shoppers do not trust e-commerce platforms and do not believe that their internet transactions are secure. Trust refers to "a consumer's willingness to accept vulnerability in an online transaction based on their positive expectations regarding an e-retailer's future behaviors" [9]. The Risk we are talking about here is the possibility of personal information and shopping behavior getting leaked out or sold to third parties. This could be an important factor for many customers including but not limited to public figures, leaders, and celebrities who are more sensitive to privacy and most of the time do not appreciate any leakage about their personal life.

3   The Idea of Anonymous on the Web

When purchasing items online, customers give out a lot of data. Names, addresses, phone numbers, date of birth, gender, emails, and many other pieces of sensitive information are usually provided for each online transaction. When such information is provided to the online platform, it just gets saved and disappears from the screen and a message saying "Order Received" pops up. The customer doesn't see where the data goes or whether it is being used for the desired purpose or not. Many questions come to mind here: "What if the website isn't trusted, and what will the merchant do with that data?" If the website isn't trusted, can they steal your data? Or even if the website is trusted, will they sell that information to a third party? Is the data just entered into that website is being kept to them? Many valid questions could come to mind when purchasing items online. With so many scams and fraudsters on the internet, how can a customer decide if a website is trusted or not? With all that being said isn't a good idea to shop anonymously?

3.1   Why Anonymous?

When talking about anonymity, we are forced to talk about reasons to be anonymous. The first reason could be the desire and the right for each individual to be anonymous while s/he is not doing anything illegal. Another good reason to stay anonymous is to shield credit card information from unauthorized users or hackers. Many clients may not feel safe inputting sensitive information into the web. Some others may not want an item to show up on their bill, or simply want to shield their identity from the internet, practice their right to privacy or anonymity!

3.2 Criminal Activities?

Anonymous purchases online are supposed to be completely safe. A lot of people see anonymous purchases as unrealistic, or only good for criminal activities, but realistically speaking they could be for everyday use. Anonymous purchases can save the client's identity, stop fraud, keep data private, and prevent unwanted things from happening to customers.

Anonymous transactions are important. Whether shopping at a large widely known website, or a smaller fishy website, privacy always matters. Big corporations are not immune to hackers, and the same goes for smaller companies. Privacy is one thing everyone deserves online, and techniques have to emerge to keep it that way.

Many will label the use of anonymity as cowardly. Without any doubt, anonymity is motivated by anxiety over possible retaliation, but that seems like a good judgment.

4   The Dark Side of Anonymity and Bitcoin-Based Operations

Besides the envy of being anonymous to protect their data, internet users have their dark side that they don't like to unveil. A study done by researchers in Carnegie Mellon University states that about 53% of interviewees admitted to malicious activities such as hacking or harassing other internet users, or engaging in "socially undesirable activities", like visiting sites that depicted violence or pornography, or downloading files illegally [10].

Money laundering is also a big concern when it comes to crypto-currency. Money laundering is the process whereby illegal funds are acquired and transferred with a false claim of legitimacy [11]. Transactions and their values on the bitcoin network are completely anonymous. Mafia would prefer to use this medium to hide their money laundering activities from governments and public. There is an innumerable mechanism by which money laundering is carried out. The emergence of virtual currencies possesses a greater risk making it easier for laundering activities to be carried out undercover without the possibility to know who is responsible for such activities [12]. Stokes (2012) in its paper titled "Virtual money laundering: the case of bitcoin and Linden dollar," highlighted the money laundering risk associated with bitcoin and Linden dollar. Their paper aimed to find similarities between online gambling and bitcoin, to recommend methods by which bitcoin and Linden dollar can be incorporated in the anti-money laundering act. Stokes stated that whatever that has value is susceptible to the crime of money laundering [13]. In the same way, any technology which encourages the transfer of value is vulnerable to being used for money laundering [14]. Bitcoin allows the transfer of money on a peer-to-peer network without disclosing the identity of the individuals involved in the transaction. The values of bitcoin to the dollar is on a steady rise. As at the time of writing this paper the value of one bitcoin is equal to $9,605.06. As of March 2012, the value of one bitcoin was $4.5. The rapid increase in value is enough to entice anyone to invest in bitcoin. Also, illegal funds are best to be invested in bitcoin, it cannot be traced to any single entity. Thus, it makes bitcoin a fertile ground for possible money launder even on a large scale. This is partly because the bitcoin network is not backed by any government or organization

[15]. Over the years, many organizations have accepted bitcoin as a medium of payment, making it even healthier for illegal activities to be carried out on the network.

It is challenging to investigate the identity of illicit bitcoin users who carry out money laundry by law enforcement agencies. This challenge is due to the anonymous nature of bitcoin transactions. Fraudsters take advantage of this character to syphon illicit funds including public funds through the bitcoin network. This makes the bitcoin network a juicy avenue attracting attackers to carry out their activities to try to redirect such funds for their benefit.

Cryptocurrencies such as bitcoin transfers are instantaneous and borderless. Users can transact in relative anonymity. These elements present a healthy ground for fraudulent activities. Many studies are linking money laundering, financing of terrorism as well as the drug industry to bitcoin [16]. It has also been linked to cybercrime activities to the dark web or dark marketplaces where consumers purchase cybercrime as a service using bitcoin. All the web or internet that are not available by regular search engines are made available only by deep well. The deep web and dark web are home for all kinds of illegal activities that go on the internet without being noticed or discovered through standard means. This includes hacking communities, illegal trade markets, and private communications. The communication on the deep web is done anonymously. Therefore, it becomes a fertile place for all kinds of illicit products sold in underground markets including drugs and guns. The services of phishing experts, hackers, and ransomware are usually advertised on the dark web. Anything illegal is possible on the deep web including the purchase of firearms, grenades, "hitman for hire", services, child pornography and many others. The deep web greatly enhances privacy by encryption. The only method of payment in the deep web or dark web is crypto-currency such as bitcoin. The dark web allows consumers to purchase details of stolen credit cards, hacking tools, usernames and passwords of compromised accounts and malware using bitcoin as the only platform for payment, making it a lucrative center for cyber attackers. The "WannaCry" ransomware that broke out in 2017 encrypts the files of their victims and makes them unavailable until the victims make a payment in bitcoin to the attackers [17]. The anonymity of the perpetrators of this kind of fraud encourages more frauds to be fired using the lucrative bitcoin network.

5   Current Anonymous e-Commerce Solutions

Online transactions usually reveal private information about customers as well as information that could be deducted as religion, health condition or even personal orientations. Most current anonymous ecommerce solutions use a Third Trusted Party (TTP) that guarantees the privacy of all information entered by the customer to finish up the transaction. In this case the TTP will be the one finishing up the transaction with both parties (seller & buyer) gets no information at all about each other [18]. TTP could be the solution to anonymity; however, who is going to protect the user's information from that "trusted" third party?

Another approach is to separate data used during the transaction [19]. When credit card, name, last name, address and other data gets separated, the attackers won't be able to track the transaction to a specific entity. However, separation of data still vulnerable to reverse engineering algorithms.

6 Blockchain Technology and Proposed System

6.1 Blockchain Technology

Blockchain is a peer to peer decentralized ledger technology [20], it was originally developed by Nakamoto [15] who invented bitcoin which allowed peer to peer cash transactions without the need for a trusted intermediary. Since then blockchain technology was used in different sectors including e-Voting [21], IoT security [22], and even in the health industry [23].

6.2 System Requirements

Our proposed solution will support three main requirements that can be illustrated as shown below:

* **Authentication:** Only people already registered in our system can sell and buy goods using "Trade Route".
* **Anonymity:** The "Trade Route" system should not allow any links between sellers and buyers. Identities should always stay private. The buyer has to remain anonymous during and after the transaction is done.
* **Accuracy:** Transactions must be accurate; No transaction can be changed or removed. If the buyer introduces the wrong address, for example, there is no way to go back and fix it. Also, no returns or refunds are allowed.

6.3 Representation of the "Trade Route" Proposed System

Our proposed system consists of a central management system that works on receiving and treating orders, as soon as the transaction information is received by the system it gets sent to miners to register them in the blockchain. When the transactions get registered in the blockchain the seller will receive the order total in "Trade Coins" and the amount will be debited to the buyer. These trade coins used in all our transactions could be used by the seller to buy other services or goods within the same platform. Trade route miners verify all transactions and save them into public "blocks" so they can be verified in the future. To motivate miners our system gives them 1 trade coins after verifying 50 blocks. Our system can process about two transactions per second and will insert information into the blockchain every 15 min (Fig. 1).

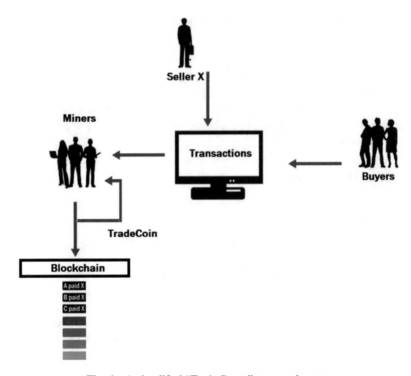

Fig. 1. A simplified "Trade Route" proposed system

7 Conclusion and Future Work

As of today, there is no legal secure way of shopping online in a private manner. In this paper, we proposed "Trade Route", a blockchain-based e-commerce solution that lets customers buy privately. However, despite all its advantages, this solution could be used by criminals to sell and buy illegal goods or simply get away from paying taxes. We suggest topping up our solution with an analyzing entity to collect and inspect transaction data to identify illegal activities and block them from using our solution. This could be done in the future using machine learning algorithms. We also assume that buyers will enter the right information when entering its order, one of the drawbacks of our system is the inability to modify or cancel any order after it has been processed.

References

1. Reddy, A., Lyer, R.: A conceptual model of the principles governing effective e-commerce. Int. J. Manag. **19**(3), 517–523 (2002)
2. Cordy, E.D.: The legal regulation of e-commerce transactions. J. Am. Acad. Bus. **2**(2), 400–407 (2003)
3. Arinto, P., Akhar, S.: Asia Pacific 2009–2010. SAGE Publications Ltd., New Delhi (2009)

4. Blazyte, A.: E-commerce revenue in Taiwan 2014–2016. https://www.statista.com/statistics/956205/taiwan-ecommerce-revenue/. Accessed 18 Feb 2020

5. Grandinetti, M.: Establishing and maintaining security on the internet. Sacramento Bus. J. **13**(25), 22 (1996)

6. Park, C., Kim, Y.G.: A framework of dynamic CRM: linking marketing with information strategy. Bus. Process Manag. J. **9**(5), 652–671 (2003)

7. Hoffman, D., Novak, T.P., Peralta, M.: Building consumer trust online. Commun. ACM **42**(4), 80–85 (1999)

8. Machrone, B.: Electronic books a horror story? Readers think so. PC Week **15**(42), 81 (1998)

9. Kimery, K.M., McCord, M.: Third-party assurances: mapping the road to trust in retailing. J. Inf. Technol. Theory Appl. **4**(2), 7 (2002)

10. Kang, R., Brown, S., Kiesler, S.: Why do people seek anonymity on the internet? In: Proceedings of the SIGCHI Conference on Human Factors in Computing Systems, pp. 2657–2666 (2013)

11. Masciandaro, D.: Global Financial Crime: Terrorism, Money Laundering and Offshore Centres. Taylor & Francis, New York (2017)

12. Campbell-Verduyn, M.: Bitcoin, crypto-coins, and global anti-money laundering governance. Crime Law Soc. Change **69**(2), 283–305 (2018)

13. Stokes, R.: Laundering: the case of Bitcoin and the Linden dollar. Inf. Commun. Technol. Law **21**(3), 221–236 (2012)

14. Demetis, D.S.: Fighting money laundering with technology: a case study of Bank X in the UK. Decis. Support Syst. **105**, 96–107 (2018)

15. Nakamoto, S.: Bitcoin: a peer-to-peer electronic cash system (2008)

16. Irwin, A.S., Milad, G.: The use of crypto-currencies in funding violent jihad. J. Money Laund. **19**(4), 407–425 (2016)

17. Mohurle, S., Patil, M.: A brief study of WannaCry threat: ransomware attack 2017. Int. J. Adv. Res. Comput. Sci. **8**(5), 1938–1940 (2017)

18. Groppe, S., Kuhr, F., Coskun, M.A.: Anonymous shopping in the internet by separation of data. Open J. Web Technol. **5**(1), 14–22 (2018)

19. AlTawy, R., ElSheikh, M., Youssef, A.M., Gong, G.: Lelantos: a blockchain-based anonymous physical delivery system. In: 2017 15th Annual Conference on Privacy, Security and Trust, 15-1509 (2017)

20. Crosby, M., Pattanayak, P., Verma, S., Kalyanaraman, V.: Blockchain technology: beyond Bitcoin. Appl. Innov. **2**(6), 7–19 (2016)

21. Ayed, A.B.: A conceptual secure blockchain-based electronic voting system. Int. J. Netw. Secur. Appl. **9**(3), 1–9 (2017)

22. Taylor, P.J., Dargahi, T., Dehghantanha, A., Parizi, R.M., Choo, K.K.R.: A systematic literature review of blockchain cybersecurity. Digit. Commun. Netw. 1–10 (2019)

23. Ayed, A.B., Belhajji, M.A.: The blockchain technology: applications and threats. In: Securing the Internet of Things: Concepts, Methodologies, Tools, and Applications. IGI Global (2020)

Computer Law and Security

Approaches to Identify Fake News:
A Systematic Literature Review

Dylan de Beer and Machdel Matthee[(⊠)] [iD]

Department of Informatics, University of Pretoria, Pretoria 0001, South Africa
machdel.matthee@up.ac.za

Abstract. With the widespread dissemination of information via digital media platforms, it is of utmost importance for individuals and societies to be able to judge the credibility of it. Fake news is not a recent concept, but it is a commonly occurring phenomenon in current times. The consequence of fake news can range from being merely annoying to influencing and misleading societies or even nations. A variety of approaches exist to identify fake news. By conducting a systematic literature review, we identify the main approaches currently available to identify fake news and how these approaches can be applied in different situations. Some approaches are illustrated with a relevant example as well as the challenges and the appropriate context in which the specific approach can be used.

Keywords: Fake news · Machine learning · Linguistics · Semantics · Syntax · Algorithms · Digital tools · Social media

1 Introduction

Paskin (2018: 254) defines fake news as "particular news articles that originate either on mainstream media (online or offline) or social media and have no factual basis, but are presented as facts and not satire". The importance of combatting fake news is starkly illustrated during the current COVID-19 pandemic. Social networks are stepping up in using digital fake news detection tools and educating the public towards spotting fake news. At the time of writing, Facebook uses machine learning algorithms to identify false or sensational claims used in advertising for alternative cures, they place potential fake news articles lower in the news feed, and they provide users with tips on how to identify fake news themselves (Sparks and Frishberg 2020). Twitter ensures that searches on the virus result in credible articles and Instagram redirects anyone searching for information on the virus to a special message with credible information (Marr 2020).

These measures are possible because different approaches exist that assist the detection of fake news. For example, platforms based on machine learning use fake news from the biggest media outlets, to refine algorithms for identifying fake news (Macaulay 2018). Some approaches detect fake news by using metadata such as a comparison of release time of the article and timelines of spreading the article as well where the story spread (Macaulay 2018).

© Springer Nature Switzerland AG 2021
T. Antipova (Ed.): ICIS 2020, LNNS 136, pp. 13–22, 2021.
https://doi.org/10.1007/978-3-030-49264-9_2

The purpose of this research paper is to, through a systematic literature review, categorize current approaches to contest the wide-ranging endemic of fake news.

2   The Evolution of Fake News and Fake News Detection

Fake news is not a new concept. Before the era of digital technology, it was spread through mainly yellow journalism with focus on sensational news such as crime, gossip, disasters and satirical news (Stein-Smith 2017). The prevalence of fake news relates to the availability of mass media digital tools (Schade 2019). Since anyone can publish articles via digital media platforms, online news articles include well researched pieces but also opinion-based arguments or simply false information (Burkhardt 2017). There is no custodian of credibility standards for information on these platforms making the spread of fake news possible. To make things worse, it is by no means straightforward telling the difference between real news and semi-true or false news (Pérez-Rosas et al. 2018).

The nature of social media makes it easy to spread fake news, as a user potentially sends fake news articles to friends, who then send it again to their friends and so on. Comments on fake news sometimes fuel its 'credibility' which can lead to rapid sharing resulting in further fake news (Albright 2017).

Social bots are also responsible for the spreading of fake news. Bots are sometimes used to target super-users by adding replies and mentions to posts. Humans are manipulated through these actions to share the fake news articles (Shao et al. 2018).

Clickbait is another tool encouraging the spread of fake news. Clickbait is an advertising tool used to get the attention of users. Sensational headlines or news are often used as clickbait that navigate the user to advertisements. More clicks on the advert means more money (Chen et al. 2015a).

Fortunately, tools have been developed for detecting fake news. For example, a tool has been developed to identify fake news that spreads through social media through examining lexical choices that appear in headlines and other intense language structures (Chen et al. 2015b). Another tool, developed to identify fake news on Twitter, has a component called the Twitter Crawler which collects and stores tweets in a database (Atodiresei et al. 2018). When a Twitter user wants to check the accuracy of the news found they can copy a link into this application after which the link will be processed for fake news detection. This process is built on an algorithm called the NER (Named Entity Recognition) (Atodiresei et al. 2018).

There are many available approaches to help the public to identify fake news and this paper aims to enhance understanding of these by categorizing these approaches as found in existing literature.

3   Research Method

3.1   Research Objective

The purpose of this paper is to categorize approaches used to identify fake news. In order to do this, a systematic literature review was done. This section presents the search terms that were used, the selection criteria and the source selection.

3.2   Search Terms

Specific search terms were used to enable the finding of relevant journal articles such as the following:

> ("what is fake news" OR "not genuine information" OR "counter fit news" OR "inaccurate report*" OR "forged (NEAR/2) news" OR "mislead* information" OR "false store*" OR "untrustworthy information" OR "hokes" OR "doubtful information" OR "incorrect detail*" OR "false news" OR "fake news" OR "false accusation*")
> AND ("digital tool*" OR "digital approach" OR "automated tool*" OR "approach*" OR "programmed tool*" OR "digital gadget*" OR "digital device*" OR "digital machan*" OR "digital appliance*" OR "digital gizmo" OR "IS gadget*" OR "IS tool*" OR "IS machine*" OR "digital gear*" OR "information device*")
> AND ("fake news detection" OR "approaches to identify fake news" OR "methods to identify fake news" OR "finding fake news" OR "ways to detect fake news").

3.3   Selection Criteria

Inclusion Criteria. Studies that adhere to the following criteria: (1) studies published between 2008 and 2019; (2) studies found in English; (3) with main focus fake news on digital platforms; (4) articles that are published in IT journals or any technology related journal articles (e.g. computers in human behavior) as well as conference proceedings; (5) journal articles that are sited more than 10 times.

Exclusion Criteria. Studies that adhered to the following criteria: (1) studies not presented in journal articles (e.g. in the form of a slide show or overhead presentation); (2) studies published, not relating to technology or IT; (3) articles on fake news but not the identification of it.

The search terms were used to find relevant articles on ProQuest, ScienceDirect, EBSCOhost and Google Scholar (seen here as 'other sources').

3.4   Flowchart of Search Process

Figure 1 below gives a flowchart of the search process: the identification of articles, the screening, the selection process and the number of the included articles.

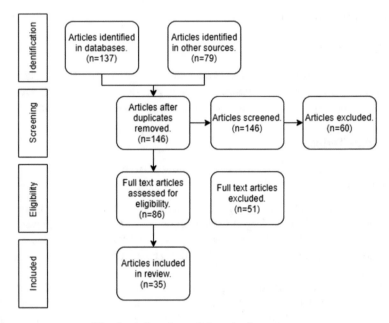

Fig. 1. A flowchart of the selection process

4   Findings

In this section of the article we list the categories of approaches that are used to identify fake news. We also discuss how the different approaches interlink with each other and how they can be used together to get a better result.

The following categories of approaches for fake news detection are proposed: (1) language approach, (2) topic-agnostic approach, (3) machine learning approach, (4) knowledge-based approach, (5) hybrid approach.

The five categories mentioned above are depicted in Fig. 2 below. Figure 2 shows the relationship between the different approaches. The sizes of the ellipses are proportional to the number of articles found (given as the percentage of total included articles) in the systematic literature review that refer to that approach.

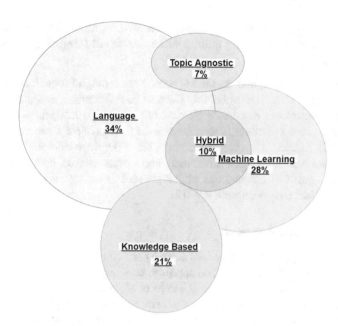

Fig. 2. Categories of fake news detection approaches resulting from the systematic literature review

The approaches are discussed in depth below with some examples for illustration purposes.

4.1 Language Approach

This approach focuses on the use of linguistics by a human or software program to detect fake news. Most of the people responsible for the spread of fake news have control over what their story is about, but they can often be exposed through the style of their language (Yang et al. 2018). The approach considers all the words in a sentence and letters in a word, how they are structured and how it fits together in a paragraph (Burkhardt 2017). The focus is therefore on grammar and syntax (Burkhardt 2017). There are currently three main methods that contribute to the language approach:

Bag of Words (BOW): In this approach, each word in a paragraph is considered of equal importance and as independent entities (Burkhardt 2017). Individual words frequencies are analysed to find signs of misinformation. These representations are also called n-grams (Thota et al. 2018). This will ultimately help to identify patterns of word use and by investigating these patterns, misleading information can be identified. The bag of words model is not as practical because context is not considered when text is converted into numerical representations and the position of a word is not always taken into consideration (Potthast et al. 2017).

Semantic Analysis: Chen et al. 2017b explain that truthfulness can be determined by comparing personal experience (e.g. restaurant review) with a profile on the topic derived from similar articles. An honest writer will be more likely to make similar remarks about a topic than other truthful writers. Different compatibly scores are used in this approach.

Deep Syntax: The deep syntax method is carried out through Probability Context Free Grammars (Stahl 2018). The Probability Context Free Grammars executes deep syntax tasks through parse trees that make Context Free Grammar analysis possible. Probabilistic Context Free Grammar is an extension of Context Free Grammars (Zhou and Zafarani 2018). Sentences are converted into a set of rewritten rules and these rules are used to analyse various syntax structures. The syntax can be compared to known structures or patterns of lies and can ultimately lead to telling the difference between fake news and real news (Burkhardt 2017).

4.2   Topic-Agnostic Approach

This category of approaches detect fake news by not considering the content of articles bur rather topic-agnostic features. The approach uses linguistic features and web mark-up capabilities to identify fake news (Castelo et al. 2019). Some examples of topic-agnostic features are 1) a large number of advertisements, 2) longer headlines with eye-catching phrases, 3) different text patterns from mainstream news to induce emotive responses 4) presence of an author name (Castelo et al. 2019; Horne and Adali 2017).

4.3   Machine Learning Approach

Machine learning algorithms can be used to identify fake news. This is achieved through using different types of training datasets to refine the algorithms. Datasets enables computer scientists to develop new machine learning approaches and techniques. Datasets are used to train the algorithms to identify fake news. How are these datasets created? One way is through crowdsourcing. Perez-Rosas et al. (2018) created a fake news data set by first collecting legitimate information on six different categories such as sports, business, entertainment, politics, technology and education (Pérez-Rosas et al. 2018). Crowdsourcing was then used and a task was set up which asked the workers to generate a false version of the news stories (Pérez-Rosas et al. 2018). Over 240 stories were collected and added to the fake news dataset.

A machine learning approach called the rumor identification framework has been developed that legitimizes signals of ambiguous posts so that a person can easily identify fake news (Sivasangari et al. 2018). The framework will alert people of posts that might be fake (Sivasangari et al. 2018). The framework is built to combat fake tweets on Twitter and focuses on four main areas; the metadata of tweets, the source of the tweet; the date and area of the tweet, where and when the tweet was developed (Sivasangari et al. 2018). By studying these four parts of the tweet the framework can be implemented to check the accuracy of the information and to separate the real from the fake (Sivasangari et al. 2018). Supporting this framework, the spread of gossip is collected to create datasets with the use of a Twitter Streaming API (Sivasangari et al. 2018).

Twitter has developed a possible solution to identify and prevent the spread of misleading information through fake accounts, likes and comments (Atodiresei et al. 2018) - the Twitter crawler, a machine learning approach works by collecting tweets and adding them to a database, making comparison between different tweets possible.

4.4   Knowledge Based Approach

Recent studies argue for the integration of machine learning and knowledge engineering to detect fake news. The challenging problem with some of these fact checking methods is the speed at which fake news spreads on social media. Microblogging platforms such as Twitter causes small pieces of false information to spread very quickly to a large number of people (Qazvinian et al. 2011). The knowledge-based approach aims at using sources that are external to verify if the news is fake or real and to identify the news before the spread thereof becomes quicker. There are three main categories; (1) Expert Oriented Fact Checking, (2) Computational Oriented Fact Checking, (3) Crowd Sourcing Oriented Fact Checking (Ahmed et al. 2019).

Expert Oriented Fact Checking. With expert oriented fact checking it is necessary to analyze and examine data and documents carefully (Ahmed et al. 2019). Expert-oriented fact-checking requires professionals to evaluate the accuracy of the news manually through research and other studies on the specific claim. Fact checking is the process of assigning certainty to a specific element by comparing the accuracy of the text to another which has previously been fact checked (Vlachos and Riedel 2014).

Computational Oriented Fact Checking. The purpose of computational oriented fact checking is to administer users with an automated fact-checking process that is able to identify if a specific piece of news is true or false (Ahmed et al. 2019). An example of computational oriented fact checking is knowledge graphs and open web sources that are based on practical referencing to help distinguish between real and fake news (Ahmed et al. 2019). A recent tool called the ClaimBuster has been developed and is an example of how fact checking can automatically identify fake news (Hassan et al. 2017). This tool makes use of machine learning techniques combined with natural language processing and a variety of database queries. It analyses context on social media, interviews and speeches in real time to determine 'facts' and compares it with a repository that contains verified facts and delivers it to the reader (Hassan et al. 2017).

Crowd Sourcing Oriented. Crowdsourcing gives the opportunity for a group of people to make a collective decision through examining the accuracy of news (Pennycook and Rand 2019). The accuracy of the news is completely based on the wisdom of the crowd (Ahmed et al. 2019). Kiskkit is an example of a platform that can be used for crowdsourcing where the platform allows a group of people to evaluate pieces of a news article (Hassan et al. 2017). After one piece has been evaluated the crowd moves to the next piece for evaluation until the entire news article has been evaluated and the accuracy thereof has been determined by the wisdom of the crowd (Hassan et al. 2017).

4.5   Hybrid Approach

There are three generally agreed upon elements of fake news articles, the first element is the text of an article, second element is the response that the articles received and lastly the source used that motivate the news article (Ruchansky et al. 2017). A recent study has been conducted that proposes a hybrid model which helps to identify fake news on social media through using a combination of human and machine learning to help identify fake news (Okoro et al. 2018). Humans only have a 4% chance of identifying fake news if they take a guess and can only identify fake news 54% of the time (Okoro et al. 2018). The hybrid model as proven to increase this percentage (Okoro et al. 2018). To make the hybrid model effective it combines social media news with machine learning and a network approach (Okoro et al. 2018). The purpose of this model is to identify the probability that the news could be fake (Okoro et al. 2018). Another hybrid model called CSI (capture, score, integrate) has been developed and functions on the main elements; (1) capture - the process of extracting representations of articles by using a Recurrent Neutral Network (RNN), (2) Score – to create a score and representation vector, (3) Integrate – to integrate the outputs of the capture and score resulting in a vector which is used for classification (Ruchansky et al. 2017).

5   Conclusion

In this paper we discussed the prevalence of fake news and how technology has changed over the last years enabling us to develop tools that can be used in the fight against fake news. We also explored the importance of identifying fake news, the influence that misinformation can have on the public's decision making and which approaches exist to combat fake news. The current battle against fake news on COVID-19 and the uncertainty surrounding it, shows that a hybrid approach towards fake news detection is needed. Human wisdom as well as digital tools need to be harnessed in this process. Hopefully some of these measures will stay in place and that digital media platform owners and public will take responsibility and work together in detecting and combatting fake news.

References

Ahmed, S., Hinkelmann, K., Corradini, F.: Combining machine learning with knowledge engineering to detect fake news in social networks - a survey. In: Proceedings of the AAAI 2019 Spring Symposium, vol. 12 (2019)

Albright, J.: Welcome to the era of fake news. Media Commun. 5(2), 87 (2017). https://doi.org/10.17645/mac.v5i2.977

Atodiresei, C.-S., Tănăselea, A., Iftene, A.: Identifying fake news and fake users on Twitter. Procedia Comput. Sci. 126, 451–461 (2018). https://doi.org/10.1016/j.procs.2018.07.279

Burkhardt, J.M.: History of fake news. Libr. Technol. Rep. 53(8), 37 (2017)

Castelo, S., Almeida, T., Elghafari, A., Santos, A., Pham, K., Nakamura, E., Freire, J.: A topic-agnostic approach for identifying fake news pages. In: Companion Proceedings of the 2019 World Wide Web Conference on - WWW 2019, pp. 975–980 (2019). https://doi.org/10.1145/3308560.3316739

Chen, Y., Conroy, N.J., Rubin, V.L.: Misleading online content: recognizing clickbait as false news? In: Proceedings of the 2015 ACM on Workshop on Multimodal Deception Detection - WMDD 2015, Seattle, Washington, USA, pp. 15–19. ACM Press (2015a). https://doi.org/10. 1145/2823465.2823467

Chen, Y., Conroy, N.J., Rubin, V.L.: News in an online world: the need for an 'automatic crap detector.' Proc. Assoc. Inf. Sci. Technol. **52**(1), 1–4 (2015b). https://doi.org/10.1002/pra2. 2015.145052010081

Hassan, N., Arslan, F., Li, C., Tremayne, M.: Toward automated fact-checking: detecting check-worthy factual claims by claimbuster. In: Proceedings of the 23rd ACM SIGKDD International Conference on Knowledge Discovery and Data Mining - KDD 2017, Halifax, NS, Canada, pp. 1803–1812. ACM Press (2017). https://doi.org/10.1145/3097983.3098131

Horne, B.D., Adali, S.: This just in: fake news packs a lot in title, uses simpler, repetitive content in text body, more similar to satire than real news. In: International AAAI Conference on Web and Social Media, vol. 8 (2017)

Macaulay, T.: Can technology solve the fake news problem it helped create? (2018). https:// www.techworld.com/startups/can-technology-solve-fake-news-problem-it-helped-create-367 2139/

Marr, B.: Coronavirus fake news: how Facebook, Twitter, and Instagram are tackling the problem. Forbes (2020). https://www.forbes.com/sites/bernardmarr/2020/03/27/finding-the-truth-about-covid-19-how-facebook-twitter-and-instagram-are-tackling-fake-news/

Okoro, E.M., Abara, B.A., Umagba, A.O., Ajonye, A.A., Isa, Z.S.: A hybrid approach to fake news detection on social media. Niger. J. Technol. **37**(2), 454 (2018). https://doi.org/10.4314/ njt.v37i2.22

Paskin, D.: Real or fake news: who knows? J. Soc. Media Soc. **7**(2), 252–273 (2018)

Pennycook, G., Rand, D.G.: Fighting misinformation on social media using crowdsourced judgments of news source quality. Proc. Natl. Acad. Sci. **116**(7), 2521–2526 (2019). https:// doi.org/10.1073/pnas.1806781116

Pérez-Rosas, V., Kleinberg, B., Lefevre, A., Mihalcea, R.: Automatic detection of fake news. In: Proceedings of the 27th International Conference on Computational Linguistics, Santa Fe, New Mexico, USA, pp. 3391–3401. Association for Computational Linguistics (2018). https://www.aclweb.org/anthology/C18-1287

Potthast, M., Kiesel, J., Reinartz, K., Bevendorff, J., Stein, B.: A stylometric inquiry into hyperpartisan and fake news (2017). arXiv Preprint arXiv:1702.05638

Qazvinian, V., Rosengren, E., Radev, D.R., Mei, Q.: Rumor has it: identifying misinformation in microblogs. In: Proceedings of the 2011 Conference on Empirical Methods in Natural Language Processing, EMNLP 2011, pp. 1589–1599 (2011)

Ruchansky, N., Seo, S., Liu, Y.: CSI: a hybrid deep model for fake news detection. In: Proceedings of the 2017 ACM on Conference on Information and Knowledge Management, CIKM 2017, pp. 797–806 (2017). https://doi.org/10.1145/3132847.3132877

Schade, U.: Software that can automatically detect fake news. Comput. Sci. Eng. 3 (2019)

Shao, C., Ciampaglia, G.L., Varol, O., Yang, K., Flammini, A., Menczer, F.: The spread of low-credibility content by social bots. Nat. Commun. **9**(1), 4787 (2018). https://doi.org/10.1038/ s41467-018-06930-7

Sivasangari, V., Anand, P.V., Santhya, R.: A modern approach to identify the fake news using machine learning. Int. J. Pure Appl. Math. **118**(20), 10 (2018)

Sparks, H., Frishberg, H.: Facebook gives step-by-step instructions on how to spot fake news (2020). https://nypost.com/2020/03/26/facebook-gives-step-by-step-instructions-on-how-to-spot-fake-news/

Stahl, K.: Fake news detection in social media. California State University Stanislaus, 6 (2018)

Stein-Smith, K.: Librarians, information literacy, and fake news. Strat. Libr. 37 (2017)

Thota, A., Tilak, P., Ahluwalia, S., Lohia, N.: Fake news detection: a deep learning approach. SMU Data Sci. Rev. **1**(3), 21 (2018)

Vlachos, A., Riedel, S.: Fact checking: task definition and dataset construction. In: Proceedings of the ACL 2014 Workshop on Language Technologies and Computational Social Science, Baltimore, MD, USA, pp. 18–22. Association for Computational Linguistics (2014). https://doi.org/10.3115/v1/W14-2508

Yang, Y., Zheng, L., Zhang, J., Cui, Q., Li, Z., Yu, P.S.: TI-CNN: convolutional neural networks for fake news detection (2018). arXiv preprint arXiv:1806.00749

Zhou, X., Zafarani, R.: Fake news: a survey of research, detection methods, and opportunities (2018). arXiv preprint arXiv:1812.00315

Digital Accounting and Auditing

Balance Sheet (Statement of Financial Position) Transformation in the Light of New Digital Technology: Ukrainian Experience

Liudmyla Lakhtionova[1]([✉]) [ID], Nataliia Muranova[1] [ID],
Oleksandr Bugaiov[1] [ID], Alla Ozeran[2] [ID],
and Svitlana Kalabukhova[2] [ID]

[1] National Aviation University, 1, Liubomyr Husar Avenue,
Kyiv 03067, Ukraine
Ludmilala@i.ua
[2] Kyiv National Economic University named after Vadym Hetman,
54/1, Peremohy Avenue, Kyiv 03057, Ukraine

Abstract. The paper contains suggestions on balance sheet (statement of financial position) transformation based on the Ukrainian reporting Form 1 in the context of approaching IFRS and European integration with the use of digital computer technologies to make it more analytical. Historical aspects of the balance sheet as an information base on the financial position of an enterprise are considered and clarified. Suggestions are put forward on transforming the assets of this reporting form in the light of digital computer technology, which is expected to improve the quality of its information content. In scientific literature, there are various ways of analyzing the financial position on the basis of the financial position statement data. Unfortunately, until recently, this issue was not sufficiently described in scientific literature with regard to the essence of the concept of financial position and a clear articulation of the organizational stages of its analysis, systematization of methods and obtained results at each stage, which necessitates the research.

Keywords: Balance sheet · Assets · Liabilities

1 Introduction

The fourth industrial revolution is introducing rapid changes into all areas of human activities. The changes are due to the use of artificial intelligence, robotics and digital technologies of various kinds. The wide use of digital computer technologies induces changes in such professions as accountant, financial analyst, auditor, etc.

The Statement of Financial Position (balance sheet) of an enterprise is the first main form of an enterprise's financial reporting in any country. The balance sheet is a accounting category representing a synthesis of accounts. We should be able to read a balance sheet and assess it critically in terms of content, structure and rational use of data. These issues have become especially important in the light of digital economics and modern computer technology.

© Springer Nature Switzerland AG 2021
T. Antipova (Ed.): ICIS 2020, LNNS 136, pp. 25–41, 2021.
https://doi.org/10.1007/978-3-030-49264-9_3

So, the issues of generating financial reports (including balance sheets) and their application are addressed by: 1) USA: L.A. Bernstain (2002) [1], T.P. Carlin (2001) [2], A.R. McMeen (2001) [2], Conrad Karlberg (2008, 2015, 2016, 2019) [3–6], R.C. Higgins (2007) [7], Charles H. Gibson (2013) [8, 9], Lyn M. Fraser, Aileen Ormiston (2010, 2013, 2016) [10], Walter Aerts, Peter Walton (2013, 2017) [11, 12], Joanne V. Flood (2015) [13]; 2) France: B. Kolass (1997) [14]; Krishna G. Palepu, Paul M. Healy (2013) [15]; 3) British scientists (London): McKenzie Wendy (2003, 2006) [16], D. Stone (1998), K. Hitcheeng (1998) [17]; P. Atrill (2007, 2015) [18, 19], E. McLaney (2007, 2015, 2017) [18–20]; 4) Ireland: Walsh Ciaran (2001) [21]; 5) Latvia: V. Paupa (2008) R. Sneidere (2008) [22]; 6) Russia: M.I. Kuter (2017, 2018, 2019, 2020) [23–26], M.M. Gurskaya (2017, 2018, 2019, 2020) [23–26]; 7) Iraq: Saoud Chayed Mashkoor Alamry (2019) [27]; 8) India: T.S. Grewa, H.S. Grewal, G.S. Grewa, R.K. Khosla (2019) [28]; 10) Scotland: A. Sangster (2017, 2020) [25, 26] and others. Many unsolved problems are discussion.

Such issues as the structure, content and assessment of the articles of an enterprise's financial position are under constant review by academic economists worldwide. Nevertheless, they do not seem to be completely solved. The problems of analytical possibilities of the balance sheet are also subjects of debate. The content and methodology of determining the finance indicators based on the balance sheet data remain unsolved problems. The current relevance of these problems is increasing in a globalized economy and with accelerated development of digital computer technology.

The aim of the article is to develop proposals to transform the Ukrainian balance sheet (financial statement) under conditions of approaching the IFRS and European integration with the use of digital computer technologies to make it more informative and analytical.

The tasks of the article are: studying the concepts of 'balance sheet' and 'statement of financial position'; presenting the history of the development of balance sheets in the world and in the independent Ukraine in the context of approaching the IFRS and European integration; describing the structure and content of the assets and liabilities of the Ukrainian balance sheet, making suggestions for their improvement under conditions of digital computer technologies; describing the analytical possibilities of the Statement of Financial Position under conditions of digital computer technologies and in conjunction with other forms of financial statements; developing organizational stages of the statement of financial position (balance sheet) taking into account the National Accounting Standard № 1 "General Requirements for Financial Statements" (Ukraine), international standards for financial statements and the EU Directive 2013/34/EU on yearly financial statements.

The solution of the tasks set will make the balance sheet more informative and analytical, which would accelerate the adoption of science-based management decisions by its users in the context of digital computer technologies.

2 Research Method

The study used: 1) general scientific methods: historical and logical approach, induction, deduction, analysis and synthesis; 2) special techniques: accounts, double entry, balance, reporting, spreadsheet, generalization.

3   The Concepts of «Balance Sheet» and «Statement of Financial Position». The History of the Development of Balance Sheets in the World and in the Independent Ukraine in the Context of Approaching the IFRS and in Compliance with the EU Directive 2013/34/EU on Yearly Financial Statements

The word «balance» comes from the Latin word «bilanx», which means two-scale. The Romans used this word in conjunction with the word 'libra' (balance) to mean two-scale balance (libra bilanx). Afterwards, words, close to «bilanx», appeared in many languages to denote «balance»: «la bilancia» in Italian, «la balance» in French, «a balance» in English, «баланс» in Ukrainian. Thus, the word 'balance' means equilibrium.

Consider the accounting balance sheet. It represents an element of the accounting method consisting in grouping the property of an enterprise by: 1) structure and allocation, 2) sources of formation. In the accounting balance sheet the word 'balance' has two meanings: 1) equality of debit and credit entries, equality of analytical and synthetic accounts, equality of the assets and liabilities of the accounting balance sheet; 2) the principal form of accounting statements that shows the condition of the assets of an enterprise and the sources of its formation in monetary terms on a certain date.

Modern American scientists Fraser Lyn M., Ormiston Aileen (2016) note: «The balance sheet shows the financial condition or financial position company on a particular date. … The balancing equation is expressed: Assets = Liabilities + Stockholders equsty [10, p. 48]. American scientists Aerts Walter, Walton Peter (2013, 2017) note: «The balance sheet (the IASB terminology is statement of financial position) gives a pictureat a given moment – the last day of the financial year – of how company has been financed and how that money has been invested in productiv capasity (plant, buildings, computers, inventories ets.)» [12, p. 4]. American scientist Joanne V. Flood (2015) також розкриває баланс [13, pp. 45–60]. French scientists Stolowy Herve, Lebas Michel, Ding Yuan (2013) note: «The very term balance sheet (now called statement of financial position) contains a message about its format. It is a set of two lists: resouces on one side (also called assets) and obligations to external parties on the side (liabilities or creditors and the net worth, coceptually owed, dy the firm as a separate entity, to Shareholders or owners)» [29].

A balance sheet is the most important accounting document, a significant source of information for the management, planning, production organization, standardizing, analysis and control. The balance sheet represents a synthesis of accounts. That's why we ought to be able to read a balance sheet and assess it critically in terms of content, structure and rational use of data.

As an accounting concept, the balance sheet has been in existence for over 500 years. The earliest reliable information concerning the usage of this word in accounting practices can be found in a ledger of the Italian banking house in 1408. In literature the word 'balance' was first used by Luca Pacioli, a famous Italian mathematician, in his work issued in 1494 and by Benedetto Cotrugli in his work written in 1458 and issued in 1573 in Venice. In legislation, the word 'balance' was first mentioned in the

Napoleonic French Commercial Code (1807). The balance sheet is the oldest form of generalizing data on the economic and financial developments of companies. More exact information on the origin of the accounting balance sheet are not available.

Luca Pacioli's work aroused a lot of debate among researchers of a history of last year's accounting. Especially much debate emerged about the structure and content of balance sheets. Fabio Besta (Besta, 1909; Sargiacomo, 2012) was the first to investigate and describe internal trial balances [30, 31]. He focused on the Balances of Andrea Babarigo, a merchant. E. Peragallo referred to Besta in his research and came to the conclusion: "Andrea Babarigo, a merchant from Venice, used such a balance in his reports in 1434, long before Pacioli" [32, p. 392]. De Roover described the balance sheets and Profits and Losses account in Francesco Datini's company in Barcelona in 1399 (Roover, 1956) [33]. The issues of Balances were highlighted by F. Melis (Melis, 1950; 1962; 1972) [34–36]. Earlier balance sheets were investigated and described by A. Martinelli (1974) [37].

The modern researchers M. Kuter and M. Gurskaya have proved that in the proprietorships of F. Datini in Pisa (the first proprietorship – 1383–1386; the second proprietorship – 1387–1392) the Profits and Losses account appeared earlier than the Trial Balance and the balance sheets, which were first produced in companies in 1394. The latter circumstance is caused by the partners' requirements to the reliability and timeliness of financial result calculation in order to distribute the profit as a reward (Kuter, 2017; Kuter, 2020; Sangster, 2017) [25, 26]. The authors mentioned above paid special attention to the fact that the Profits and Losses account had always been in the Ledger, and nobody had ever made a copy of the Profits and Losses account as it was very detailed and occupied several folios. When studying medieval balance sheets, M. Kuter and M. Gurskaya and co-authors discovered the earliest examples of Asset Impairment and Depreciation assessments. Using archive sources, they have proved that only the Asset Impairment was assessed at companies of that time whereas the Depreciation was assessed only once in 1399 in Barcelona (Kuter, 2018) [23]. And, which is especially important, it was shown how medieval "accountants" established reserves from the profit before its distribution for improving the accounting reliability (Kuter, 2020) [23].

In those early years, discussions were already held about the content and structure of balance sheets. The theory of a single row of accounts became widespread. The first yearly accounting balance sheet was compiled at Francesco Datini's trading company in the 1390s. The sheet was the prototype of today's balance sheet. In XIX century joint stock companies began to appear in Europe, and they published their balance sheets in newspspers. The weired structure of the balance sheets drew attention of unbiased users who started criticizing the existing balance sheet form. The French accountants Eugène Léautey and Adolphe Guilbault were the first to respond. In their work issued in the middle 1880s, they remarked that the assets of the balance sheet contained fictitious assets. Similarly, the liabilities of the balance sheet contained fictitious liabilities.

In the first half of the XX century, Johann Friedrich Schehr, the founder of the German cameral accounting, suggested reforming the balance sheet compiling procedure and renaming the balance sheet headlines: PROPERTY – for the left side and CAPITAL – for the right. The theory of two rows of accounts (this is what the theory of Schehr is called) substituted for the theory of a single row of accounts. The balance sheet became more understandable.

In the second half of the XIX century, financial markets and banking actively developed in the USA and Great Britain, which brought about presenting financial statements in order to get loans or participate in a stock market. As a consequence, well structured and quite uniformed financial statements appeared. They were compiled based on the interests of various users of these statements including managers. Now balance sheets compiled in accordance with the IFRS requirements includes not two but three components: assets, capital and liabilities.

Balance studies is a field of knowledge dealing with the economic substance of accounting balance, principles of its design, rules of assessing the entries and the use of balance information for managing an enterprise. In the second half of the XIX century the following accounting schools were created: Italian, German, French, Anglo-American (pragmatic approach). Each of them had its own approaches to studying balance sheets.

Balance sheets have come a long evolutional way to form. Each stage of social development has been characterized by its own structure of balance sheets. In Ukraine, the most typical features (since 1925) have always been: sources of compiling, time frames for compiling, amount of data, content, form. The transformation of the content and structure of accounting balance sheets in Ukraine has been dependent on the development of accounting (Table 1).

Table 1. Stages of the development of accounting and financial statements in Ukraine

Years	Stages of the development of accounting and financial statements
1917–1932	Accounting during the transition from capitalism to socialism. Development of the socialist type of accounting principles
1932–1945	Development of the methodology and organization of accounting that was aimed to monitor the implementation of plans, conservation of the social property, implementation of cost-effectiveness measures, creation of new forms of accounting, creation of new methods and techniques of cost accounting and product costing, establishment of internal controls and of sectorial accounting
1945–1965	Improvement of the unified accounting system on a national scale. Primary documentation content regulation, introduction of the regulatory method for accounting and calculation
1965–1991	Implementation of the economic reform aimed to empower enterprises. Introduction of economic self-sufficiency. Methodological change in planning. Improvements in planning accounts, development of general and sectorial regulations on planning, accounting and product costing. Creation of automated control systems. Intensification of research in the methodology and organization of accounting
1991–2000	Transition period caused by the political transformations in Ukraine. Formation of the socialist-to- market economy transition accounting. Ukraine emerges as an independent country. Emergence of preconditions for development of the Ukrainian national accounting and reporting system Adoption of the Programme of accounting system reformation using international standards (28.10.1998)

(*continued*)

Table 1. (*continued*)

Years	Stages of the development of accounting and financial statements
2000–2013	Emergence of the national accounting system, with international standards taken into account. Implementation (on 01.01.2000) of the Law of Ukraine "On Accounting and Financial Reporting in Ukraine" (16.07.1999). Issuance of Accounting Standards (AS) and National Accounting Standards (NAS) complying with IFRS and IAS. Adoption of the new plan of accounts, new accounting registers and financial reporting forms. Improvement of the methodology of evaluating assets, liabilities and capital. Development of auditing. The Cabinet of Ministers of Ukraine approves the strategy of IFRS application (Decree № 911-p of 24.10.2007)
2013–up to now	Improvement of the plan of accounts, accounting registers and financial reporting forms. Intensification of bringing the national accounting system to IAS Compiling financial statements in compliance with IFRS (for enterprises representing the public interest). Development of activities aimed to fulfil the Directive 2013/34/EU. Implementation of the national standards: NAS 1 and NAS 2

As Ukraine used to be an integral part of the USSR, the history of Ukraine's accounting development is viewed accordingly. In 1991 Ukraine became an independent country and began to develop its accounting system independently. During these stages the balance sheets of enterprises were changed.

In 1998 was adopted the Program of accounting system reformation using international standards according to which all Ukraine's enterprises were to implement the new accounting system within 2000–2001. The duration of the transition period was established 2 years.

On January 1, 2000 the Law of Ukraine "On Accounting and Financial Statements in Ukraine" was implemented (16.07.1999) [38]. The law defines the legal framework for accounting regulation, organization and implementation and compiling financial statements in Ukraine. In 1999 the first Ukrainian Accounting Standards (hereinafter AS) regarding financial statements and the methodology of their compilation were designed and introduced with effect from 01.01.2000: AS 1 «General Requirements for Financial Statements», AS 2 «Income Statement», AS 3 «Statement of Cash Flows», AS 4 «Report on equity» (invalid since 19.03.2013).

During 2000–2013 the Ukrainian balance sheet was in constant transformation and improvement. But we will not study the retrospective changes. Instead, we will consider in detail the balance sheet form (financial statement) provided by Form № 1 now in force in Ukraine. This balance sheet form has changed its content and title in compliance with the National Accounting Standard (hereinafter NAS) 1 «General Requirements for Financial Statements» approved by the Ministry of Finance of Ukraine (Decree № 73 of 07.02.2013) [39]. During 2013–2020 the content and structure of the balance sheet were also changed. However, we will criticize the form of the Ukrainian balance sheet now in force and give our proposals regarding it.

In 2017 the Law of Ukraine «On Changes to the Law of Ukraine «On Accounting and Financial Reporting in Ukraine» [40] was adopted to bring the national legislative norms to the provisions of the Directive 2013/34/EU [41]. The law came in force on 01.01.2018, with some of its provisions implemented on 01.01.2019. The changes introduced in accounting and financial statements are supposed to have the following influence on enterprise operation: 1) introducing an electronic format of financial reporting (the year 2019 was the first reporting period) and its publication on a company's website together with the audit findings (for financial institutions, large and medium-sized enterprises that represent the social interest). This enables increasing the transparency and comparability of financial reporting figures; 2) compulsory use of IFRS [42] by enterprises representing the social interest. It will increase the responsibility of Ukrainian companies to the society and allow minimizing the users' information and economic risks (in particular, investment risks) for their financial statements; 3) introduction of the management report containing both financial and non-financial information on the state of an enterprise and its development prospects, major risks and operational uncertainties. This report allows the enterprise to develop an effective risk-oriented management system that meets the company's all strategic objectives and tasks; 4) gradation of domestic enterprises: micro-, small, medium-sized and large. Such differentiation defines the requirements for compiling financial statements – large enterprises must use IFRS, small and micro-enterprises are exempt from submitting the management report; medium-sized enterprises are allowed not to show the non-financial information in their management reports; 5) introduction of accessible and public financial reporting information allows prompt responses to requests for information from individuals and entities on the whereabouts of the enterprise. Thus, the role of the financial report, including the balance sheet (financial condition statement), has significantly increased. The legislative framework list for compiling financial reports in Ukraine can be found on the website of the Ministry of Finance of Ukraine.

In accordance with the Law of Ukraine «On Accounting and Financial Reporting in Ukraine» and NAS 1 "The National Accounting Standard is a regulatory act establishing principles and methods of accounting and of compiling financial statements by enterprises (except the ones which, as provided for by law, compile financial reports in compliance with IFRS and National Accounting Standards in the public sector), which is designed based on IFRS and EU legislation for accounting and approved by the central executive body responsible for the development and implementation of public policy in accounting» [38]. NAS 1 «General Requirements for Financial Statements» defines the purpose and structure of financial statements, the principles of their compilation and the requirements for recognition and description of their elements. The provisions of NAS 1 apply to financial statements and consolidated financial statements of entities (enterprises) of all forms of ownership (except banks and budgetary institutions) that are obliged to submit financial statements in accordance with the law. According to NAS 1, the balance sheet (statement of financial position) is a statement of financial position of an enterprise that shows its assets, liabilities and equity on a certain date [39].

A financial statement in Ukraine is composed of: a balance sheet (statement of financial position) (hereinafter balance sheet), an income statement (statement of comprehensive income) (hereinafter далі income statement), a statement of cash flows, a report on equity, and notes to the financial statements [39]. The balance sheet of an enterprise is prepared as at the end of the last day of the reporting period [39]. The form and structure of financial statement items are established by NAS 1 and presented in appendices 1 and 2. Information in a balance sheet is shown in an appropriate item if it meets the criteria: the information is essential; the item can be reliably assessed. According to the Methodological Recommendations on How Financial Statements Are to Be Filled Out, «balance sheet compilation is aimed to provide users with complete, accurate and unbiased information on the financial situation of the enterprise at the reporting date» [43]. Part II of the Recommendations describes the balance sheet structure, which presents the enterprise's assets, liabilities and equity. The total assets in the balance sheet shall equal the total liabilities plus equity. The assets are shown in the balance sheet if they can be reliably assessed and they are expected to entail economic benefits in the future. The liabilities are shown in the balance sheet if they can be reliably assessed and the expected economic benefits may decrease as a result of their repayment. The equity is shown in the balance sheet simultaneously with the assets or liabilities that influence its change [43].

The financial situation of an enterprise is not defined in regulatory instruments. However, in IFRS, IAS and NAS the concepts of balance sheet and of statement of financial position are equated. French scientists Stolowy Herve, Lebas Michel, Ding Yuan (2013) note: «The very term balance sheet (now called statement of financial position) contains a message about its format» [29]. Information given in a balance sheet describes the financial position of an enterprise. It includes the assets, liabilities and equity. Thus, assets, liabilities and equity are the elements directly related to financial position assessment. Now there are various opinions on the structure of enterprise financial position indexes. The American scientists Fraser Lyn M., Ormiston Aileen (2016) [10, pp. 204–238]; Aerts Walter, Walton Peter (2013, 2017) [11, pp. 231–260: 12, pp. 450–483]; Joanne V. Flood (2015) [13, pp. 29–44] analyze the financial condition of a company in detail.

We will consider financial position assessment indexes such that can be defined by the balance sheet data only: 1) dynamics of the content and structure of the balance sheet assets, of the current assets, non-current assets, receivables and inventories; 2) dynamics of the content and structure of the balance sheet liabilities, of the equity, creditor's equity, long-term liabilities and ensuring, current liabilities and ensuring and accounts payable; 3) indexes of the financial stability of an enterprise; 4) enterprise liquidity indexes; 5) balance sheet liquidity and its critical value; 5) enterprise solvency. The complex of the indexes above represents the core of the financial position of an enterprise and its assessment.

The substantive content of a balance sheet is defined by its structure. Here we present Form 1 of the enterprise balance sheet as part of a financial statement now valid in Ukraine (Table 2).

Table 2. Balance sheet content and structure according to AS 2 NAS 1

AS 2 «Balance» [44] (annuled)		NAS 1 «General requirements for financial position» [39] (actual)	
Balance sheet (Form 1)		Balance sheet (statement of financial position) (Form 1)	
Assets	Equity and liabilities	Assets	Equity and liabilities
I. Non-current assets II. Current assets III. Future expenses	I. Equity II. Ensuring of future expenses and payments III. Long-term liabilities IV. Current liabilities V. Future incomes	I. Non-current assets II. Current assets III. Non-current assets kept to be sold and disposal units	I. Equity II. Long-term liabilities and ensuring III. Current liabilities and ensuring IV. Liabilities associated with non-current assets kept to be sold and to disposal units
Balance (Line Code 280)	**Balance** (Line Code 640)	**Balance** (Line Code 1300)	**Balance** (Line Code 1900)

As seen from Table 2, the balance sheet structure has changed significantly. A new approach has been introduced to the numeration of the balance line codes: three-digit codes have changed to four-digit ones. Part III of the actual balance sheet assets has been changed, and future expenses are incorporated into Part II «Current Assets». The separate sub-account 286 «Non-current assets and disposal units kept to be sold» has been introduced into the plan of accounts, and Part III of the balance sheet shows «Non-current assets kept to be sold and disposal units». Part II «Ensuring of future expenses and payments» has been excluded. In the actual form of the balance sheet (statement of financial condition) there are Part II «Long-term liabilities and ensuring» and Part III «Current liabilities and ensuring». Part V «Future incomes» has been removed and the separate Part IV «Liabilities associated with non-current assets kept to be sold and to disposal units» has been formed. The annulled form contained more parts for Equity and Liabilities. The content of each part of the balance sheet (statement of financial condition) has changed. The number of compulsory items has reduced whereas that of additional ones has increased. The additional items are not compulsory to use if their related information is absent and was absent in the previous periods. The methodology of compiling Form 1 is presented in the Methodological Recommendations on How Financial Statements Are to Be Filled Out [43]. Changed the march to line code numbering balance. There were three-digit codes, four-digit line codes became. Thus, Form 1 has changed greatly. We will consider the changes in the balance sheet in terms of assets and liabilities.

4 Structure and Content of the Assets and Liabilities of the Balance Sheet. Suggestions for Improving the Statement of Financial Position in the Light of New Digital Technology

The actual form of the balance sheet (statement of financial position) is presented in Appendix 1 of NAS 1 "General Requirements for Financial Statements" [39]. Suggestions for improving the assets and liabilities of the balance sheet (statement of financial position) concern the changes in the number, headlines and list of their items (Table 3).

Table 3. Balance sheet content and structure according to NAS 1 (left) and to the suggested balance sheet form (right)

NAS 1 «General requirements for financial statements»		Suggestions	
Balance sheet (statement of financial position) (Form 1)		Statement of financial position (balance sheet) (Form 1)	
Assets	Equity and liabilities	Assets	Equity and liabilities
I. Non-current assets II. Current assets III. Non-current assets kept to be sold and disposal units	I. Equity II. Long-term liabilities and ensuring III. Current liabilities and ensuring IV. Liabilities associated with non-current assets kept to be sold and to disposal units	I. Non-current non-financial assets II. Non-current financial assets III. Current non-financial assets IV. Current financial assets	I. Equity II. Long-term liabilities and ensuring III. Current liabilities and ensuring IV. Liabilities associated with non-current assets kept to be sold and to disposal units
Balance (Line Code 1300)	**Balance** (Line Code 1900)	**Balance** (Line Code 1300)	**Balance** (Line Code 1900)

The balance sheet in the information base of financial analysis occupies a central position. For that reason its structure and content must be directed at strengthening the logic of displaying information, improving comprehension, raising analytics, simplifying for the use of information in further analytical studies in the light of new digital technology. This is due to the requirements of approaching to IFRS and fulfilling requirements of the EU Directive № 2013/34/EU, accelerating European integration processes, developing digital economics, globalizing economical processes and harmonizing financial statements. The changes made to the name of the reporting Form 1 are quite correct. The expression «Statement of financial position» added in parentheses is a correct specification of the name. We accept this name. It sets the Ukrainian balance sheet form closer to the recommendations of IFRS. Due to the discussion among scientists from different countries of the world about the concept of financial condition and the area of its analytical research it is assessed rather positively. In the specified name of the given reporting form the attention is focused on the purpose of the balance sheet as an information base for assessing financial position of an enterprise. But the word «balance sheet» in the name is appropriate to be put second.

The balance sheet asset reflects the economic resources of the enterprise as to their content and allocation. Balance sheet asset reflects the nature (direction) of the use of capital, so it is divided into fixed and working. Balance sheet assets are divided into non-current and current assets. This is also stated in IFRS. According to IFRS, each enterprise individually decides on the feasibility of dividing assets into current and non-current assets.

According to NAS, all assets of an enterprise are necessarily divided into non-current and current assets. In order to further improve the presentation of source information in the balance sheet to assess the financial position of the enterprise, it is considered appropriate to make changes to the structure of the asset reporting form No. 1.

Suggestions are made to: 1) divide the existing part I «Non-current assets of the balance sheet asset into two separate parts: part I «Non-current non-financial assets» and part II «Non-current financial assets»; 2) divide the existing part II of the balance sheet asset «Current assets» into two separate parts: part III «Current tangible assets» and part IV «Current financial assets»; 3) the existing third part «Non-current assets kept to be sold and disposal units» of the balance sheet asset is recommended to include in the proposed part III «Current tangible assets» as a specific commodity.

Such changes in the content of the balance sheet asset will make it more detailed and informative. The information provided is more open, simple, accessible, and understandable to users. This will facilitate and speed up the financial position analysis process. There is no need to further group and select source information to analyze the absolute and relative indexes of asset, liability, financial stability, solvency and liquidity. This is especially important in the context of artificial intelligence and new computer technology.

In addition, it is generally accepted and understood that balance sheet parts are formed by combining individual items. Thus, each part of the balance must distinguish separate items (indexes). But in the current form of balance sheet (both in assets and liabilities) separate parts are presented, although no items are provided in them.

It is incorrect to enlarge the item «Inventories» and include production inventories, low value and non-durable items, unfinished production, finished goods and commodities in its content. According to the information in these items, it is visually that the user can make a conclusion about the type of activity of the enterprise. It is the content of inventories that is determined by the peculiarities of the activity of a manufacturing, trading, or service enterprise. Combining the information in this item reduces the information content of the balance sheet, and confuses users. It further complicates the work of financial analysts and managers. In accordance with the Methodological Guidelines for Completing the Financial Reporting Forms, the list of balance sheet items is additional. We consider items «Production inventories», «Unfinished production», «Finished goods», «Commodities» (excluding non-current assets kept to be sold and disposal units), «Non-current assets kept to be sold and disposal units» should be made mandatory in the balance sheet. As to their nature and in economic terms, non-current assets kept to be sold and disposal units are commodities. Commodity is a current asset. In Ukraine information on this type of commodities in the Chart of Accounts is displayed in the sub-account of the account «Commodities». Therefore, it is logical to show this type of product in the content of current financial (tangible) assets in the balance sheet too.

According to NAS 1, the item «Accounts receivable for products, goods, works, services» is shown at net realizable value only. The information about provision for doubtful debts is removed. This is irrelevant and ambiguous. The calculation of the provision for doubtful debts is borrowed from foreign accounting practices, information on it is important for management decisions. Therefore, the «Provision for doubtful debts» item should be made mandatory again and reflected in the balance sheet. It is worth dwelling on the complex item «Cash and cash equivalents». In order to eliminate the additional work of users in obtaining the transcript of this complex item, it is obligatory to select the item «Cash», «Cash at banks in national currency», «Cash at banks in foreign currency». For countries where national currency (in Ukraine - hryvnia) is used in domestic turnover, it is important to show in the statement of financial condition cash in different types of currencies. The amount of cash equivalents should be included in the item «Current financial investments». Indeed, they are a type of current financial investments and are reflected in the Chart of Accounts in a single control account with division into sub-accounts. The availability of these items in the Reporting Form 1 will accelerate the analysis of cash, solvency and liquidity of the enterprise, the liquidity of the balance in new digital technology. If you refer to IFRS, they do not provide for the display of non-current assets and sales groups in a separate part of the statement of financial position. Therefore, the proposed recommendation to remove this part in the balance sheet asset approximates the Ukrainian balance sheet to IFRS. Research on balance sheet liability items will be a matter for future research.

Thus, the proposed recommendations expand the source information and enhance the analytical balance. This is important for users. They will be able to quickly obtain financial analysis results and make timely management decisions through the use of advanced computer technology. It should be noted that the proposed recommendations for the balance sheet (statement of financial position) increase the usefulness of its information. We emphasize that users have access to the information about the activities of the company only through financial statements. According to the Law of Ukraine «On Accounting and Financial Reporting in Ukraine» financial statements (including balance sheets) are not a trade secret [40]. IFRS developers can take into account our suggestions and supplement the classification of non-current and current assets and improve the information content of this reporting form.

5 Analytical Capabilities of the Balance Sheet in Conjunction with Other Forms of Financial Report. The Organizational Stages of the Analytical Study

Analysis of a balance sheet (financial position) is an issue of financial analysis and analysis of financial statements of enterprises. The history of balance sheet analysis as well as the methodology and organization of its analytical study has long been researched in the academic world. Lately, balance sheet analysis (financial report analysis) has been covered by:: 1) USA: Conrad Karlberg (2008, 2015, 2016, 2019), Lyn M. Fraser, Aileen Ormiston (2010, 2013, 2016), Walter Aerts, Peter Walton (2013, 2017); 2) France: Krishna G. Palepu, Paul M. Healy (2013); 3) British scientists: P. Atrill (2007, 2015), E. McLaney (2007, 2015, 2017); 4) Iraq: Saoud Chayed Mashkoor Alamry (2019); 5) India: T.S. Grewa H. S. Grewal G. S. Grewa, R. K. Khosla (2019) and others. The balance sheet represents a clear picture of the financial condition of an enterprise. In order to assess and predict the financial condition of an enterprise one should be able to read the balance sheet and have good proficiency in the methodology of its analysis. To be able to read the balance sheet means to know the content of each its item, ways of its assessment and its relationships with other balance sheet items, the character of possible changes in each item and the influence of these changes on the financial position of the enterprise. Consider the analytical capabilities of a balance sheet (statement of financial position) and the organizational stages of analytical processes (Table 4).

Table 4. Analytical capabilities of the financial report (balance sheet) and organizational stages of the study

Analytical capabilities of the balance sheet (statement of financial position)	Analysis stages	Methods of analysis	Analysis results
Assessing the dynamics of the content and structure of the balance sheet liabilities by its parts, of the equity, creditor's equity, long-term liabilities and ensuring, current liabilities and ensuring, commercial and other accounts payable	I	Horizontal, vertical; absolute and relative indexes	Absolute and relative indexes of the content and structure of the balance sheet liabilities by its parts, of the equity, creditor's equity, long-term liabilities and ensuring, current liabilities and ensuring, commercial and other accounts payable
Assessing the dynamics of the content and structure of the balance sheet assets by its parts, non-current non-financial assets, non-current financial assets, current non-financial assets, current financial assets, inventories, commercial and other receivables	II	Horizontal, vertical; absolute and relative indexes	Absolute and relative indexes of the content and structure of the balance sheet assets by its parts, non-current non-financial assets, non-current financial assets, current non-financial assets, current financial assets, inventories, commercial and other receivables
Assessing the financial stability of the enterprise and the reserve of the main operating activity (safe operation area)	III	Horizontal, vertical; absolute and relative indexes	A set of absolute and relative indexes of the financial stability of the enterprise and of the reserve of the main operating activity запасу (safe operation area)

(*continued*)

Table 4. (*continued*)

Analytical capabilities of the balance sheet (statement of financial position)	Analysis stages	Methods of analysis	Analysis results
Assessing the balance sheet liquidity	IV	Horizontal; absolute indexes	Indexes of the balance sheet liquidity or non-liquidity and of the balance sheet liquidity critical value
Assessing the solvency and liquidity of the enterprise	V	Horizontal, vertical; absolute and relative indexes	A set of absolute and relative indexes of the solvency and liquidity of the enterprise
Involving the data of Form 2 "Income Statement (Statement of comprehensive income)"			
Assessing the turnover of various kinds of the enterprise's capital	VI	Horizontal; relative indexes	A set of relative indexes of the capital turnover
Assessing the profitability of the enterprise, assets and equity	VII	Horizontal; relative indexes	A set of relative indexes of the profitability of the enterprise, assets and equity
Generalization and summing up of the obtained results on the enterprise's financial position and financial results	VIII	Horizontal, vertical; absolute and relative indexes	Generalized conclusions and recommendations on improving the financial position and financial results of the enterprise

As seen from the table above, we can speak about 8 organizational stages of balance sheet analysis. The table is a representation of the logical analytical study of the statement of financial position. The methods of the balance sheet analysis at each stage of the study are specified in it. The results of the analysis are displayed for each of its stages. Special attention should be paid to the last stage of the analysis. In addition to the correct calculations at the previous stages, financial analysts and other users should be able to correctly generalize the conclusions and develop recommendations on improving the financial position of an enterprise for the future and enhancing its competitiveness.

Highlighting the proposed four sections of the balance sheet asset is expected: 1) to facilitate and accelerate, without additional processing of the initial information, the determination of the following financial position indicators: financial risk coefficient based on the net debt; coefficient of liabilities associated with non-current assets kept to be sold and with disposal units; six assets liquidity groups (absolute liquidity, high liquidity, accelerated liquidity, quick liquidity, slow liquidity, hard liquidity); balance liquidity and its degree depending on the enterprise status (large, medium, small, microenterprise), business type and industry sector; availability of permanent working capital; high, accelerated (strict), quick (critical), current liquidity indexes; liquid cash flow, etc.; 2) to accelerate the calculation of the financial needs of the enterprise; 3) to facilitate the determination of a minimum sum of money required for further successful activity of the enterprise. Recommendations on improving the financial position of an enterprise for the future and enhancing its competitiveness.

Appropriate analytical work performed by financial analysts and financial managers will be helpful in making sound decisions on the timely choice of enterprise activities financing methods and on enhancing its competitiveness.

6 Conclusion

In the course of the study, the following results have been obtained: the conclusions on the history of the development of accounting and balance sheets have been clarified; an approach to defining the essence of the financial position is suggested; suggestions have been put forward on clarifying the name of the form of the Ukrainian statement of financial position and on balance sheet assets transformation; the logical structure (organizational stages) of the enterprise performance analysis on the basis of the statement of financial position data has been described. The paper represents a significant contribution to modern thinking concerning the issues of the statement of financial position structure and content, understanding the concept of financial position and analytical capabilities of the balance sheet at all stages of its study.

The recommended transformation of the statement of financial position (balance sheet) will help to ensure a more rapid implementation of a thorough analytical study of the financial position of the enterprise in the digital technology. Analytical calculations will be accelerated on the basis of the source information correctly grouped according to the balance sheet data. There is no need for users to further process the balance data. It will increase efficiency, timeliness, economy, quality of information processing of through the application of modern computer technologies. The suggestions made will facilitate the calculation of the totality of all indicators of financial position. Many scholars studied the nature and content of the balance sheet, analyzed the statement of financial position differently, but we do not have information on the precise formulation of the organizational stages of analysis, systematization of methods and its results in the context of each step. This research is an important contribution to theoretical and practical knowledge in the fields of accounting, financial reporting, financial analysis and financial statement analysis. The existing discussion issues on the concept of financial stability, solvency, balance sheet and enterprise liquidity, critical value of balance liquidity, a set of financial position indexes in terms of their content, grouping, methods of determination are the subject of future scientific research.

References

1. Bernstain, L.A.: Analiz Finansovoy Otchyotnosti: Teoriya, Praktika I Interpretatsiya (Financial Statement Analysis: Theory, Practice And Interpretation). Financy i Statistika, Moscow (2002)
2. Karlin, T.R., Makmin, A.R.: Analiz Finansovykh Otchyotov (Na Osnove Gaap) (Financial Report Analysis (Based On Gaap)). INFRA-M, Moscow (2001)
3. Carlberg, C.: Analiz Finansovoy Otchyotnosti S Ispolzovaniyem Excel (Financial Statement Analysis Using Excel). Dialektika, Kiev (2019)
4. Carlberg, C.: Regressionny analiz v Microsoft Excel (Regression Analysis Microsoft Excel). Wiliams, Moscow (2016)

5. Carlberg, C.: Biznes-Analiz S Pomoshchyu Microsoft Excel (+Cd-Rom) (Business Analysis With Microsoft Excel). Wiliams, Moscow (2008)
6. Carlberg, C.: Biznes-Analiz S Ispolzovaniyem Excel (Business Analysis: Microsoft Excel 2010. Wiliams, Moscow (2015)
7. Higgins, R.S.: Finansovy analiz: instrument dlya prinyatiya biznes-resheniy (Financial Analysis: a Tool for Business Decision Making). OOO "I.D. Wiliams", Moscow (2007)
8. Gibson, C.H.: Financial Statement Analysis, 13th edn. Cengage Learning EMEA, Boston (2013)
9. Gibson, C.H.: Finance analysis. In: Financial Reporting and Analysis, 11th edn. Univ. of Toledo, Toledo (2009)
10. Fraser, L.M., Aileen, O.: Understanding of Financial Statements, 11th edn. Pearson, Boston (2016)
11. Walter, A., Peter, W.: Global Financial Accounting and Reporting: Principles and Analysis, 3rd edn. Cengage Learning EMEA, Boston (2013)
12. Walter, A., Peter, W.: Global Financial Accounting and Reporting: Principles and Analysis, 4th edn. Cengage Learning EMEA, Boston (2017)
13. Flood, J.M.: Wiley GAAP 2015: Interpretation and Application of Generally Accepted Accounting Principles. Wiley, New York (2015)
14. Kolass, B.: Upravleniye finansovoy deyatelnostyu predpriyatiya. Problemy, kontseptsii i metody (Managing the Financial Activity of an Enterprise. Problems, Conceptions and Methods). Finansy, UNITI, Moscow (1997)
15. Palepu, K.G., Healy, P.M.: Business Analysis Valuation Using Financial Statements, 5th edn. Cengage Learning EMEA, Boston (2013)
16. Wendy, Mc.K.: Ispolzovaniye i interpretatsiya finansovoy otchyotnosti (Financial Statement Usage and Interpretation). Balans-Klub, Biznes Buks (2006)
17. Stone, D., Hitching, K.: Buchgalterskiy otchyot i finansovy analiz (Accounting and Financial Analysis). "Sirin", "Biznes-Inform", Moscow (1998)
18. Atrill, P., McLaney, E.: Accounting and Finance: An Introduction, 8th edn. Pearson, London (2015)
19. Atrill, P., McLaney, E.: Accounting and Finance for Non-Specialists, 9th edn. Pearson Education Limited, London (2015)
20. Eddie, M.: Business Finance: Theory and Practice, 11th edn. Pearson, London (2017)
21. Ciaran, W.: Kliuchovi finansovi pokaznyky. Analiz ta upravlinnia rozvytkom pidpryiemstva (Key Financial Indexes. Analysis and Management of Enterprise Development). Vseuvyto; Naukova Dumka, Kyiv (2001)
22. Paupa, V., Sneidere, R.: Uzdevumu krajums finansu analize. Baltimoras konsultaciju centres, Riga (2008)
23. Kuter, M., Gurskaya, M., Andreenkova, A., Bagdasaryan, R.: The early practices of financial statements formation in Medieval Italy. Acc. Hist. J. **44**(2), 17–25 (2018)
24. Kuter, M., Gurskaya, M., Bagdasaryan, R.: The structure of the Trial Balance. In: Antipova, T. (ed.) ICIS 2019. LNNS 78, pp. 103–116. Springer, Basel (2020). https://doi.org/10.1007/978-3-030-22493-6_11. Accessed 09 Feb 2020
25. Kuter, M., Sangster, A., Gurskaya, M.: The formation and use of a profit reserve at the end of the 14th century. Acc. Hist. **25**(1), 69–88 (2020). https://doi.org/10.1177/10323732 19870316. Accessed 09 Feb 2020
26. Sangster, A., Kuter, M., Gurskaya, M., Andreenkova, A.: The determination of profit in medieval times. In: Antipova T., Rocha Á. (eds.) Information Technology Science, MOSITS 2017. Advances in Intelligent Systems and Computing, vol. 724. Springer, Cham (2017). https://doi.org/10.1007/978-3-319-74980-8_20. Accessed 09 Feb 2020

27. Alamry, S.C.M.: Overview of financial statements analysis. In: Analysis of Financial Statements. Al-Alalamia for Printing and Designs Sammawa, Iraq (2019)
28. Grewal, T.S., Grewal, H.S., Grewal, G.S., Khosla, R.K.: Analysis of Financial Statements. Sultan Chand & Sons Private Limited, New Delhi (2019)
29. Herve, S., Michel, L., Yuan, D.: Financial Accounting and Reporting: A Global Perspective, 4th edn. Cengage Learning EMEA, Boston (2013)
30. Besta, F.: La Ragioneria, 2nd edn (1909). (In 3 Volumes. Facsimile Reprint. Rirea, Rome (2007))
31. Sargiacomo, M., Servalli, S., Andrei, P.: Fabio besta: accounting thinker and accounting history pioneer. Acc. Hist. Rev. **22**(3), 249–267 (2012)
32. Peragallo, E.: Origin of the trial balance. Acc. Rev. **31**(3), 389–394 (1956)
33. De Roover, R.: The development of accounting prior to Luca Pacioli according to the account-books of Medieval merchants. In: Littleton, A.C., Yamey, B.S. (eds.) Studies in the History of Accounting, pp. 114–174. London (1956)
34. Melis, F.: Storia della Ragioneria. Cesare Zuffi, Bologna (1950)
35. Melis, F.: Aspetti Della Vita Economica Medievale (Studi Nell'archivio Datini Di Prato). Monte dei Paschi di Siena, Siena (1960)
36. Melis, F.: Documenti Per La Storia Economica Dei Secoli Xiii—Xvi. Leo S. Olschki, Firenze (1972)
37. Martinelli, A.: The Origination and Evolution of Double Entry Bookkeeping to 1440. ProQuest Dissertations & Theses Global (1974)
38. Ukrainy, Z.: "Pro buchgalterskyi oblik ta finansovu zvitnist v Ukraini" № 996-XIV vid 16.07.1999 p. (ostanni zminy № 2545-VIII vid 18.09.2018 r.) (The Law of Ukraine "On Accounting and Financial Reporting in Ukraine" № 996-XIV of 16.07.1999. (latest changes № 2545-VIII of 18.09.2018). https://zakon.rada.gov.ua/laws/show/996–14. Accessed 09 Feb 2020
39. Natsionalne polozhennia (standart) buchgalterskoho obliku 1 "Zahalni vymohy do finansovoi zvitnosti": zatverdzheno nakazom Ministerstva finansiv Ukrainy № 73 vid 07.02.2013 r. (National Accounting Standard 1 "General Requirements for Financial Statements": approved by the Decree of the Ministry of Finance of Ukraine № 73 of 07.02.2013. http://www.minfin.gov.ua. Accessed 09 Feb 2020
40. Ukrainy, Z.: Pro vnesennia zmin do Zakonu Ukrainy "Pro buchgalterskyi oblik ta finansovu zvitnist v Ukraini" № 2164-VIII vid 5.10.2017 r. (The Law of Ukraine "On Changes to the Law of Ukraine "On Accounting and Financial Reporting in Ukraine") № 2164-VIII of 5.10.2017. https://zakon.rada.gov.ua/laws/show/2164–19. Accessed 09 Feb 2020
41. Directive on the annual financial statements, consolidated financial statements and related reports of certain types of undertakings of the European parliament and of the council of 26 June 2013 N 2013/34/EU. https://eur-lex.europa.eu/legal-content/EN/ALL/?uri=CELEX% 3A32013L0034. Accessed 09 Feb 2020
42. Mizhnarodni standarty finansovoi zvitnosti (International Financial Reporting Standards). https://zakon.rada.gov.ua/laws/main/929_010. Accessed 09 Feb 2020
43. Methodichni recommendacii do zapovnenniy form finansovoy zvitnosti № 433 vid 28.03.2013 (Methodical recommendations for completing the financial statements № 433 of 28.03.2013. https://zakon.rada.gov.ua/rada/show/v0433201-13/conv. Accessed 09 Feb 2020
44. Polozhennia (standart) buchgalterskoho obliku 2 "Balans": zatverdzheno nakazom Minis-terstva finansiv Ukrainy № 87 vid 31.03.1999 r. (Accounting Standard 2 "Balance": approved by the Decree of the Ministry of Finance of Ukraine № 87 of 31.03.1999). https:// zakon.rada.gov.ua/laws/show/z0396-99. Accessed 09 Feb 2020

Risk-Based Internal Control over Formation of Financial Reporting

Elena Dombrovskaya(⊠) ⓘD

Financial University under the Government of the Russian Federation,
Leningradsky pr., 49, Moscow 125993, Russian Federation
den242@mail.ru

Abstract. Currently, definite trends have been outlined for a change in the role of internal control in the corporate management system. Previously, core competency was determined by evaluating conformance with certain standards, in particular when preparing financial reporting. Today, priorities shift towards the need to assess and identify risks, create measures to minimize them, and ensure the reliability and completeness of financial reporting. Considering the ever-increasing list of different risks, as well as their increasing influence, the internal control system should have a risk-oriented focus. In the course of internal control, organizations assess risks of various operations and facilities. Control over the formation of financial reporting should be aimed at presenting relevant and truthful information in the reporting. Herewith, the risks of material misstatement of the reporting information should be minimized. Disregard of the risks of material misstatement reduces the truthfulness of financial reporting and creates favorable conditions for the operation of accounting and reporting data. The article presents the problems of generating a system of internal control over the formation of financial reporting based on risk factors and their minimization.

Keywords: Financial reporting · Risk-based · Internal control

1 Introduction

The truthfulness of the financial reporting of companies is a major concern that is debated a lot by the global community. Thus, the U.S. Securities and Exchange Commission (SEC) has announced its investigations related to the financial misstatement and accounting fraud as a key priority. In total, the SEC took 821 control actions in 2018 (754 – in 2017), following which for law violators in the field of regulation of securities markets and disclosure of information to investors (including violations related to misrepresentation of financial reporting), more than 3.9 billion dollars of misappropriated property and penalties were collected. Despite the fact that in 2018 the number of control actions taken in the area of financial accounting fraud and information disclosure decreased by more than 35% compared to 2017 (from 76 to 49), their occurrence still indicates the existence of potential threats to the investor. Major violations were recorded in areas such as inaccurate accounting, inappropriate bookkeeping, and artificial overstatement of profit [1].

© Springer Nature Switzerland AG 2021
T. Antipova (Ed.): ICIS 2020, LNNS 136, pp. 42–52, 2021.
https://doi.org/10.1007/978-3-030-49264-9_4

In FY 2019, the most common complaint categories reported by whistleblowers were Corporate Disclosures and Financials (21%), Offering Fraud (13%), and Manipulation (10%) [2].

A large number of cases of misrepresentation and falsification of financial accounting indicators requires the development of a control system for its formation. A lot of research is focused on the follow-up control of financial reporting, aimed at identifying signs of its misrepresentation and falsification by external users of information. It stands to mention the work of M. Beneish [3] and M. Roxas [4], who found dependence between the falsification of financial reporting and particular financial indicators. Based on this dependence, models have been developed to assess the risk of falsification of the company's reporting. Obviously, the use of these models will not allow avoiding falsification but will only reduce the risks of using incorrect financial information.

A condition ensuring the reliability and truthfulness of the accounting information is the availability of an effective system of preliminary internal control over the formation of financial reporting. It should be part of the corporate governance system and be based on the principles of the corporate culture of the company.

Organizational and methodological aspects of internal control are considered in the works of L.I. Kulikova and D.R. Satdarova [5]. The authors note that financial statement fraud results in financial and nonfinancial losses for the company. Thereby, the goodwill and image of the company degrade, its investment and consumer appeal decreases, and relations with business partners collapse. Internal control is one of the effective methods of management and control of the company. To prevent the falsification of financial reporting, the method of external and internal accounting compliance control is recommended.

Akhmetshin E.M. pays great attention to the regulatory environment of supervisory activities. He uses an institutional approach to characterize in his work the control system for the formation of financial reporting in Russia. The author believes that in the present context, the activities of the company should be coordinated with external factors that determine the correctness of economic decisions. For each specific situation, the correct models and algorithms should be developed for the most effective behavior. Akhmetshin E.M. has revealed that effective control is hindered by noncompliance with the principle of substance over form in accounting practice, differences between the rules of book-keeping and tax accounting, as well as between Russian accounting principles and international financial reporting standards [6].

Knechel W.R. and Salterio S.E take a broadside approach to assessing the company's business risks by an auditor. In the United States, when auditing private companies and PCAOB standards, the emphasis is upon the risk of material financial misstatement and auditors' activities to gain reasonable assurance that there is no such misstatement. The authors consider internal control over the formation of financial reporting as an element of a comprehensive audit of the company [7].

Valaskova K., Kliestik T., and Kovacova M. have discussed the practical aspects of financial risk management in their study. As part of the financial risk monitoring, the authors have assessed the financial risks of Slovak enterprises by identifying the most significant factors. As a result of multiple regression analysis, statistically significant determinants have been identified that affect the future financial development of the company, as well as a regression model for predicting bankruptcy. The most significant indicators identified allow performing internal control of the reliability of financial reporting indicators. Statistical analysis tools can form the basis of the internal control procedure in the company [8].

Studies based on the use of modern digital technologies and considering their capabilities to improve the quality and reliability of reporting information are of great interest. Therefore, in their work Efimova O., Rozhnova O., and Gorodetskaya O. have examined the capabilities of XBRL technology to improve the system of financial and nonfinancial reporting of companies. Based on the identification of the XBRL advantages for the formation of various types of reporting, the authors have shown that the use of XBRL provides effective management of companies and their business processes as well as improves the quality of reporting information [9].

A risk-based approach can be applied as a tool directly during the formation of financial reporting. Demina I. and Dombrovskaya E. believe that in order to disclose information about the financial state and results of operations fully and reliably, the company should take into account the existing and potential risks during the formation of financial reporting and disclose the existence and nature of the impact of such risks on the financial reporting [10]. There are no requirements for generating a risk-oriented accounting system and forming financial reporting in either national or international laws and standards. Nevertheless, the standardization of the issues under consideration is required to provide the basis for clear and informative reporting information.

It is possible to carry to number of researches of the early reporting [11–13].

2 The Primary Approaches to Generate a Risk-Oriented System of Internal Control over the Formation of Financial Reporting

The purpose of the internal control system over the formation of financial reporting based on the risk-based approach is to control the implementation of all accounting policies and other regulatory documents in relation to risks by the responsible agencies.

Considering the organizational aspect, the main purposes of the internal control service are to increase the efficiency of units, ensure the reliability of financial reporting data, provide compliance of the company with the current legislation, detect violations of regulations and rules as well as identify risks and respond adequately. Thus, two main purposes of the internal control system over the formation of financial reporting based on the risk-oriented approach can be distinguished:

1) Detection of violations of regulations and requirements of accounting policies and other documents regarding risks;
2) Identification and response to the identified risks.

The accounting system and the internal control system should be two interrelated components of the risk-oriented system for the formation of financial reporting since neither one nor the other component is functioning in the organization on its own but is the stage of achieving the organization's purpose of generating reliable financial reporting.

The system of internal risk-based control over the formation of financial reporting should be developed taking into account the principles set forth in the COSO «Guidelines for Internal Control». Table 1 presents the model of interaction between the internal control system and the accounting system as part of the risk-based approach to the formation of financial reporting:

Table 1. Model of the risk-oriented system of internal control and accounting

Internal control system	Accounting system
Control environment risk assessment	Recognition, assessment, and recording of the accounting events
Internal control procedures	
Information and communication monitoring	Formation of financial reporting

The control environment is an important component of the internal control system, since this component, which contains a set of standards and requirements developed by the organization, evaluates the importance and significance of internal control in the organization.

The control environment is based on the corporate governance system adopted by the company, supported by the philosophy and style of management, as well as the organizational structure. The control environment is based on compliance with the requirements of laws and standards, as well as established ethical norms and principles, by the company and its employees. The control environment for the formation of financial reporting is determined by the accounting policy of the company, the adopted procedure for the formation of financial reporting, the implementation of the professional code of ethics and company policy regarding internal control.

The risk-oriented system of internal control over the formation of financial reporting involves the identification and assessment of risks that may lead to misrepresentation of the accounting information. In this regard, it should be noted that the risk-oriented approach cannot be effective if used during control procedures only. The accounting system should also have a risk-oriented focus. Reliable and truthful reporting information can be obtained only with a comprehensive approach to accounting and control systems.

At the stage of presentation of accounting events in the accounting system, the following risks occur: legitimacy and validity of recognition of an asset or liability, income or expenditure; timeliness and completeness of recognition of accounting events; the accuracy of recognition assessment; correctness of the presentation or classification of accounting events.

Internal control procedures are aimed at identifying all risks associated with the formation of financial reporting:

- procedures for control of the actual presence and state of objects, inventory;
- control procedures for the presentation of the organization's assets and liabilities during a valid assessment;
- control procedures for the presentation of the impact of risks on the financial reporting, including through the generation of allowances;
- control procedures for the operation of information and analysis systems of the company;
- control procedures to prevent deliberate accounting falsification;
- control procedures for the identification, recording, assessment, and management of the company's risks;
- control procedures for compliance with accounting policies regarding the formation of financial reporting and compliance with all requirements for the impact and recording of the identified risks.

Information and communication as components of the internal control system are aimed at the timely exchange of information required for organization units or management to make a relevant decision, taking into account the identified risks.

The last component of the internal control system for the formation of financial reporting is monitoring. It is aimed at establishing the efficiency of all components of the system. Permanent monitoring procedures at various stages of business processes allow receiving timely feedback on the operation of the system and the need for further actions.

The system of internal control over the formation of financial reporting should be as formalized as possible. The process of this formalization involves the description of all significant business processes that affect the formation of financial reporting. Obligations and responsibilities should be differentiated among services and employees responsible for control procedures. All components of the internal control system should be subjected to interval updating. In addition, the internal control system should be regularly tested for the effectiveness of control procedures.

3 Peculiarities of the Internal Control System for the Formation of Financial Reporting in a Construction Company

During the study, the organization of the internal control system in a construction company has been considered. The company implements a complete range of works from design and construction and installation works to commissioning of various facilities. Construction companies are heavily marketed, have a broad range of services and continued production cycle. Many of them are public, and reliable information about their activities is of interest to a large number of users. The construction company considered during the preparation of the article is a part of the group and is a subsidiary of the joint-stock company, which is the leader in construction engineering in the Russian market.

The organizational structure of the company is built on a functional basis, which provides clear differentiation between the authorities and responsibilities of each organization unit and employee. The corporate governance system is based on the articles of association and the code of corporate governance. The profitability performance profiles of assets and economic efficiency of the company are significantly lower compared to the industry average indicators, which calls for special attention when determining its strategy and development prospects.

That is, the control environment is formed in the company, which determines the perception by all employees of the internal control system. For its effective implementation, it is required to identify and analyze risks that arise in the course of achieving the company objectives, and which are the basis for determining control procedures and risk management measures. The study has shown the importance of macro-level risks and industry risks in the company's activities. They are listed in Table 2.

Table 2. The basic macro-level risks and industry risks in the construction company activities

Macro-level risks	Industry risks
1. Country and regional risks	1. Production risk
2. Regulatory risks	2. Technological risk
3. Financial risks:	3. Organizational risk
– liquidity risk	4. Commercial risk
– investment risk	5. Competition aggravation risk
– advancing risk	6. Risk of hazardous job
– interest rate risk	7. Climatic risk
– currency risk	8. Price volatility risk
– inflation risk	9. Operational risk
	10. Social risk

The company pays serious attention to risk assessment of its activities and discloses information about uncertainties and risks in the annual report. Their significance and degree of impact on the company's activities are determined based on the analysis and assessment of risks (Table 3).

Table 3. Forms and level of impact of risks on the construction company activities

Risk type	Influence level	Risk impact	Control methods
Political risk	Low	Cuts in public expenditures for construction	Search for individual customers
Sanction risk	Low	Introduction of international sanctions for the procurement of construction materials	Search for different vendors
Regional risk	Low	Transportation problems	–
Legal risk	Low	Expenses on payment of additional tax liabilities	–
Currency risk	Low	Increase in cost of supply from external suppliers	–
Interest rate	Low	Increase in cost of debt	–
Inflation risk	Low	Reduction of cost of the performed works and receivables; increase in price for materials, services, energy	–
Liquidity risk	High	Lack of sufficient monetary resources and liquid assets	Inventory of working capital, improvement of receivables management policy
Credit risk	High	Insolvent debtors	Provision creation, the activity of the counterparty evaluation commission, tightening the advance payment procedure for contractors
Price volatility risk	High	Continuing increase in material prices	A competition-based centralized procurement system
Competition aggravation risk	High	Redistribution of the scope of works in progress within the group of companies	Search for a new customer
Production risk	High	Downtime of equipment and labor power	Optimization of the payment schedule to prevent interruptions in water and electricity supply
Commercial risk	High	Dishonesty of suppliers and contractors	The activity of the counterparty evaluation commission
Organizational risk	High	Schedule overrun for completed work due to improper design specifications and estimates	Search for skilled personnel, aimed at the correct implementation of documentation
Risk of hazardous job	Low	Accidents on the construction site	–
Climatic risk	Low	Steady precipitation, wind, abnormal temperatures that impede construction	–

(*continued*)

Table 3. (*continued*)

Risk type	Influence level	Risk impact	Control methods
Social risk	Low	Shortage of skilled personnel	–
Operational risk	Low	Additional expenses for maintenance of equipment and machinery	–
Risk to goodwill	Low	Loss of goodwill for suppliers and contractors, buyers and customers	Support at the level of interaction with economic entities and regulatory authorities

Thus, the company's activities are highly exposed to several types of risks. Most risks are common, caused by the influence of external factors that are difficult to manage and predict their impact accurately.

The form of appearance of each type of risk is unique to the company. Each type of risk at the macro level and industry level, interpreted under the conditions of the company's operation, can be considered specific to it.

To implement the risk-oriented system of control over the formation of the company's financial reporting, it is reasonable to act in the areas shown in Fig. 1:

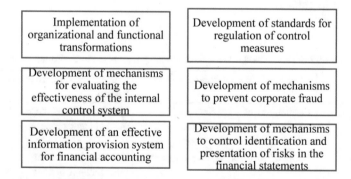

Fig. 1. Areas of implementation of the risk-oriented system of internal control over the formation of financial reporting

The business mechanism for introducing the risk-oriented control over the formation of financial reporting includes the following:

– organizational and functional transformations aimed at the clear separation of roles and responsibilities for risk management between the company departments. Although the internal audit department should be appointed responsible for the compliance of the formed financial reporting with the requirements of the risk-

oriented system, the audit function should be assigned to each employee of the company;
- generation of mechanisms for analysis, evaluation, and monitoring of the reliability and effectiveness of the internal control system;
- creation of a modern and effective information provision system for all levels of management, which provides for timely monitoring of compliance with the established tasks, contract terms, standards, norms, and procedures during the implementation of business processes;
- development of standards determining common requirements for the implementation of control measures, the formalization of results thereof, as well as the level of personnel qualification of internal control units;
- development and implementation of a policy to prevent, detect, and eliminate corporate fraud, deliberate falsification of accounting events in order to gain self-interest, abusive acts, and corrupt practices, and to ensure compliance with the economic interests of the company;
- the development and implementation of requirements and policies on procedures for identifying risks, assessing their impact, recording them in accounting and financial reporting within the framework of accounting policies.

Thus, the company's internal control system should change its focus towards the risk-oriented system. The basic differences between the risk-oriented system and the traditional control system are presented in Table 4.

The differences between the risk-oriented system and the traditional control system are presented in Table 4.

Table 4. The differences between the risk-oriented system and the traditional control system

Criterion	Traditional internal control system	Risk-oriented internal control system
Ultimate goal	Verification of compliance with regulatory requirements and reliability of financial reporting	Improvement of the risk management process
Stakeholders	Financial director, chief accountant	Organization owners and top executive management
Periodicity	Periodic	Continuous
Advantages	Increases confidence in the correctness of the formation of financial reporting in accordance with the developed regulations	Allows presenting data, taking into account the impact of all types of risks for assessing and planning the activities of the organization
Disadvantages	Does not always identify significant risks timely due to a focus on financial activities	Difficult to use because it is expensive and not regulated by the legal requirements

4 Results

Under current conditions, financial reporting cannot be formed without taking into account risk factors that affect the activities of any company. The process of control over the formation of financial reporting should also be risk-oriented.

Each company should develop a list of risk management tools. For this purpose, accounting data can be used. However, they are applicable for monitoring certain types of risks, for example, credit, currency, or liquidity risk. Each economic entity is exposed to a large number of risks, the impact of which is not related to the processes inside it. The used risk management tools are specific for different companies and should be disclosed in the financial reporting along with the detailed information about the identified risks.

The mechanism for generating the risk-oriented system of control over the formation of financial reporting is not regulated at the legislative level. Therefore, the use of such a system can be called voluntary. However, according to the experience, owners do not want to incur additional expenditures for the formation of an effective risk management system, therefore, the risk-oriented system of control over the formation of financial reporting is not commonly used.

Risk management and disclosure of information about them in the financial reporting is part of the company's internal control system; therefore, the internal control system should also be risk-oriented. The system of a risk-based internal control over the formation of financial reporting should be developed taking into account the principles set forth in the COSO «Guidelines for Internal Control». The effectiveness of the risk-oriented internal control system is determined by the resistance to misrepresentation of the reporting data, capable of ensuring compliance with laws and regulations and contributing to the soundness of assets of an economic entity.

5 Conclusion

Currently, there are many options for manipulating financial reporting. The risk that the company embellishes its financial indicators for the purpose of debt financing is extremely high. To minimize the risk of deliberate misrepresentation of financial reporting, a developed legal framework in the field of fraud prevention and high information transparency of companies is required. Unfortunately, many public companies do not have audit reports and notes to the reporting in the public domain, which results in distrust of the financial reporting and also indicates the lack of important information required for the reliable assessment of the company's financial standing.

The risk of deliberate misrepresentation of financial reporting arises in the case of an ineffective corporate governance system in a company since the responsibility to control the preparation of financial reporting lies within the internal control service, which is a part of the corporate governance system. Thus, the higher the quality of corporate governance, the lower the risk of financial reporting misrepresentation. The quality of corporate governance depends on the composition, structure, and responsibilities of internal services and committees as well as on the level of information transparency provided by the company. Assessing the quality of corporate governance enables indirect garnering the insights into the reliability of the prepared financial reporting.

References

1. SEC announces enforcement results for FY (2018). https://www.sec.gov/files/enforcement-annual-report-2018.pdf. Accessed 20 Jan 2020
2. SEC announces enforcement results for FY (2019). https://www.sec.gov/files/enforcement-annual-report-2019.pdf. Accessed 20 Jan 2020
3. Beneish, M.: The detection of earning manipulation. Financ. Anal. J. **55**(5), 24–36 (1999)
4. Roxas, M.: Financial statement fraud detection using ratio and digital analysis. J. Leadersh. Account. Ethics **8**(4), 56–66 (2011)
5. Kulikova, L.I., Satdarova, D.R.: Internal control and compliance-control as effective methods of management, detection and prevention of financial statement fraud. Acad. Strateg. Manag. J. **15**, 98–109 (2016)
6. Akhmetshin, E.M., et al.: Institutional analysis of the regulatory and legal framework for financial reporting control in Russia. Eur. Res. Stud. **21**, 130–141 (2018)
7. Knechel, W.R., Salterio, S.E.: Auditing: Assurance and Risk. Routledge, New York (2016)
8. Valaskova, K., Kliestik, T., Kovacova, M.: Management of financial risks in Slovak enterprises using regression analysis. Oecon. Copernic. **9**(1), 105–121 (2018)
9. Efimova, O., Rozhnova, O., Gorodetskaya, O.: XBRL as a tool for integrating financial and non-financial reporting. In: Antipova, T., Rocha, A. (eds.) Digital Science. DSIC 2019. Advances in Intelligent Systems and Computing, vol. 1114, pp. 135–147. Springer, Cham (2020)
10. Demina, I., Dombrovskaya, E.: Generating risk-based financial reporting. In: Antipova, T., Rocha, A. (eds.) Digital Science. DSIC 2019. Advances in Intelligent Systems and Computing, vol. 1114, pp. 387–402. Springer, Cham (2020)
11. Gurskaya, M., Aleinikov, D., Kuter, M.: The early practice of analytical balances formation in F. Datini's companies in Avignon. In: Antipova, T. (ed.) Integrated Science in Digital Age, ICIS 2019. LNNS, vol. 78, pp. 91–102. Springer, Cham (2020). https://doi.org/10.1007/978-3-030-22493-6_10
12. Kuter, M., Gurskaya, M., Andreenkova, A., Bagdasaryan, R.: The early practices of financial statements formation in Medieval Italy. Account. Hist. J. **44**(2), 17–25 (2017). https://doi.org/10.2308/aahj-10543
13. Kuter, M., Gurskaya, M., Bagdasaryan, R.: The structure of the trial balance. In: Antipova, T. (ed.) Integrated Science in Digital Age, ICIS 2019. LNNS, vol. 78, pp. 103–116 (2020). https://doi.org/10.1007/978-3-030-22493-6_11

Development of Accounting in Digital Economy Era

Olga A. Frolova[1] , Irina V. Milgunova[2] , Natalya P. Sidorova[1] ,
Nadezhda S. Kulkova[1] , and Elena N. Kitaeva[1(✉)]

[1] Nizhny Novgorod State Engineering and Economic University,
Knyaginino, Russia
kitaeva_72@inbox.ru
[2] Southwest State University, Kursk, Russia

Abstract. With the emergence of economic activity, a person also needed to register the facts of this side of his life. Accounting was born, which with the growth and expansion of production activity of the person gradually, due to the requirements of time, was formed in three main areas of account: accounting financial, management and tax. Accounting, as a living system, has evolved and acquired modern characteristics. At present stage, automation and digitalization are firmly embedded in our lives, changing not only our habits and preferences, but also other aspects of our lives. Accounting will not disappear, just as economic life itself will not disappear, but it will be transformed along with it, and it will continue in a different, more advanced and progressive form. Based on the analysis of the economic literature, it is concluded that the activity of accountants is reduced to the formation of accounting policies, interpretation of events, and classification of the facts of the company's economic activity. The profession of "accountant" in our usual sense will be a thing of the past, replaced by more qualified specialists with the functions of financial supervisor.

Keywords: Digital economy · Economics · Digitalization · Accounting · Systematization of accounting · Implementation of new technologies

1 Introduction

The implementation of modern information technologies is rapidly and significantly changing our lives: our everyday life, our communication, our professional environment. New opportunities and prospects are opening up for our entire society, which is the "digital economy."

The definition of digital economy is contained in Presidential Decree № 203 from 9th of May 2017 "On the Strategy for Development of Information Society in the Russian Federation on 2017–2030 years": "digital economy is an economic activity in which the key factor of production is digital data, the processing of large volumes and the use of the results of analysis of which, compared to traditional forms of economic management, significantly increase the efficiency of various types of production, technologies, equipment, storage, sale, delivery of goods and services" [1].

The government program "Digital Economy of the Russian Federation," approved by the Order of the Government of the Russian Federation from 28th of July, 2017 №

© Springer Nature Switzerland AG 2021
T. Antipova (Ed.): ICIS 2020, LNNS 136, pp. 53–59, 2021.
https://doi.org/10.1007/978-3-030-49264-9_5

1632-r, based on the decree of the President, is based on the assumption that digital economy is "a key factor of production in all spheres of social and economic activity and in which effective interaction, including cross-border, of business, scientific and educational community, state and citizens is ensured" [2].

The definition of the digital economy as an economy based on digital technologies clearly and not ambiguously indicates that the solution of the problems of digitalization of the Russian economy is essentially reduced to implementation of digital technologies in all sections of the economic activity of the country and - in regions, and in industries, and on macro-, meso-, micro-levels, and in all border sections of the domestic economic complex. Accordingly, strategic management of processes of digitalization of the economy of the Russian Federation is, first, management of processes of development of digital technologies, bringing their practical readiness to full use in a real reproduction cycle; and secondly, management the processes of "embedding" of digital technologies into real economic activity, their 100% involvement in economic life of the country [3].

Acceptance of digitalization conditions (high-speed and digital communication channels and electronic storage of information) will lead to:

- significant increase in labor productivity;
- centralized management of the economic sphere;
- global automation, standardization;
- openness of the economy to the population [4, p. 58].

2   Impact of Digitalization on Accounting in Russia

Digitalization has also affected such system as accounting, which can (in fact) be characterized as a complete microeconomic model of the digital economy: exists in the system of numbers, uses the "language of numbers," applies its own, special method of recording numbers (double recording) and so on.

Accounting, as a system, evolved together with society and economy: from rock inscriptions, records on wooden tablets, paper registers to electronic document management. The processing of digital information has also changed: sticks, abacus, counting machines, arymometers, calculators; we remember manual records in T-shaped accounts, now we are talking about application computer programs, modern IT-technology products.

The high efficiency of digital processing of accounting information is beyond doubt. At the same time, a number of questions arise about a significant change both in the Russian accounting and reporting system and in the accounting systems of other countries. There is a problem of uniformity of accounting methods of different countries, computer programs of accounting information processing, as well as integration of accounting data and reporting.

Digitalization is the introduction of transformation process that is innovation, which will change accounting; digital accounting will offer new segments of services and products.

The traditional view of accounting information is being replaced by innovative ones: online accounting, cloud storage of accounting information, reference and information systems. In turn, this will encourage global change in the companies themselves. With emergence of economic activity, a person also needed to register the facts of this side of his life.

"Accounting modeling is one of the ways of conceptual reconstruction of business life facts, business processes, business situations and business structures, value chains" [5, p. 126].

The problem of rethinking and developing the methodology for cost accounting and calculating processes and products is particularly relevant. Solving the problem is caused by the need to analyze past activities and plan for the future. "One of the most important parts of accounting system is the information component of two subsystems - financial and management accounting. There is a reorientation from control function to informative one based on the organization of points of digital transformation of the enterprise. It is necessary to develop new indicators, methods for collecting and processing not only financial information, but also the sufficiency of its integration with information about other aspects of the business and the external environment [6, p. 53].

The development of digital accounting also leads to the development of its components: it is also the introduction of additional accounts, sub-accounts, as well as changes in accounting objects. New forms of assets, liabilities, capital are emerging. Thus, there is a need for new principles of systematization of accounting, addition of criteria for reflection of economic information. The role and functionality of accounting system becomes different, as dynamically and aggressively developing multifunctional digital information systems are able to absorb it.

A striking example of the possibility of introducing new methods in solving accounting problems is mobile management accounting, in which it is necessary to reflect as traditional functions, as well as methodology, planning, modeling, analysis, forecasting. The information component of accounting comes to the fore, that is, the control function is replaced by the information function.

At the same time the profession of accountant is also being transformed. Accounting workers have always been in demand in the labour market, together with the name of the profession its functionality has changed: from a simple scribe to a bookkeeper and accountant.

In the conditions of modern digitalization of economy, the accounting service may face the problem of human capital availability and adequate knowledge of IT workers and technical aspects. It is therefore necessary to rethink the role of the accounting staff. This means understanding processes, IT-proximity, knowledge of data analysis, on which accounting and controlling staff should be focused [7].

Today qualified accountant should know not only the accounting methodology, but also various tax schemes, tax planning methods, civil and administrative legislation, as well as has practical experience as an accountant in one or more industries (production, construction, wholesale and retail trade, services, entertainment business, public catering, insurance business, etc.). An accountant must know one or more specialized accounting programs. The whole accounting department fits on the hard disk of the computer and you can manage it with just one click [8, p. 54].

The work of modern accountant has become more multidimensional and multi-disciplinary. It is obvious that it is possible to meet the needs of the modern world only by seeking new knowledge, skills to analyze and monitor changes in the external environment. An accountant who is not able to respond quickly to changes in tax, civil, and accounting legislation will not be useful to his employer.

Most states, represented by economists, representatives of IT organizations and the government, solve the task of how to convert the profession of accountant from "paper" to electronic one. This is the most important goal of the digitalization process itself [9].

3 Results

Further steps towards digitalization of accounting, and economy as a whole, will make it possible to fully automate accounting and tax accounting, which will free the accountant from some already obsolete functions. Conducting an analysis of company's financial position will remain one of the most important functions of the accountant, time for its conduct will become more, and therefore, the quality will increase.

An example of the use of digital technologies in the field of accounting is the blockchain technology.

Blockchain is a multifunctional and multilevel information technology designed to reliably account for various assets. It may be a means of registering, recording and exchanging any financial, tangible and intangible assets. In fact, blockchain is a new organizational paradigm for coordinating any type of human activity [10].

This system is endowed with the most important property from the point of view of accounting work: information created in the blockchain is reliable and protected from further changes.

Another area of application of digital technologies in accounting is introduction of software robots and artificial intelligence for automation of business processes in enterprises, which is called RBA technology [11].

As the term "distributed registry technology" implies, blockchain technology is based on the ability to create and share unique digital records without a centralized confidant party. By means of clever combination of cryptography and peer-to-peer networks, this technology guarantees the accuracy and transparency of the information stored and transmitted in the system, providing some additional advantages, such as the ability to see all previous recording states and create programmable records, so-called "smart contracts" [12].

The implementation of new technologies in the field of recording, accounting and controlling of the facts of economic life of modern enterprises, as well as promising changes in the economic and financial aspects of the activities of various companies, implies a change in the accounting system itself. Proposals are made to include perspective accounts, differential accounts, control accounts, and non-financial information accounts in the existing system of accounts along with active, passive, and active-passive accounts.

The double record is no longer exhaustive to reflect the full economic nature of accounting. We are already talking about the possibility of using triple and quadruple recording systems. Progress in technology of processing of accounting data has a significant impact on the change in the accounting objects.

New forms of automation, including robots and algorithms, in which the latest progresses in artificial intelligence have been applied, replace not only workers in production, but also accountants, lawyers and other specialists. In 2000 there were 600 stockbrokers in the New York office of finance group "Goldman Sachs". In 2017 there are only two brokers left, and the main work is performed by automatic exchange trading programs. This trend can be observed in dozens of Wall Street stock firms. Such change is likely to lead to further concentration of wealth in the hands of owners of capital and intellectual property [12].

The ideas and problems of digitalization capture various spheres of our lives. Today it is difficult to find a sector of the economy that would not be affected by automation. The most high-tech industries, such as software production and distribution, banking, service, and others, are being transformed at the fastest pace. In other industries, such as agriculture, digitalization is slower, but there are also significant shifts. The process of digitalization is aimed at maximizing the profitability of production, as the introduction of new technologies allows to reduce the cost.

The problem of digitalization is gaining momentum, the relevance is due to a large number of authors who study this issue. We can note the works of Shamina O. V. [14], Frolova O. A. [15], Zubenko E. N. [16], Makarychev V. [17].

The use of blockchain technology may be relevant to public sector of economy, for example in state audits conducted by state auditors. It is necessary to emphasize research of Antipova T. in this field, who highlights the most relevant property of blockchain technology, as an ideal way to prevent fraud with budget money [18].

4 Conclusion

Based on the results of the analysis of the economic literature and legal acts, the authors of this article have formed the opinion that the activities of accountants in modern conditions of the digital economy should cover the following functions:

- professional judgment on issues of economy, accounting, legal and social aspects of society activity;
- work on the formation of not only accounting, tax, but more management reports;
- analysis and expert opinion on issues of business development, as well as potential results of certain management decisions;
- analysis of current and future activities of the enterprise;
- formation and presentation of any type of reporting in electronic form;
- use of the most advanced domestic and foreign information technologies;
- search for new theoretical aspects, development of methods for developing and improving accounting.

Thus, in our opinion, the activity of accountants is reduced to the formation of accounting policies, interpretation of events, and classification of the facts of the company's economic activity. These processes will be carried out through professional judgment by specialists who understand the algorithm and economic content of the essence of modern data processing methods. The profession of "accountant" in our usual sense will be a thing of the past, replaced by more qualified specialists with the functions of financial supervisors. Such specialists can recommend strategic solutions for business. A person of a new profession needs to combine the functions of financier, accountant, analyst and auditor.

So, in response to the new needs of the digital economy, accounting is being transformed, and the digitalization of accounting is a necessary step in its development. The era of universal digitalization leads to changes (i.e. improvements) in the accounting system and methods, the relevance of which will not be lost.

The future of the profession we see in the transformation of accounting workers from low levels of qualification to higher ones. This itself is already a positive trend not only for the preservation, but also for the development of the profession.

The latest technologies are being created to solve emerging human problems related to the need for evolution. But technology itself can exacerbate some of the problems and even create new ones. Along with the advantages of the digital economy, there may be problems quite trivial. These are high costs that are inevitable in the creation of the latest communication systems.

The competitiveness of the Russian economy largely depends on the transition to digital format. We believe that the technologies of the future should only expand human capabilities, not define them.

References

1. Presidential Decree № 203 from 9th of May 2017 On the Strategy for Development of Information Society in the Russian Federation on 2017–2030 years. http://www.consultant.ru/document/cons_doc_LAW_216363/
2. Order of the Government of the Russian Federation from 28th of July, 2017 № 1632-r "About the approval of program "Digital Economy of the Russian Federation"". http://www.consultant.ru/document/cons_doc_LAW_221756/
3. Yakutin, Yu.V.: Russian economy: digital transformation strategy (to constructive criticism of the government program "Digital Economy of the Russian Federation"). Management and business administration № 4, pp. 27–52 (2017). https://lomonosov-msu.ru/archive/Lomonosov_2019/data/16731/93175_uid107396_report.pdf
4. Pchelinceva, M.R.: Harmonization of accounting and environmental reporting in the context of digitalization. Cent. Sci. Bull. 9S(50S), 58–59 (2018). https://elibrary.ru/item.asp?id=34980911
5. Eremenko, V.A., Stepanova, R.M., Alpatieva, S.R.: Application of modeling in accounting. Collection: Construction. Architecture. Economy. Materials of the International Forum "Victory May 1945": collection of articles. Ministry of Education and Science of the Russian Federation, Don State Technical University, Trade Union of Workers of National Education and Science of the Russian Federation, pp. 126–130 (2018). https://elibrary.ru/item.asp?id=34996909
6. Karpova, T.P.: Directions of accounting development in the digital economy. News St. Petersburg State Econ. Univ. 3(111), 52–57 (2018). https://elibrary.ru/item.asp?id=35085010

7. Debus, C., Schilling, R.: Digitalization: Accounting vs. Treasury. KPMG Corporate Treasury News, 93rd edn. (2019). https://home.kpmg/de/en/home/insights/2019/08/digitaliza tion-accounting-treasury.html
8. Korzhova, O.V., Filimonov, A.A.: Digital economy in accounting. Materials of XI international student scientific conference "Student Scientific Forum 2019". Sci. Rev. **4**, 53–55 (2019). https://scienceforum.ru/2019/article/2018015461
9. Minkovskaya, A.: Digitalization of accounting and reporting, and the possibility of its development in the Republic of Belarus. Phys. Math. Educ. **4**(18), 112–114 (2018). https://doi.org/10.31110/2413-1571-2018-018-4-018
10. Svon, M.: Blockchain: Scheme of New Economy (English translation). Publishing House "Olympus – Business", Moscow (2017)
11. Arears of applications of blockchain. https://mycryptocurrency24.com. Accessed 06 Mar 2019
12. Shvab, K.: Technologies of the Fourth Industrial Revolution. Eksmo, Moscow (2018). (English translation)
13. Decree of the President of the Russian Federation from 1st of December 2016 № 642 "On the Strategy of Scientific and Technological Development of the Russian Federation". http://www.consultant.ru/document/cons_doc_LAW_207967/
14. Shamina, O.V.: Statistical assessment of informatization of economy of Nizhny Novgorod region. Bull. NGIEI **4**(71), 93–100 (2017). https://elibrary.ru/item.asp?id=29028534
15. Frolova, O.A., Shamin, A.E., Shavandina, I.V., Kutaeva, T.N.: Smart village. Problems and prospects in Russia. In: Digital Science. Scopus Database, pp. 400–486 (2019). https://doi.org/10.1007/978-3-030-37737-3_4
16. Mizikovsky, E.A., Zubenko, E.N., Sysoeva, Yu.Yu., Ilicheva, O.V., Frolova, O.A.: Production accounting system at breweries. In: Antipova, T. (eds.) Integrated Science in Digital Age, ICIS 2019. Lecture Notes in Networks and Systems, vol 78. Springer (2020). https://doi.org/10.1007/1-4020-0612-8_324
17. Frolova, O., Makarychev, V., Yashkova, N., Kornilova, L., Akimov, A.: Digital transformation of agro-industrial complex. In: 5th International Conference on Agricultural and Biological Sciences (ABS) WGD Conference Series: Earth and Environment Science, vol. 346 (2019). https://doi.org/10.1088/1755-1315/346/1/012029
18. Antipova, T.: Using blockchain technology for government auditing. In: 13th Iberian Conference on Information Systems and Technologies (CISTI), Caceres, Spain, 13–16 June IEEE, Portugal Sect (2018)

Beneish Model as a Tool for Reporting Quality Estimation: Empirical Evidence

Maksim Volkov[(✉)]

Shevchenko st. 8, r. 6, 300026 Tula, Russia

Abstract. In this research, a number of key approaches to fraud identification has been reviewed. A need for a quantitative model of reporting quality estimation, in particular from investment decision analysis standpoint, has been discussed. A Beneish model – a quantitative model of estimation the probability of financial result manipulation – has been tested on a wide scope of international companies from metals and mining sector in order to test the empirical efficiency of such model and analyze the most important factors that contributes to the reporting quality. Each factor of the model has been separately analyzed in order to outline the tendencies and possible reasons for financial result manipulation.

Keywords: Beneish model · Reporting quality · Financial results manipulation · Financial reporting manipulation

1 Introduction

Development of modern technologies has led to the increase in popularity of quantitative methods for financial analysis purposes. Nowadays, around 93% of data is stored digitally [1], which largely contribute to such popularity. Investors, analysts and researchers utilize quantitative tools not only for common financial ratio analysis and modelling, but also for qualitative means, such as evaluation of reporting quality. Due to insufficient amount and poor quality of data presented in annual reports, financial analysis quite often results in questionable conclusions, which in turn has led to the fact that more than 30% of M&A deals result in value destruction [10]. Usually, low quality accounting data can be characterized as insufficient (in terms of amount of data presented), of low reliability or of low value for the decision-making process (irrelevant). Obviously, those attributes are interrelated. It is usually up to the expert judgement of an analyst to decide, whether the data is of sufficient quality and whether it can be used for financial analysis or not.

Even though it is possible to decide, if the amount of data is sufficient or if it is relevant for financial analysis, it is usually impossible to judge on the reliability of the data presented in financial reports. This is why due diligence and audit procedures are suggested almost for any deal or decision related to investment or financing. By executing financial, operational, technical and legal due diligence procedures one can verify the reliability of information presented and hence make decisions that are more appropriate.

© Springer Nature Switzerland AG 2021
T. Antipova (Ed.): ICIS 2020, LNNS 136, pp. 60–68, 2021.
https://doi.org/10.1007/978-3-030-49264-9_6

In cases, when an entity is trying to raise additional financing, it might have an intention to utilize what is called a framing bias: trying to play the way the data is presented in order to make its asset look more favorable for investors or creditors. Usually the amount of data that is required to be disclosed in financial statements is not sufficient for decision making; hence, the manner, the amount and the structure of additional information disclosed is usually not regulated, which makes such behavior from entities possible, leading to no legal consequences. Also, the numbers in financial reports can also be legally played by utilizing creative accounting techniques: capitalizing expenses, forming reserves when it is needed etc. However, entities can also misrepresent information by disclosing false data, which in many jurisdictions can be viewed as crime.

Apart from investment or financing decisions, such data might be presented for other purposes: for management remuneration calculations, for tax reporting, for providing data to workers and public entities etc. However, from the author's point of view, misrepresentation of data reflected in financial statements has the most crucial consequences as it violates the very foundations of financial statements: to give investors sufficient and appropriate data to make an investment decisions.

The most popular indicators for misrepresentation are therefore profit measures. Low quality of profits is usually associated with aggressively recording profits and conservatively recording losses. This policy usually leads to a large gap between cash flow and profits, which can be viewed as a sign of poor reporting quality by some investment specialists.

Sources of income also influence the quality of profits: for instance, income, generated by selling fixed assets, or profits, recorded due to reserve being written off, - any one-off operation would lead to low quality of earnings. These techniques could also be used to perform creative accounting, adjusting figures to the level required. And as it was described before, such techniques do not violate reporting standards – either GAAP or IFRS.

Due diligence procedures certainly lead to higher reliability of data presented which would contribute to efficiency increase of any financial decision in general. Nowadays well-established risk-oriented approach to due-diligence procedure [2] can be viewed as a necessary tool for fundamental value estimation. However, such procedures might be quite expensive, especially when it comes to deals and decisions of smaller size. Quite often, people seek due diligence procedures that are cheaper, rather than of higher quality. It is also quite costly in terms of time to perform a high-quality due diligence. Considering these, the quality of data presented, including the quality of financial reporting, should also be considered as a factor of value creation for decision-making process.

There is a number of fundamental researches that focus on fraud identification and explanation. One of those is Financial Shenanigans by Schilit and Perler [11]. In their research, seven financial "sins" related to earnings manipulations has been outlined [11]. In general, those "sins" are related to two areas: timing or size of income, expense or liability recognition.

Another approach is also based on warning signals of financial statements, which are often referred to as "red flags". The red flags are usually related to unusual increase of decrease in profitability, cash flow, assets, liabilities, equity accounts etc. [4] Being more empirical and practice-oriented, this approach is quite similar and intuitive, yet there is a lack of clear quantitative rules on what can be viewed as a "red flag".

A need for a tool that can be used for preliminary brief evaluation of reported earnings quality on one hand that is also quite fast and not overcomplicated on the other hand has stimulated the development of a number of quantitative methods that allow for probability estimation of financial reporting manipulation. One example of such a model is Beneish Model, that allow to quantify the probability of financial result manipulation only by using public financial reports. Also, by using this model, it is possible not only to calculate the probability of financial reporting manipulation, but also to decompose it by factors and perform analysis of factors that contribute such probability. This functional, in turn, makes it possible to perform benchmarking, relative peer-to-peer analysis, dynamics evaluation of each factor that contributes to financial reporting quality, - in other words, it allows for more in-depth analysis in order to provide relevant information for efficient financial decision-making.

The goal of current research is to test Beneish model for relative analysis purposes on Metals and Mining sector companies, as well as to identify the most important factors that contribute the reporting quality. Also, it should be stated that the accuracy of model presented, as well as its suitability for empirical analysis of reporting quality has not been tested in current research.

2   Materials and Methods

Beneish model has been developed in 1999 by professor Messod Beneish from the University of Indiana in his research «The Detection of earnings manipulation» [8]. Professor Beneish has used observations from 1982 to 1992 for model construction, sourcing the data from Standard&Poors Compustat database.

The model is represented by a probit model, with a Z-statistics as an output (or, according to professor Beneish, so-called M-statistic, - an acronym for manipulation). The result of the model makes sense regardless of its sign or absolute value.

$$M = -4,84 + 0,92 * DSRI + 0,528 * GMI + 0,404 * AQI + 0,892 * SGI + 0,115$$
$$* DEPI - 0,172 * SGAI - 0,327 * LVGI + 4,679 * TATA \qquad (1)$$

This model includes eight exogenous variables, each of which can be taken or can be calculated from the financial reports, which contributes significantly for the model suitability. Each variable and the algorithm for its calculation is shown at Table 1.

Table 1. Exogenous variables of Beneish model and its formulas.

Variable	Acronym	Calculation formula
Days' sales in a receivable index	DSRI	$(\text{Account Receivables}/\text{Sales}_t)/(\text{Account Receivables}_{t-1}/\text{Sales}_{t-1})$
Gross margin index	GMI	$[(\text{Sales}_{t-1} - \text{COGS}_{t-1})/\text{Sales}_{t-1}]/[(\text{Sales}_t - \text{COGS}_t)/\text{Sales}_t]$
Asset quality index	AQI	$[1 - (\text{Current Assets}_t + \text{PP\&E}_t + \text{Investments}_t)/\text{Total Assets}_t]/[1 - ((\text{Current Assets}_{t-1} + \text{PP\&E}_{t-1} + \text{Investments}_{t-1})/\text{Total Assets}_{t-1})]$
Sales growth index	SGI	$\text{Sales}_t/\text{Sales}_{t-1}$
Depreciation index	DEPI	$(\text{Depreciation and Amortisation}_{t-1}/(\text{PP\&E}_{t-1} + \text{D\&A}_{t-1}))/(\text{D\&A}_t/(\text{PP\&E}_t + \text{D\&A}_t))$
SG&A Expenses Index	SGAI	$(\text{SG\&A}_t/\text{Sales}_t)/(\text{SG\&A}_{t-1}/\text{Sales}_{t-1})$
Leverage index	LVGI	$[(\text{Short-term Debt}_t + \text{Long-term Debt}_t)/\text{Total Assets}_t]/[(\text{Short-term debt}_{t-1} + \text{Long-term debt}_{t-1})/\text{Total Assets}_{t-1}]$
Total accruals to total assets	TATA	$(\text{Income from main activity}_t - \text{Cash inflow from main activity}_t)/\text{Total assets}_t$

Each variable presented also has a fundamental value, apart from being used for the purposes of current model, as each variable can be used for analysis purposes separately.

According to professor Beneish, a decrease in Account receivables turnover may occur due to the increase in competitive environment and, therefore, better payment terms provision for customers. However, in most cases an increase is Account Receivables growth relative to Sales growth can clearly be viewed as a sign of revenue manipulation, which is why DSRI indicator has been included in the model.

Negative gross profit margin dynamics, according to the research, can be viewed as a sign of decreased efficiency of company's activities. In case of gross margin decrease, there is a motivation for the management to manipulate the financial result, which may lead to aggressive revenue recognition.

In current model, by asset quality the author considers the relation of tangible fixed assets to total assets. An increase in such indicator may be the sign of company's intention to capitalize expenses, which might also influence the quality of financial results.

An increase or decrease in sales itself is not a sign of manipulation, however according to professor Beneish, many professionals believe that the risk of financial reporting manipulation is higher among high-growth companies. The reason is, companies are trying to maintain high growth rate, which is why they are trying to aggressively record additional profits. Therefore, there is a clear motivation for financial result manipulation for the company.

Decrease in D&A growth rate may be the consequence of more conservative costs recognition, by reviewing the useful life or depreciation policies. It may lead to the decrease in profit quality.

A decrease in SG&A share to total revenue may be the sign of more aggressive revenue recognition or more conservative costs recognition, which also influence the profit quality.

All else being equal, the more debt the company has, the more motivation there is to manipulate profit figures. This justify the inclusion of LVGI index in the model.

Total accruals to total assets index represent the quality of recognized revenue by subtracting actual cash inflow from accrued income. The larger the share of such an accrual to overall assets, the higher the probability of financial results manipulation.

According to the research, DSRI, GMI, AQI, SGI and TATA indicators are statistically significant. Hence, there is a variation of the Beneish model that excludes SGI, DEPI and TATA metrics. In this research, the basic model has been used, with results verified by using the simplified version.

One of the features of the model is that the probability of financial result manipulation increases as exogenous variables increase, and at some point, the probability growth decreases, which represents approximation to 100% probability. This justifies the output of the model as an M-Score, a measure of cumulative distribution.

There is an issue with identifying the critical value for M-Statistic comparison. In general, there are two types of errors: the first one is to classify the company as a company that doesn't manipulate financial results, when, in fact, it does; and the second – to classify the company as a financial result manipulator when it is not true. The first type of error might lead to up to 40% losses of capitalization of portfolio; the second one might lead to hidden losses in a form of lost opportunities. By balancing between those two types of errors, professor Beneish outlines an optimal M-Score being equal to −1,78, which corresponds to 4% probability of type 1 error. However, in current research more conservative critical value of M-Score has been considered, which equals to −2,22, - corresponding to 2% probability of type 1 error.

Despite the fact that the accuracy and suitability of the model is not the subject of current research, the main assumptions of the model should be considered in order to perform a proper analysis of results acquired:

1. **Model has been constructed as a probit model.** As well as Altman's Z-score model that is used for bankruptcy probability identification, the model is a probit model. The model might produce results of highest accuracy on a sample that has been used for its construction, but its accuracy might be flawed when used on other samples. That is why it is crucial to consider the sector, business-model, geography and other fundamental factors when using the model.
2. **Decrease in accuracy due to the model being outdated.** As any model that has been constructed using statistical methods, without its coefficients being updated on a regular basis, its accuracy might be flawed.
3. **Significance level.** According to professor Beneish, there is a known minimal probability of error that depends on a predetermined significance level.
4. **Preliminary nature of the results.** As it was stated before, for proper verification of financial reporting reliability, a due diligence procedures are required. Current model can be viewed as a tool for preliminary reporting quality evaluation for large samples and can't be considered to be used as an alternative to proper due diligence.

Current research, apart from assumptions mentioned, also assumes a number of assumptions related to the sample of observations, as well as to the critical values of M-Statistics. It is also suggested that even though the M-Score for a certain company might be higher than the corresponding critical value, it can't be an evidence of company manipulating its financial reporting. Rather, it would mean that this company has higher chances on doing so.

The sample for current research includes 434 metals and mining companies from 41 countries; the average market capitalization of those is approximately USD 2 bln. The Bloomberg Database has been used as a source for observation, with the data taken as of 31.12.2016. In addition, Chinese companies (158 entities) represent the highest share of companies in the sample. Russian companies are represented by Poly-metal PLC., EVRAZ PLC., Severstal JSC., MMK JSC., Rusal JSC., VSMPO-AVISMA JSC., Raspadskaya JSC., Mechel JSC.

The results of the model are shown at Fig. 1.

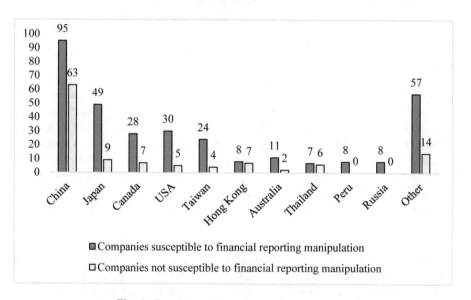

Fig. 1. Beneish model testing results by country.

As a result, 117 companies that could be susceptible to financial reporting manipulation has been identified. It should be noted that the majority of companies that are susceptible to financial reporting manipulation – both in absolute and relative terms – are the companies from China (63 entities). None of the Russian companies from the sample tested is susceptible to financial reporting manipulation. However, there is a −2,2205 value of M-Score for Polymetal PLC., which can be considered as a boarder value when the critical M-Score is −2,22.

In most cases, for companies susceptible for financial reporting manipulation, a DSRI coefficient is usually higher than the same coefficient for non-susceptible com-panies. Mostly it was due to the higher increase in account receivables with relatively stable growth of revenue. This can't be viewed as a positive sign, as an increase in revenue is outset by increase in working capital.

There is also a substantial increase in TATA indicators for companies that a susceptible in financial reporting manipulation. The same indicator is negative for non-susceptible companies, which means that the accrued income is in fact lower that the operating cash inflow.

SG&A expenses are higher for non-susceptible companies. On one hand, this might be the sign of more motivated and efficient management, or more reliable internal control systems. On the other hand, insufficient amount of management remuneration might lead to motivation for higher financial result recognition in order to receive additional bonuses. It is also worth mentioning that the majority of companies that was included in the sample are Asian companies, with its specifics and management quality.

The majority of companies susceptible to financial reporting manipulation are the companies operating in countries with high competitiveness level, which conforms the overall tendency and is theoretically and practically sound.

The detailed results for Russian companies are shown in Table 2.

Table 2. Beneish model output results for Russian M&M companies.

Company	DSRI	GMI	AQI	SGI	DEPI	SGAI	LVGI	TATA	M-score
Polymetal PLC.	2,88	0,60	1,08	0,99	0,89	0,78	1,52	−0,24	−2,22
MMK JSC.	1,06	0,68	1,19	0,97	0,79	0,83	1,22	−0,17	−3,39
Rusal JSC.	1,15	0,60	0,84	0,96	1,07	0,93	1,23	−0,12	−3,27
VSMPO-AVISMA JSC.	1,00	0,91	0,86	1,01	0,79	1,16	1,30	−0,15	−3,44
EVRAZ PLC.	0,81	0,79	1,20	0,91	0,84	0,92	1,19	−0,28	−4,14
Mechel JSC.	0,70	0,84	1,20	0,75	0,69	1,02	1,63	−0,54	−5,76
Raspadskaya JSC.	0,69	1,37	1,86	0,81	0,71	0,70	1,71	−0,44	−4,67
Severstal JSC.	0,55	0,87	0,79	0,88	0,48	0,87	1,26	−0,38	−5,08

The profile of Russian companies matches the average profile of companies that are not susceptible to financial reporting manipulation, which is a good sign in terms of overall investment attractiveness of Russian companies. In terms of critical values, for Polymetal PLC. there is a high share of DSRI value, which was due to the high growth of account receivables in relation to revenue growth. The account receivables has increased from USD 12,25 mln to USD 34,95 mln. There is no explanations on this matter in the annual report.

It is also worth mentioning that for Polymetal PLC. as well as for other Russian companies cash inflow from operating activities is higher than the revenue recorded, which is a good sign for investors in terms of revenue quality. However, this leads to another additional matter of liquidity management efficiency.

In current research, a hypothesis of whereas M-score is connected to the overall return of a company for the next two periods has also been tested. According to the results acquired, there is no statistical evidence of such correlation. However, with multiple regression being tested, where separate indicators of Beneish model has been used as exogenous variables and total return per share as endogenous variable, there was a clear indication that GMI, LVGI and TATA indicators significantly contribute to overall return forecasting.

3 Conclusion

We can conclude, that M-statistic can be viewed as an efficient tool for preliminary financial reporting quality evaluation, especially when it comes to large samples of companies. However, it should not be treated as a substitute to common due diligence procedures.

As a result of model testing, the following thesis has been proved:

- The main indicator that influences the financial reporting quality would be the relation of Account receivables growth to overall revenue growth – DSRI metric of Beneish model;
- Total return to shareholders and financial reporting quality are not necessarily related in a short-term perspective;
- The competitiveness level is directly related to the provability of financial reporting figures manipulation;
- SG&A expenses in relation to sales are inversely related to the probability of financial results manipulation.

However, it is strongly suggested that the thesis mentioned should be tested on different samples of companies from different regions and industries.

The results acquired can be used for preliminary evaluation of reporting quality and hypothesis construction for further analysis. Companies itself might find it useful to utilize this model in order to adjust the list of indicators and the nature of data disclosed in financial reporting. This would contribute to the overall increase in the quality of information disclosed and decrease the degree of information asymmetry.

The model tested is quite popular among researches and analysts all over the world. It is obvious that outcomes obtained through current research, may be used as an input for investment analysis of companies. It is also suggested to update the coefficients and assumptions according to the research goals in order to make current model suitable for practical purposes. Lastly, the model can and should be further improved by assuming non-linear functional forms, stochastic nature of financial reporting manipulation and other quantitative matters that can be accounted for by using modern technologies of financial modeling and analysis.

References

1. Aris, N.A., Othman, R., Arif, S.M.M., Malek, M.A., Omar, N.: Fraud Detection: Benford's Law vs Beneish Model (2013)
2. Antonova, N.: Due diligence planning as technology for business risk assessment. In: ICIS 2019, LNNS, vol. 78, pp. 60–71 (2019)

3. Barilenko, V.I.: Complex Analysis of Business Activities: Textbook and Practice Book. URAIT, Moscow (2019)
4. Buljubasic, E., Halilbegovic, S.: Detection of financial statement fraud using Beneish model (2017)
5. CFA®: Program Curriculum. Level 2 Volume 2 Financial reporting and analysis. CFA ® Institute, p. 295 (2019)
6. Efimova, O.V.: Analytical aspects of information disclosure in notes to financial statements. In: Auditorskiye Vedomosti, vol. 3, pp. 38–50. Unity-Dana, Moscow (2015)
7. Beneish, M.: Earnings Quality and Future Returns: The Relation between Accruals and the Probability of Earnings Manipulation, May 2005. 60 p.
8. Messod Beneish the Detection of Earnings Manipulation, June 1999. 21 p.
9. Roshchektaev, S.A., Roshchektaeva, U.Y.: Detection of fraud in financial statements: the model M. Beneisch. In: Nauchniy Vestnik Yuzhnogo Instituta Menedzhmenta, vol. 2, pp. 37–43. UIM, Krasnodar (2018)
10. Sharkov, D.A.: Due diligence: fundamental procedure for M&A deal structuring. In: Problemy sovremennoy Economiki, vol. 1, no. 53, pp. 123–125. NPK Rost, Saint-Petersburg (2015)
11. Schilit, M.H., Perler, J.: Financial Shenanigans: How to Detect Accounting Gimmicks & Fraud in Financial Reports. McGraw Hill (2010)

Peculiarities of the Capital Account Formation in the Period of Reorientation of Accounting to Double-Entry Method

Mikhail Kuter⊙, Marina Gurskaya(✉)⊙, and Dmitry Aleinikov⊙

Economy Department, Kuban State University,
Stavropolskaya st., 149, 350040 Krasnodar, Russia
marinagurskaya@mail.ru

Abstract. This study allows tracing the procedure of capital account formation on which the corpo and sovracorpo of Francesco di Marco Datini's company in Avignon are reflected during the reorientation period from the single-entry method and mingled accounts to the double-entry method and bilaterian account (1398–1401). Examples are given of the reflection of the capital account both in the form of a paragraph and its continuation when the balance is carried overusing the double-entry method. An example of assigning the indicator to the Trial Balance prepared in the Secret Book due to changes of the company's partners is shown. The study is based on archival sources preserved in the Fondo Datini of the State Archive of Prato (Italy).

Keywords: Corpo and sovracorpo · Medieval accounting · Digital technology in accounting history

1 Introduction

Francesco di Marco Datini di Prato (1335–1410) was a famous merchant, whose commercial activities are carried out in the many places of Medieval Europe. But not only successful trading promoted to save memory about him and his sole proprietorships and companies. Thanks to thorough reflected every transaction in the accounting books and different Ricordanze and Quaderni, thanks to extensive and regular corresponding, thanks to a lot of saved documents (bills of exchange, insurance policies, business letters, etc.) scholars of different areas of sciences can make their research by using the real documents of XIV–XV centuries for that.

Avignon in France became the beginning of the long-term path for Francesco di Marco, who was born in Prato (approximal 20 km from Florence), where he moved when he was fifteen. Avignon was at that time the residence of Popes ("Avignon Captivity" or "Avignon Papacy" 1309–1378) and one of the most important trading-cities of Europe, it attracted a lot of foreigners, who strived to a new and better life. Datini had to learn a lot of things before he started his own business, which in time, turned out in the vast trade network is included branches and companies around Europe.

During the period 1350 to 1363, we have not enough information about what Datini did at that time in Avignon [1, p. 135]. The first his company was established in 1363 with Nicola di Bernando as a partner.

T. Antipova (Ed.): ICIS 2020, LNNS 136, pp. 69–80, 2021.
https://doi.org/10.1007/978-3-030-49264-9_7

This study is focused on the more later period of activity of this company. The accounting books of that period are survived almost well, they allow to study the features of the commercial life of the region, its requirements and trading connecting. In the survived accounting books the researchers can find the information about the features of medieval accounting, methodic and practice, some of which we use currently.

The main purpose of this study is to trace the procedure and identify peculiarities of the capital account formation in the period of reorientation of accounting to double-entry method.

2 Research Method

The principal research method adopted in this study is archival. It uses material found in the State Archive of Prato. This research team has been working with the material in this archive for the past 14 yearr and many of the records have been recorded and linked together using logical-analytical modelling. This is an approach that we developed for the purpose of enabling entries in the account books to be traced visually between accounts and books and from page to page. By adopting this approach, we are able to see the entire accounting system electronically, making entries and their sources clear in a way that is not possible if all that you have is the original set of account books. This enables us to consider each transaction in detail, trace its classification, and so explain the bookkeeping and accounting methods adopted without misinterpretation. This approach allows for a new analysis and interpretation of accounting practices for periods when there was no concept of a standard method or a single approach to accounting or financial reporting. In this article for periods when there was no concept of either a standard method or a unified approach to either financial recording or financial reporting. In this paper, we present the bookkeeping method adopted by the accountant in Avignon. As far as we are aware, the entries in the account books included in this study have never previously been analyzed.

3 A Brief History of Research on the Datini's Sole Proprietorships and Companies and Their Accounting System

The most comprehensive describing of Datini companies' activity give, of course, Italian scholars. Primarily should mention the contribution of Federigo Melis – a founder and director of International Institute of Economic History "F. Datini", who presented to contemporary researchers in detail with a commercial activity of Medieval Merchant. In his publications [1, 2] described carefully the structure and organization of the activities of sole proprietorships and companies of Francesco Datini, and also, he did a huge job to inventory and saving of survived accounting books and documents.

But the Datini's Archive was mentioned by historians of economics and accounting before Federigo Melis. The first who studied documents of Francesco Datini and who mentioned the importance of them, were two citizens of Prato: Marino Benelli and Cesare Guasti. According Melis [2, p. 89] Marino Benelli was the first who started to sort and to list survived documents and Cesare Guasti, who was a director of the State Archives and Superintendent of the archives of Tuscany, he studied some of them [3].

But the first to publish the results of a study describing the life, business, and documents of Francesco Datini was Isidoro del Lungo [4] and Giovanni Livi [5].

Fabio Besta [6], contribution of who in development of accounting history, was described in detail by Massimo Sargiacomo, Stefania Servalli and Paolo Andrei [7] was the first who mentioned accounting books of Datini's companies in the accounting history literature [6, pp. 317–320].

In 1914, Sebastiano Nicastro [8] published an inventory of saved documents in the Datini archive, compiled by Giovani Livi. In 1922, Gaetano Corsani [9] published a work on the history of accounting, where he mentioned accounting books stored in the Archive.

In 1928, Enrico Bensa [10] published the results of his research on documents stored in the Datini archive. The book contained many photographs and descriptions of documents that reveal the features of medieval business in Tuscany. The book has been used as a reference tool by many historians of Economics and accounting during the 20th century.

Of course, it is necessary to note the most profound research conducted in the second half of the 20th century by T. Zerbi [11] and A. Martinelli [12].

Iris Origo, in 1957 presented her book "The Merchant of Prato" [13] which provided a surprisingly complete study of the life and time of an Italian entrepreneur. This book looks at more than just Datini as a merchant and manufacturer but also tells about his everyday environment.

But not only Italian researchers were interested in Francesco Datini, but Raymond de Roover also wrote a lot about him [14].

If we talk about modern authors, first, we should mention the team of the International Institute of Economic History "F. Datini" under the direction of Giampiero Nigro. The published monograph [15] allows us to consider the features of the life and character of Francesco Datini, his relationship with his family, the development and individual aspects of his business, the characteristics of creating and operating in sole proprietorships and companies.

It is worth noting the Chapter in [15] dedicated to the company in Avignon since the activity of this company is the main interest of this study.

An equally exciting and essential source for archival research is the Inventory of Cecchi, Elena, published in 2004. This book entirely dedicated to the company of Francesco Datini in Avignon [16].

Many authors continue to write about the Datini archives today, about the history of the archive, its contents, Datini's personality, and his life. However, insufficient research has been devoted to the development of accounting practices in enterprises and companies Francesco Datini.

4 Organization of Activity in the Company in Avignon

Activity of the sole proprietorship and companies of Francesco Datini in Avignon can be divided into periods: before returning to Italy and after that returning (1363–1382; 1382–1410); by organizational forms of business entities (companies – 1363–1373, Datini's proprietorships – 1373–1382, companies – 1382–1410); accounting methods (on the mingled accounts – 1363–1398, on the bilaterian accounts – 1398–1410) [17].

The company was reorganized in 1382 when Datini left Avignon and returned to Prato. Until 1398, the company's partners included Francesco di Marco Datini da

Prato, Boninsegna di Matteo, and Tieri Benci. Boninsegna di Matteo, who was responsible for accounting, kept the accounts "in a (mingled) column." According to the partner's contract, one of the main tasks for him was to make the audit and to prepare a detailed analytical report on each item once a year, most of which, as we know, was conducted by inventory. In the end, he did his job well.

On April 30, 1392, Tommaso di ser Giovanni from Lucca joined the company as a Manager. It seems that Tommaso di ser Giovanni got good practice at Avignon, including in accounting. In 1396, Datini directed him to strengthen the company in Milan.

Boninsegna di Matteo died on 25 December 1397. Datini immediately returned Tommaso di ser Giovanni to Avignon. The company needed a person who could keep accounts.

Francesco Datini made a decision that sets Tommaso di ser Giovanni on probation. As a result, from January 1, 1398, to October 31, 1401, the composition of the partners remained unchanged, including the deceased Boninsegna di Matteo. And only on November 1, 1401, Tommaso di ser Giovanni di Lucca was included in the company's partners.

From the first day you joined the company (January 1, 1398) Tommaso di ser Giovanni, who is entrusted with accounting functions, begins to reorient accounting in the company in Avignon to double entry. The subject of our research is the accounting system of Datini's company in Avignon during the period 1398–1401. Let's focus on the accounting books that are preserved and stored in the archives of Francesco Datini in the State archive in Prato and relate to the analyzed period.

First, we should mention Libro Segreto (Prato, AS. D. No. 161), which contains information for the period 1398–1404. Libro Segreto is a "secret book" (record book or private ledger containing the partnership agreement, the shares of capital contributed by each partner, the final balance, and the division of profits and all assets at the dissolution of the partnership) [18]. The content of this book is surprising. Information for 1398–1401 (October 31) is located until page 86r. data for the end of 1401 is placed until page 114r. records for 1401 (November 1) – 1404 are located from page 131r. Then the recordings are interrupted, and from page 133v, quite unexpectedly, the second half of 1410 is recorded.

It is also necessary to mention Libro Segreto (Prato, AS. D. No. 162), which contains information for the period 1401–1408.

Cash registers (Entrata e Uscita) have been preserved from this period:

- Prato, AS. D. № 126–including entries for 1396–1398;
- Prato, AS. D. № 127 – including entries for 1399;
- Prato, AS. D. № 128 – including entries for 1400–1401;
- Prato, AS. D. № 129 – including entries for 1401–1404;
- Prato, AS. D. № 130 including entries for 1408–1412.

As for Memoriale, they are Archived under the numbers Prato, AS. D. No. 82–87, and cover the period from 1398 to 1407[1].

---

[1] http://datini.archiviodistato.prato.it/.

5 Corpo, Sovracorpo and Their Accounting

From the beginning of the thirteenth century, a new model of Business Society was appeared in the Europe: the *company (compagnia)* or *partnerships*. The emergence of companies was based on an earlier partner practice – *commenda*. Commenda was a so-called contract with a sleeping partner, which provided some capital, but did not participate in the activity personally. In over time, it has evolved into a sit-down organization based in an established location, managed by partners, employees, or agents abroad.

The people united in Compagnia drafted a charter according to which they agreed to invest capital in a commercial enterprise, which was usually to last from three to five years. The partnership contract specified each partner's contribution to capital and how their share of profits was to be determined; they specified the name of the firm, its duration, its place of business, and its activities in General terms; they designated the partner with managerial authority; they obligated the partners not to engage in any other business during the partnership period, and they prohibited the withdrawal of capital from the partnership until the partnership was terminated.

The capital that partners were to invest in the business was primarily authorized capital for initial commercial or financial investments and to provide liquidity for subsequent transactions that may take a long time [19, p. 65].

At the Medieval time, in addition to the authorized capital, which consists of amounts awarded by members (*corpo di compagnia*), additional costs paid by the participants themselves or from third parties for investment, funds called "*sov-racorpo*" or "*sopraccorpo*" or "*fuori corpo*" flocked to the company – money loaned to a partnership by a partner in addition to his share of the capital, upon which interest was paid. Most often, the company's capital was only a small percentage compared to the share of funds invested as excess capital, which could even exceed in several times the corporate equity.

The authorized capital of the company, established in Avignon in 1383 and replaced the sole proprietorship of Datini, was 3.000 florins. It needs to be mentioned that the total amount of the capital was f. 3866 s. 0 d. 8 and included f. 866 s. 0 d. 8 as "sovracorpo". Boninsegna di Matteo, and Tieri di Benci were working partners and they did not pay a part of the capital but pretended for the part of the profit. According to the contact of first period (from 1382 to 1385), the profit was shared as 1/2 for Datini, 1/3 for Tieri di Benci and 1/6 for Boninsegna di Matteo [1].

In 1386 Andrea di Bartolomeo became a working partner of the company and Datini increased the capital on 500 florins. But, as it was before, only he is who invested money for the company. In 1388 the structure of corpo was changed again. Datini disinvested part of the capital (800 florins), but the same amount invested Tieri di Benci (400 florins) and Andrea di Bartolomeo (400 florins). And only the main partner loaned money to the company, how it can see in the capital's structure (Sovracorpo was f. 1008 s. 6. d. 11).

Once again, the capital (corpo) structure will change in 1389 when Andrea di Bartolomeo leaves the company. The amount of capital will be reduced by its share of 400 florins and by the end of the company's existence will be 3100 florins, of which 2700 florins is the share of Datini and 400 florins is the share of Tieri di Benci.

The size of sovracorpo changed annually and was formed both at the expense of part of the profit left by the partners and at the expense of additional amounts contributed.

As mentioned above, the period of the company's activity since 1398 is of interest in this paper. From that time on, the reorientation of the company's accounting to double-entry method begins. Libro Segreto (Prato, AS. D. №161), begun in 1398, allows us to consider the process of this reorientation.

The Fig. 1 shows a photocopy of the first page of this book, whose entries are reflected in the column (mingled account form - credit at the top, debit at the bottom). It is worth noting the presence of cross-references, which is one of the main elements of the double-entry method in accounting. We can see from which book and from which page the information has been transferred to the credit of the account.

The picture on the Fig. 1 presents the capital account of this company. This can be determined by the first entry in the account. In this region, a rule was followed in creating account entries: the first entry on each side of the account bears the name under which the account is opened and the following operations: the entered date phrase to write off, or credits – de'Dare or dev'avere ('must give' or 'must have' or their plural), the amount of the transaction and then any additional information the accountant deemed necessary [20]. The first line of the account on page 2r indicates the presence of accounts payable to three partners of the company: "Francesco di Marco and Boninsegna di Matteo and Tieri di Benci must have", this designation is typical for the capital account of the company. The translation of the capital account (2r) is shown in Table 1.

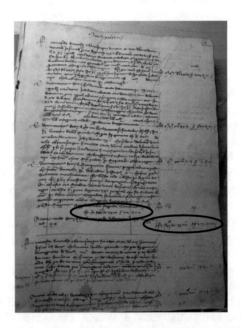

Fig. 1. The page in the Libro Segreto Prato, AS. D. №161, c. 2r

As you can see in the photocopy, the indicators are placed on this page vertically. The page contains two accounts-Prato, AS. D. №161, c. 2r(1) and Prato, AS. D. №161, c. 2r(2), separated by a solid horizontal line. The year 1398 is written at the top of the page. In the upper right corner – page number "2". Also, in the upper right corner of Federico Melis wrote "1" with pencil. That means the page is Prato, AS. D. №161, c. 2r - the first saved page in the book.

Table 1. Translation of the first account on the page Prato, AS. D. №161, c. 2r

1398	
Francesco di Marco and Boninsegna di Matteo and Tieri di Benci **must have** until January 1st 1385, f. 3103 s. 13 d. 10 *provenzali* we found in cash and debtors at the Quaderno di Cassa yesterday December 31st at the first account of three years passed along with Francesco and they were assigned in cash to this *ragione* and the sum, applying the exchange at s. 24 each, is worth f. 2586 s. 9 d. 10 *provenzali* are in part from the profit made during the aforementioned three years as in the Quaderno de Ragionamento B at c. 3 of those time which we have assigned to this *ragione*	f. 2586 s. 9 d. 10 *provenzali*
And they **must have** on this day f. 2175 s. 17 d. 6 of which we found ourselves indebted, once that on this day we deduced the creditors, of the account of the first three years we assigned to this *ragione* that is: 65 debtors counting f. 7045 s. 10 d. 4 *provenzali* and creditors 25 counting f. f. 4869 s. 16 d. 10 *provenzali* as written further in this book. What remains is the said f. 2175 s. 17 d. 6 *provenzali* as it can be seen comparing the total amount of debtors at the Libro Nero A at c. 2 and creditors at the same book at c. 2 and recalculated in that Quaderno de Ragionamento	f. 2175 s. 17 d. 6
And they **must have** on this day f. 1362 s. 1 d. 3 *provenzali* we took from the Libro Grande covered in parchment [*pergamino*] B from c. 2 where they had to have on December 31st 1386, we got them from Francesco di Marco and Boninsegna di Matteo and Tieri di Benci and Andrea di Bartolomeo, they are for the surplus we found of merchandise from f. 3000 up of the said account of three years. They remain in the profit of the three years in question and are together with the sum already mentioned in the capital of this *ragione*	f. 1362 s. 1 s. 3
And they **must have** on this day f. 500 *denaro corrente* for which we stated that Francesco di Marco himself must give in this book at c. 82 they are for as many goods of this account and of the profit of the first three years, and we put the sum in the capital [*corpo*] of the partnership on his behalf for f. 3000 when Andrea di Bartolomeo da Siena joined the partnership, that is January 1st 1385. And he **must have** the said f. 3000 which become 3500 in the Secret Book of the Partnership A at c. 102, and for this reason and for this Secret Book we wrote	f. 500
The total amount is f. 6624 s. 8 d. 7	
They had on this day as we put they **must have** in this book at c. 44	f. 6624 s. 8 d. 7

In this account, the entries contain cross-references to accounts held in other Libri Grande and Libri Segreti, which belong to prior periods and where prior profits were reported. This account shows the profit earned in the period 1382–1385 – f. 2586 s. 9 d. 10 and f. 2175 s. 17 d. 6, and the profit earned in 1386 – f. 1362 s. 1 s. 3. The last entry reflects the increase in capital (corpo) by 500 florins that Datini contributes at the time Andrea di Bartolomeo was added to the number of partners. After four points of the account is calculated the total amount that sums up the credit of the account. The total credit amount of the account is f. 6624 s. 8 d. 7 (first ellipse in Fig. 1).

The last entry is made in the debit of the account, according to which the amount f. 6624 s. 8 d. 7 (the second ellipse on Fig. 1) is transferred in this book to the credit of the account Prato, AS. D. №161, c. 43v-44r(3), which, like all accounts starting from the second folio, is kept in the bilateral (Venetian) form (Fig. 2) and is located third on the folio (marked with the second ellipse). Account Prato, AS. D. №161, c. 44r(3) (Fig. 2) is crossed from top to bottom (from right to left) with two slashes. This means that the account is closed and data from it is transferred to another account (accounts). This account is an extension of the capital account.

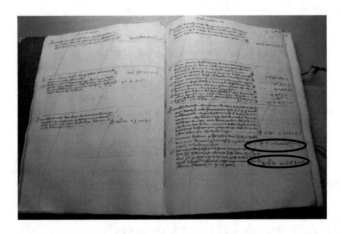

Fig. 2. Formation of the capital account in bilateral form, Prato, AS. D. №161, c. 43v—44r(3)

As already mentioned, when switching from a Florentine (mingled) account "in the paragraph") to the bilateral account, there was an entry from debit one account to credit another. The accountant, in this case, saw not an elementary transfer from one form of information medium to another, but the transaction in the double-entry method.

The translation of the third credit entry on the account reads as follows:

"And they **must have** on this day f. 6624 *denari correnti* s. 8 d. 7 *provenzali* we put they had in this book before at c. 2, they are for the profit made during the first three years before Francesco went away, this money was assigned to the capital of this *ragione* as it can be seen in this book at c. 2 in 4 entries, sum is f. 6624 *denari correnti* s. 8 d. 7".

Let's look at the first two entries on this account. The first entry says:

"Francesco di Marco and Boninsegna di Matteo and Tieri di Benci must have on this day f. 1702 denari correnti s. 4 d. 1 provenzali, we took from the Libro Grande coperto di perghamena (covered with parchment) D from c. 187 where they must have on December 31st 1388, we had from Francesco di Marco and Boninsegna di Matteo and Tieri di Benci and Andrea di Bartolomeo, for money we got from them the year before as sovraccorpo of partnership for the profit made of 2 years ending by December 31st 1387".

The second credit entry on the account (marked with the first ellipse in Fig. 2.) says:

"And they must have as we put, they must give in this book at c. 36 where they should have to have for one account in 2 entries took from the Secret Book of the partnership, sum is f. 1279 s. 9 d. 5".

This entry is formed as the amount of two credit entries and transferred from the debit account of Prato, AS. D. No. 161, pp. 35v–36r(4). Translation of two credit entries allowed a complete representation of the capital account formation procedure at the moment of transition to bilaterian accounts and double-entry method (Table 2).

Table 2. The translation of the fourth account (Prato, AS. D. №161, c. 35v–36r(4))

Francesco di Marco himself, Boninsegna di Matteo himself, Tieri di Benci himself **must have** on this day as we took from the Libro Grande covered in parchment labeled C from c. 124 where they should have on April 25th 1387 they had them from the account of the warehouse, that is from Francesco, Boninsegna and Tieri and Andrea di Bartolomeo for part of the profit made during the last year in the account of the warehouse below belonging to them	f. 671 s. 2 d. 6 *correnti*
And they **must have** on this day f. 608 *correnti* s. 6 d. 11 *Piccioli* we took away from the Libro Grande covered in parchment labeled C from c. 137 where they should have had on 31st December 1389, and they had them from Francesco and Boninsegna and Tieri and Andrea di Bartolomeo, and they **must have** the aforementioned money for a *ragione* written in the Secret Book B at c. 40 where they must have the said amount for balance of the profit made in 1386 and 1387. And this *ragione* was put in the said book and here is reckoned together with the aforementioned partners	f.608 s. 6 d. 11 *correnti*

This entry shows that the capital account also reflects the profit of previous periods of the company's activity (1387–1389). Thus, it can be argued that in the reorientation to double entry, the Tommaso di ser Giovanni method has formed a capital account, where it collected the profit indicators of previous periods.

Figure 3 shows the chart of capital account formation at the moment of closing the company and preparing the Trial Balance (October 31, 1401). As it can be clearly seen on the diagram, the amount f. 6624 s. 8 d. 7 (transferred from the first info page), the third on the account of settlements with the owners of the company (on the photocopy shown in Fig. 2, circled by the ellipse).

On October 31, 1401, Tommaso di ser Giovanni closed the accounts in the Secret Book and built in this book an internal Trial Balance, which consisted of two accounts placed on two folios: Prato, AS. D. №161, c. 84v-85r and Prato, AS. D. №161, c. 85v-86r.

On 31 October 1401, the partnership agreement expired and there was a change in the composition of partners. Boninsegna di Matteo, who died in 1397, was excluded from the partners of the company and Tommaso di ser Giovanni was included instead. It was therefore necessary to close the accounts and prepare the Trial balance.

Amount f. 9605 s. 22 d. 1, accumulated in the credit account of Prato, AS. D. №161, c. 43v-44r(4), is transferred to the fourth entry of the "Creditors" section of the first folio of Trial Balance. The debit entry on the account being closed reads as follows:

"Francesco di Marco and Boninsegna di Matteo and Tieri di Benci themselves must give on October 31th 1401 as we assigned them as creditors to Francesco di Marco and Co. of Avignon in this book at c. 85 they must have the balance of this account above, sum is f. 9605 s. 22 d. 1".

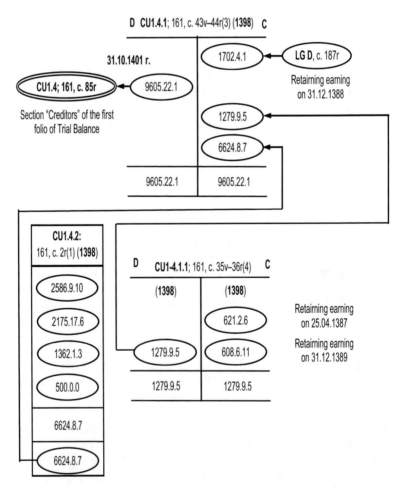

Fig. 3. Scheme of capital account formation in preparing the Trial Balance (Prato, AS. D. №161)

The contra entry in the section "Creditors" of Trial Balance indicates: "And they must have on this day as Francesco and Boninsegna and Tieri are assigned to them as creditors as in this book at c. 43 and put at c. 89", where the capital account in the old structure will be re-opened as of November 1, 1401, and it includes information about accumulating previous Corpo and Sovracorpo.

6 Conclusion

The main purpose of the article was to describe the procedure of capital account formation during the period of accounting reorientation to the double-entry method based on archival materials. The paper gives examples of capital account reflection both in the form of a paragraph (mingled) form and in a bilateral form when its balance is transferred by the double-entry method. The procedure of capital indicator formation for reflection in the trial balance, which was compiled in 1401 in the secret book, when the company's partners were changed, is considered. The research allows noting that the capital account included both the basic authorized capital and the part of profit left by partners, as sovracorpo.

References

1. Melis, F.: Aspetti della vita economica medievale (studi nell'archivio Datini di Prato). Monte dei Paschi di Siena, Siena (1962)
2. Melis, F.: The archives of a fourteenth century merchant and banker: Francesco Di Marco Datini, of Prato. PSL Q. Rev. 7(28), 29 (1954)
3. Guasti, C.: Ser lapo Mazzei: lettere di un notaro a un mercante del secolo XIV con altre lettere e documenti. Le Monnier, Firenze (1880)
4. del Lungo, I.: Florentia: uomini e cose del Quattrocento. G. Barbèra, Firenze (1897)
5. Livi, G.: Dall'archivio di Francesco Datini, mercante pratese: celebrandosi in Prato addì XVI d'agosto MDCCCCX auspice la pia casa de'Ceppi il V centenario della morte di lui. F. Lumachi, Firenze (1910)
6. Besta, F.: La Ragioneria, seconda edizione. Volume I, parte prima Ragioneria generale. Vallardi, Milano (1909)
7. Sargiacomo, M., Servalli, S., Andrei, P.: Fabio Besta: accounting thinker and accounting history pioneer. Account. Hist. Rev. 22(3), 249–267 (2012)
8. Nicastro, S.: L'archivio di Francesco di Francesco di Marco Datini in Prato. L. Cappelli, Rocca S. Casciano (1914)
9. Corsani, G.: I fondaci ei banchi di un mercante pratese del Trecento: contributo alla storia della ragioneria e del commercio; da lettere e documenti inediti. La Tipografica, Prato (1922)
10. Bensa, E.: Francesco di Marco da Prato: notizie e documenti sulla mercatura italiana del secolo xiv. Fratelli Treves, Milan (1928)
11. Zerbi, T.: Le Origini della partita dopia: Gestioni aziendali e situazioni di mercato nei secoli XIV e XV. Marzorati, Milan (1952)
12. Martinelli, A.: Notes on the origin of double entry bookkeeping. Abacus 13(1), 3–27 (1977)
13. Origo, I.: The Merchant of Prato, Francesco di Marco Datini. J. Cape, London (1957)

14. de Roover, R.: The development of accounting prior to Luca Pacioli according to the account-books of Medieval merchants. In: Littleton, A.C., Yamey, B.S. (eds.) Studies in the History of Accounting, pp. 114–174, London (1956)
15. Nigro, G.: Datini Francesco di Marco: the man the merchant. Firenze University Press, Firenze (2010)
16. Cecchi Aste, E.: L'Archivio di Francesco di Marco Datini fondaco di Avignone: inventario, vol. 1. Ministero Beni Att. Culturali, Istituto Poligrafico e Zecca dello Stato, Roma (2004)
17. Aleinikov, D.: What the study of the early accounting books in F. Datini's companies in Avignon has given. In: Antipova, T., Rocha, Á. (eds.) The 2018 International Conference on Digital Science. Disc 2019. Advances in Intelligent Systems and Computing, vol. 1114, pp. 374–386. Springer, Cham (2019)
18. Edler, F.: Glossary of Mediaeval Terms of Business, Italian Séries, 1200–1600, vol. 18. Mediaeval Academy of America, Waverly Press, Baltimore (1934)
19. Goldthwaite, R.: The Economy of Renaissance Florence. JHU Press, Baltimore (2009)
20. Goldthwaite, R.: The practice and culture of accounting in Renaissance Florence. Enterp. Soc. 16(3), 611–647 (2015)

Accounting Policy as the Key Factor of the Interaction of Various Types of Accounting in the Context of Digitalization of the Economy

Ekaterina Olomskaya⬤, Ruslan Tkhagapso$^{(\boxtimes)}$⬤, and Fatima Khot⬤

Kuban State University, 149, Stavropolskaya Street, 350040 Krasnodar, Russian Federation
rusjath@mail.ru

Abstract. The article discusses the emergence and development of the concept of accounting policy in Russian economic practice. A comparative characteristic of the category is presented in accordance with IFRS, as well as Russian accounting and tax regulations. The fundamental differences of the considered economic categories are highlighted. The structure of the internal regulatory document on accounting policies from the perspective of digitalization of economic relations is studied in detail. The aspects of accounting policy are disclosed. A detailed and reasoned comparative analysis of accounting policies for the purposes of financial, managerial and tax accounting in the digital economy is presented.

Keywords: Digital economy · Accounting · Management accounting · Managerial accounting · Tax accounting · Formation and disclosure of accounting policy

1 Introduction

The concept of "digital economy" was first introduced by one of the world-famous authorities in the field of business strategies – Canadian scientist Don Tapscott in 1995 when he described the features of network intelligence [1]. Today, a digital or virtual economy is understood as a combination of relations in the field of economics, culture and social life, based on the implementation of electronic technologies. The digital economy, as a new type of business with the prevailing value of information and methods of managing it in production and consumption cycles, serves as a basic element of the development of the economy as a whole. Its impact on the financial sector, accounting, insurance, commerce, medicine, education and others is undeniable. Advanced technologies contribute to the emergence of new ways of cooperation of market participants aimed at joint problem solving.

The areas of activity of firms and business entities available for remote management include: planning, management and control; business analysis; accounting and audit; reporting; service delivery; goods delivery; logistics; marketing.

T. Antipova (Ed.): ICIS 2020, LNNS 136, pp. 81–92, 2021.
https://doi.org/10.1007/978-3-030-49264-9_8

The international practice of accounting has brought significant changes to the scientific literature, reflecting the growing interest of the scientific and business communities in non-standard approaches in the field of accounting, expanding its conceptual foundations. Financial, managerial and tax accounting are now distinguished as the main types of accounting. At the same time, a lot of researchers [2–5], referring to the requirements of the modern economy, suggest distinguishing other types of accounting. These include: creative accounting, social accounting, forward accounting, functional accounting, corporate accounting, multidimensional accounting, environmental accounting, etc.

In some papers, an attempt to distinguish individual objects of accounting supervision is made, which is being justified with the need for effective management of the company. The latest cloud technologies have influenced the emergence of such a concept as virtual accounting. Modern interactive information processing technologies have led to the emergence of the concept of "virtuality" in accounting. The emergence of virtual accounting is explained by the need to process any business operation in various multi-level positions, which gives a significantly more detailed outlook on information.

The issues of the development of new types of accounting are closely related to the development of the theory and methodology of accounting science as a whole, which undergoes a detailed analysis and description of accounting types in the context of globalization and digitalization. Regardless of what types of accounting emerge in practice, the main role in the organization and cooperation of various types of accounting is played by an accounting policy of the organization. Even in the context of modern digitalization, the relevance of accounting policies is growing.

2 Research Results

In the modern economy, partial or full implementation of the business activities of business entities through online resources is actively promoted. For companies, the main advantages of using the capabilities of digital technologies are: increased productivity of workers and a simultaneous reduction in costs due to freelance; reduction of bureaucracy and corruption; "transparency" of operations; nullification of the human error – the possibility of an error caused by emotions, physical condition, as well as the real possibility of expanding the circle of potential buyers. The essence of the digital economy is to accelerate the processing of information, reducing routine work and making employees available for other tasks.

When building a data security system at the level of an economic entity, it is necessary to take into account information from different perspectives: a list of information resources; those responsible for security, administering these resources; interconnection of information systems; employee access to information resources.

At the present stage of development of the digital economy in organizations, due attention should be paid to the issues of forming the accounting policy of the company, as a starting point in the development of the accounting information system of any organization.

The accounting policy, being a self-sufficient document, contributes to improving the economic security of business entities on the basis of predetermining, controlling and comprehensively solving their problems in the field of accounting (financial, managerial) and tax accounting to ensure the sustainability of the business and the fact that contractors, the state, owners and personnel trust it.

In order to ensure the economic security of a business entity and to manage the risk when choosing a particular accounting method (financial, managerial) and tax accounting, an economic entity must take into account the influence of each method on the formation of financial results of an economic entity and taxation.

The progressive development of the Russian accounting system through adaptation in the environment of International Financial Reporting Standards sets the tone for accounting regulation on the basis of generally recognized principles.

The establishment of general principles in the context of digitalization allows one to ensure the unity of the accounting methodology, while not neglecting the features of the functioning of economic entities. The latter is implemented through such important accounting tools as an accounting policy and professional judgment of an accountant.

Thus, the active digitalization of the Russian economy in the context of ongoing adaptation to IFRS and isolation of tax accounting as an independent information subsystem in Russia, have put forward the substantiation of theoretical and practical aspects of the formation and interaction of relevant accounting policies. It is well known that accounting policy is part of the organization's financial policy, manifested in its voluntary choice of one of several possible options for accounting and reporting.

In the modern economy, three types of accounting policies can be distinguished based on the coverage of economic entities (Fig. 1).

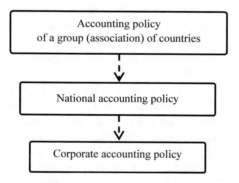

Fig. 1. Classification of accounting policies by coverage

The emergence of large international associations determines the need for inter-action of countries at various levels of public life. An important role in these processes is played by ensuring a common understanding of economic information with signif-icant differences in national accounting standards. Nevertheless, it is almost impossible to objectively achieve complete unity of accounting rules in all countries. The reasons for this are the differences in the applied management models, the degree of development of economies, the degree of freedom of regulation of the accounting system. Therefore, the category of *national accounting policy* is of great importance.

In Russian accounting practice, the concept of "accounting policy" first appeared with the adoption of the Accounting Policy Regulation, approved by order of the Ministry of Finance of the Russian Federation No. 10 of 20.03.1992 (item 6 of Sect. 2). This document stated that an enterprise should "comply with the adopted accounting policy (methodology) during the reporting year to reflect certain business transactions and valuation of property, determined on the basis of established rules and business conditions. Changes in the accounting policy compared to the previous year should be explained in the annual financial statements."

At that time the accounting and tax accounting were not yet imperatively separated and, therefore, all those elements that the organization had the right to choose when preparing financial statements automatically extended to its tax system. During the transition period of the Russian economy, accounting policies served as a legal means of minimizing the amount of income and profit. However, in this case, the organization could have problems with attracting investments, bank lending, counterparty trust, etc.

It became possible to talk about the emergence of an independent accounting policy, formed for the purpose of calculating taxes, with the adoption of part two of the tax code, the chapters on value added tax and income tax specifically. As a result, business entities had the opportunity to increase accounting profits and at the same time reduce tax revenues: present their financial situation in the best light and at the same time reduce the tax burden. The only limitation in this case was the increase in labor costs.

The Accounting Regulation "Accounting Policy of an Enterprise", published in 1994, played the role of a fundamental document in the field. With the adoption in 1996 of the Federal Law "On Accounting", accounting policy as one of the most important elements of accounting for a market economy was fixed at the legislative level. Article 6 of the law determined the need for its formation and documentation, the basic elements, and also states the need for its consistent application. In the current Federal Law No. 402-FZ "On Accounting", article 8 defines the accounting policy as "the totality of the methods used by an economic entity for accounting". Moreover, "an economic entity independently forms its accounting policy, guided by the legislation of the Russian Federation on accounting, federal and industry standards" [6]. And here the orientation of the Russian accounting legislation to the Anglo-Saxon model is manifested, in accordance with the Law on Accounting, documents in the field of accounting regulation include: federal standards, industry standards, regulatory acts of the Central Bank of the Russian Federation, recommendations in the field of accounting from professional communities and standards of an economic entity, which are designed to put the organization and its accounting in order.

At this point, for the first time in three decades of reforming accounting in Russia, economic entities have been given the right to independently choose the procedure for developing, approving, amending and repealing standards (item 12 [6]). At the same time, the standards of an economic entity must not contradict federal, industry standards and the regulations provided in part 6 of article 21 of the Central Bank regulations of the Russian Federation (item 15 [6]).

   The currently in force Regulation "Accounting Policy of an Organization" sets out the principles for formation of an accounting policy, the rules for introducing amendments and addenda, and the specifics of disclosure in reports. According to item 2 of this provision, "the accounting policy of an organization refers to the adopted set of methods for conducting accounting: initial observation, cost measurement, current grouping and final generalization of facts of economic activity" [7].

   Based on this definition, the main objective of the accounting policy is recognized as the choice of accounting methods and their consistent application in practice. It is noteworthy that neither the definition itself nor the text of the standard indicates the purpose of its development, which is to ensure the reliability of financial statements.

   In the International Financial Reporting Standards, there is another emphasis: "accounting policy is a set of specific principles, foundations, conditions, rules and practices adopted by the company for the preparation and presentation of financial statements" (IFRS 1 "Presentation of Financial Statements"). Organizations should formulate and apply accounting policies in such a way that the financial statements comply with the requirements of all standards and interpretations.

   Full disclosure of accounting policies allows users to objectively assess the financial condition of the organization and make informed economic decisions. The organization must describe the selected principles and methods of accounting.

   The existence of many types of accounting (financial, managerial, tax, statistical, strategic, etc.) is designed to satisfy the interests of various groups of users of accounting information to the maximum extent, which, in turn, predetermines accounting policies of organization (Fig. 2). Since the principles and rules of their maintenance can differ significantly, the assertion about the existence of an independent accounting policy with respect to each type of accounting is obvious. In other words, nowadays we can talk about the existence of accounting policies for managerial, financial and tax accounting.

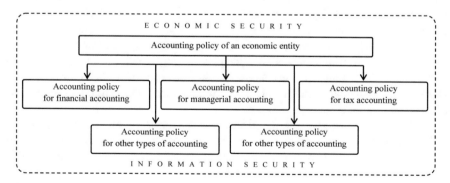

Fig. 2. Structure of the accounting policy of an economic entity in the context of digitalization of the economy

   Accounting policy for managerial accounting (management accounting policy) implies the maximum number of options. This type of accounting is not regulated in any way, and an accountant can choose among an unlimited number of elements, up to determining such global positions as the use of double entry, a single money meter, primary documents, etc. There are also no requirements regarding registration and disclosure of such an accounting policy.

The accounting policy for financial accounting (financial accounting policy) is formed by organizations on the basis of accounting legislation that defines the final list of elements among which a choice is allowed. It is registered by a decree and disclosed in an explanatory note.

The accounting policy for tax accounting (tax accounting policy) uses the rights provided by tax legislation regarding the choice of possible options for calculating the tax base. Compared with the accounting policy for financial accounting, the list of freedoms granted by the Tax Code is smaller, since the priority is to satisfy the informational interests of the only user – the state represented by tax authorities.

Here we elaborate on the issues of developing accounting policies for management accounting. When introducing a management accounting system in organizations for monitoring and managing its activities, it should be remembered that the stages of development and applying management accounting are almost identical to the stages of development and implementing financial accounting. Regarding the regulatory regulation of each type of accounting, it is necessary to take into account the needs of the end users of information. If we are talking about financial accounting, the accounting and reporting are regulated at the legislative level in accordance with generally accepted rules and norms, but if we are talking about managerial accounting, due to the fact that the information user is only an internal user, it is formed according to the requirements and requests of the organization's managers in accordance with the internal company standards. Accordingly, this leads to serious differences in the requirements and scale of accounting for the accounting policies of both systems – financial and managerial accounting.

In scientific practice, researchers provide different definitions of accounting policies for management accounting [8, 9]. Some authors consider the accounting policy of an organization as the totality of accounting methods adopted by it, calculating the cost of production (work, services) and compiling internal reports for the purpose of monitoring and managing the organization [10]. Others claim it is some internal regulations of an organization, which defines the principles and organizational foundations for the construction and management accounting, as well as a set of methods for its maintenance and compilation of management reporting, focused on the information requests of the organization's managers [11].

It is difficult to disagree with the statement as the formation of an accounting policy entirely depends on the requirements and requests of the organization's management personnel and, accordingly, the accounting policy of managerial accounting, unlike other types of accounting policies, is subjective.

Thus, it is possible to conclude that accounting policies for management accounting can be understood as a set of methods for maintaining an organization's management accounting and providing information about its production and business activities as well as its structural divisions for accounting, planning, control and effective management of both the organization as a whole and its individual structural divisions, responsibility centers, types of activities, products, operations, accepted by the economic entity itself.

It should be noted that all authors agree that the accounting policy for management accounting should cover and include not only a description of the methods and techniques of accounting, but also form management reporting.

The composition, content, structure and stages of development of accounting policies are purely individual for each economic entity, it is a kind of "unidentifiable intangible asset" of the organization and are considered to be a trade secret of the organization.

The choice and justification of the main aspects of accounting policies for both management accounting and financial accounting are influenced by the same factors. The structure of the accounting policy for management accounting may also include organizational, technical and methodological aspects. Various authors identify the following factors that influence the formation of a management accounting system in an organization (Fig. 3).

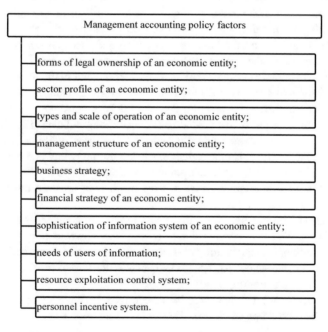

Fig. 3. Factors which influence the formation of a management accounting policy of an organization

Having compared financial and management accounting in the process of formation of accounting policies, the following conclusions can be made [10].

The Russian national standard "Accounting policies of an organization" [7] establishes the rules for the formation (selection or development) and disclosure of accounting policies of legal entities under the laws of the Russian Federation. Accordingly, if an organization selected accounting methods from those proposed by law and documented them in accordance with its internal standards within the framework of financial accounting, then the goal of creating an accounting policy in financial accounting is considered achieved. As for managerial accounting, it is not enough to limit oneself to the steps mentioned above, it is also necessary to take into account the level of competence of accountants and analysts and, of course, the accumulated experience in managing managerial accounting both in our country and in Western countries.

It leads to the fact that the rules of management accounting, its methods and techniques can differ significantly from those used in financial accounting – one can use any of the methods proposed in theory and practice. In this case, it is crucial not to contradict to the legislation of the country in which the economic entity operates.

It is also not enough to develop a financial accounting policy and approve an appropriate set of documents for it. It is necessary to develop a whole system of internal organizational documents of an economic entity in the form of instructive, methodological, reference, regulatory and other documentation.

Compilation of accounting policy for management accounting is of great importance for a company which intends to enter a new level of a market or uphold its current position. Management will not be able to make decisions that will lead the company to growth in the long term on the basis of accounting and financial reports only.

At the same time, the role of modern technologies in managing the economic processes of an economic entity should be taken into account in the form of automation of the decision-making process; analysis and forecasting capabilities; receiving information directly from users for creation digital models of consumers, processes.

Next, we consider the issues related to tax accounting policies [12]. Some authors identify the concept of "accounting policies for tax purposes" with the concept of "tax policy", which, we believe, is unacceptable. As accounting policy is recognized as part of an organization's overall economic policy, tax accounting policy is one of the areas of tax policy, which, in addition to accounting mechanisms, uses a range of financial instruments.

It is commonly known that there are two important prerequisites for the emergence of accounting policies: economic interests and the conditionality of accounting. Modern accounting, by its nature, objectively provides the opportunity to choose options for its maintenance. An accountant, under the influence of many economic interests, forms an accounting policy.

These interests, we believe, can be categorized in three enlarged groups that gradually reduce the scope of choice. These are the interests of:

1) the state (formed with the account of the opinions of state bodies regulating accounting and tax accounting, the business community as a whole (creditors, banks, insurers, etc.), the public, and professional organizations). Under their influence, a national (international) accounting policy is created;
2) owners and top managers (actual owners of the organization and persons whose interests they promote). They determine the most important priorities of the accounting policies of the organization. Unlike countries with developed market economies, in Russia this block of interests is in its infancy. Only a very small percentage of owners understand the very fact that accounting affects their income and wants to take a direct part in managing this process;

3) accounting employees (accountants, economists, etc.). Taking into account the limitations established above, accountants, on the basis of their own professional judgment, specify the accounting policies of the organization, make final decisions regarding the reflection of a fact of economic life in accounting.

The groups listed above form factors which determine the accounting policies of the organization. These are:

1) restrictions established by national accounting policies;
2) the owner's requirements for the distribution of financial results over time; and
3) professional judgments of the accountants when they choose a way to reflect the facts of economic life.

The listed factors allow us to trace the process of formation of accounting policies according to the top-down approach.

This circumstance is very important in the context of ongoing changes in accounting related to the adoption of IFRS. The category "accounting policy" receives a deeper content. This is not just a choice between the alternatives provided by IFRS and its documenting in a decree for the upcoming fiscal year. Due to the high abstractness of IFRS, their focus on accounting principles does not take into account (and incapable of doing so) all possible situations that accountants may encounter. Every day they are faced with the whole variety of financial and economic practices, and they often need to determine how to reflect this or that fact of economic life. At the same time, IFRS focuses on compliance with principles and the application of professional judgment.

Thus, the choice of accounting policies (within the framework of a single concept on global issues, which should be determined before the beginning of the year) becomes an everyday concern. Therefore, we can talk about the existence of two levels of accounting policies of the organization connected with:

1) determining the main accounting methods for the upcoming fiscal year;
2) choosing a way of recording specific business transactions in the framework of the accepted methods of accounting during the reporting period.

Accounting policies currently being developed by Russian organizations should include at least the financial, tax, and, if necessary, management aspects. Each of them, as a rule, implies a common (introductory), organizational, and methodological part.

The "General Provisions" section contains information about an organization, the regulatory framework that served as the basis for the development of accounting policy, the basic principles of accounting and reporting, possible cases of deviation from them.

In the "Methodical Section" the elements of the accounting methodology are set out, which determine the options for reflecting information about production, business and financial activities, proceeding from alternative techniques and methods.

The "Organizational section" describes the elements that determine the forms of maintaining and organizing accounting by an economic entity.

We believe that, taking into account the stated provisions, the number of elements of the tax accounting policy should be, at least, substantially reduced. Moreover, we consider it highly doubtful whether there is an accounting policy for tax purposes as such. The reasons for this are as follows:

a) in theory, it is not needed: there is only one user of information generated in tax accounting – the state, there is no conflict of interests and the need to search for compromises;

b) the purpose of tax accounting is not providing accurate information, as proclaimed in the Tax Code, but just one of the necessary technical conditions. In financial accounting, this is indeed an aim, since there an objective ambiguity, variability in the understanding and interpretation of facts of economic life by an accountant, called professional judgment, exist. In tax accounting professional judgment is not necessary – it is unacceptable, the same circumstances related to the payment of taxes cannot be interpreted differently by different taxpayers or tax authorities. They should be unambiguously perceived by all subjects of tax legal relations. Unlike financial accounting, which was placed by Western and, more recently, Russian methodologists, outside the legal field and even above it, the tax one has a pronounced legal characteristic and cannot exist without close connection with the law;

c) from an economic (financial) point of view, the presence of a tax accounting policy is extremely unprofitable for the state, since it allows to minimize taxes legally, without receiving anything in return. These are direct losses of the state;

d) if IFRS find it possible and necessary to reduce the number of alternatives (in financial accounting!), then in tax accounting, there are at least three more (mentioned above) reasons to do so.

There may be a concern that a significant reduction in the choice of methodological elements of tax accounting policies will lead to the formation of such accounting policies for financial accounting, which will not be based on professional judgment and the desire to ensure the reliability of financial statements, but the requirements of tax legislation. In fact, this is not entirely true (or rather, not true at all). This circumstance is only a "necessary" condition, but by no means a "sufficient" one for such a scenario. It all depends, first of all, on the accountant's desire to make reliable reports, and if it is not there, then within the framework of the current methodology, one can quite successfully bring financial and tax accounting policies closer, giving priority to the latter. What today can be observed in practice (especially in small business).

Thus, the reduction of alternatives provided to economic entities in the formation of their tax accounting policies is more than justified. This measure will significantly increase the "transparency" of the information generated in tax accounting, provide effective control over the correct calculation and payment of taxes, and will help to reduce the number of organizations looking for ways to improve financial results not at the expense of production, business and financial activities, but through the creation and application of various schemes for optimizing tax payments. The recommendations made will contribute to the solution of the tasks of the accounting policy, while simplifying the methodology and at the same time improving the quality level of tax accounting.

3 Conclusion

Thus, the complex of methodological, organizational and technological issues of accounting policy contributes to the systematic formation of information about its environment, which allows us to expand the scope of possible solutions and increase the level of economic and information security of an economic entity.

The main aim of the paper is to define the ways of transformation of the accounting policy of a company in the context of digitalization of the economy. Abstracting, generalization, idealization, and historical and logical methods of scientific research are applied in the paper.

Based on the results of the study, we can conclude that the digitalization of the economy requires the development of automation systems of accounting in order to reduce the number of errors by improving and optimizing practical recording and the amount of information in information systems, and using built-in algorithms of internal automated control of accounting data. All this requires the development of regulations and cloud technologies which can guarantee complete confidentiality and security of data and exclude the possibility of data loss or unauthorized access.

The authors express hope that the issue of formation of an accounting policy and search for modern accounting tools aimed at digitalization of economic processes will be reflected in the works of researches.

Growing system of telecommuting imposes certain limitations and special features, which pose new challenges for the society. Managers have to make management decisions from the outside of their companies, taking into account empirical data from the employees who, also, work remotely. All the aforesaid factors emphasize the research rationale.

References

1. The Age of Networked Intelligence: On "The Digital Economy" by Don Tapscott. Inf. Soc. **2**, 67–70 (2001)
2. Baruch, L.: Toward a theory of equitable and efficient accounting policy. Acc. Rev. **63**(1), 1–22 (1988)
3. Ali, M., Ahmed, K.: Determinants of accounting policy choices under international accounting standards: evidence from South Asia. Acc. Res. J. **30**(4), 430–446 (2017)
4. Pontoppidan, C.A., Brusca, I.: The first steps towards harmonizing public sector accounting for European Union member states: strategies and perspectives. Publ. Money Manage. **36**(3), 181–188 (2016)
5. Kuter, M., Lugovsky, D.: Accounting policies, accounting estimates and its role in the preparation of fair financial statements in digital economy. In: Antipova, T. (eds.) Integrated Science in Digital Age. ICIS 2019, vol. 78, pp. 165–176. Springer, Cham (2020)
6. "On Accounting": Federal Law No. №402-FZ, 6 December 2011
7. Russian Statutory Accounting "accounting policies of an organization" (RSA 1/08): Ministry of Finance Decree RF No. 106n, 06 October 2008
8. Denisov, U., Ionina, M.: Digital economy: concept and nature of the phenomenon. Context Reflection: Philos. World Hum. Being **8**(1A), 199–206 (2019)

9. Safarov, A.: Accounting Policy. Internal Standards of Accounting. Managerial Accounting 5 (2006)
10. Mukhina, E.: Characteristics of accounting policy for managerial accounting concepts. Hum. Sci. Res. **8** (2015). http://human.snauka.ru/2015/08/12344
11. Vakhrushina, M.: Management Accounting. National Education, Moscow (2013)
12. The Tax Code of the Russian Federation: Part Two. Federal Law No. 117-Fz, 5 August 2000

Digital Business and Finance

The Main Approaches to Assessing Efficiency of Tax Administration and Control in the Context of Digitalization

Larisa Drobyshevskaya⬤, Elena Vylegzhanina$^{(\boxtimes)}$ ⬤,
Vera Grebennikova⬤, and Elena Mamiy⬤

Kuban State University, 149, Stavropolskaya Street, Krasnodar 350040, Russia
e_vylegzhanina@mail.ru

Abstract. The subject of the study is the development of tax administration and tax control. The development of the digital economy and information space makes new demands on tax control and methods of its assessment. In this regard, the article highlights global trends in the development of tax administration and determines the principles of constructing an effective system of tax administration and control. The article reveals the essence of the risk-oriented approach in conducting control measures, highlights the key elements of tax authorities' risk-oriented model of control and supervision and the introduction of information systems in the Russian Federation. The authors state that it is necessary to improve the methodology for evaluating the effectiveness of tax control as an important scientific and practical task for the development of the state's fiscal policy. A comparative analysis of the main approaches to the formation of methods for evaluating the effectiveness of tax administration process and control is carried out. Taking into account the given scientific and methodological provisions, the development directions of tax authorities control activities effectiveness assessment are formulated.

Keywords: Taxation · Tax administration · Tax audit · Effectiveness of tax control · Desk audit · Field audit · Risk oriented tax control

1 Introduction

Aims and goals. The purpose of the study is to develop the main approaches to the formation of models for determining the effectiveness of tax administration and control based on a critical assessment of the theoretical views of Russian and foreign researchers and established practice, both local and global.

Methodology. Dialectical and formal logic were used in the course of the study. Tax control performance evaluation methodologies were analyzed combining monographic, logical, statistical methods of analysis and comparative method.

Results. As a result of the study, various approaches to assessing the performance of tax authorities in general and tax audits in particular were identified and further vectors of model development were identified.

© Springer Nature Switzerland AG 2021
T. Antipova (Ed.): ICIS 2020, LNNS 136, pp. 95–111, 2021.
https://doi.org/10.1007/978-3-030-49264-9_9

Conclusions. An integrated model for assessing the effectiveness of tax authorities control work should include the aspect of control measures preventive effectiveness, which can be implemented on the basis of organizational, technological and preventive-disciplining effectiveness of control checks.

2 Relevance of the Task to Improve Tax Administration and Control

The challenges of ensuring economic growth and development are inextricably linked to improving tax systems. From the state's point of view, taxation is an important economic tool for ensuring economic growth and attracting financial resources for national development programs [1]. Statistical data indicate that the value of the tax system as the main source of budget revenue in any state is undeniable [2]. At the same time, the actual tasks of maximizing tax collection and minimizing tax evasion are difficult to achieve in practice. A holistic assessment of the effectiveness and optimality of the tax system, identifying problem areas, is a complex multi-factor topic for scientists since there is no single system for evaluating the tax system's indicators. Tax administration is distinguished by a special subsystem in the structure of the tax system, where tax administrations exert a controlling influence on tax revenues in the built-in tax control subsystem [3].

When building and improving tax systems, minimizing administrative and compliance costs is crucial for both the state and taxpayers. Adam Smith called this principle the "economy of collecting" [4]. Similarly, the benefits of any reform in the revenue administration of the budget system should outweigh the costs associated with its adoption, including transition and implementation costs. The positive global trend of the last two decades has become the increase of tax administration efficiency and level of informatization. The 2019 OECD report on tax administration indicates an increase in tax authorities' investment in digitalization in order to:

1. facilitate the filing and payment of taxes;
2. automation of data analysis of taxpayers and third parties;
3. improving communication between taxpayers and tax authorities.

Most foreign jurisdictions digitalize their tax systems. In OECD countries, about 90% of taxpayers have switched to e-filing tax returns and e-payment of taxes, which reduces the time spent by taxpayers.

Various types of electronic billing systems are used all over the world. Tax authorities can use platforms that allow trading partners to exchange electronic documents over a specific network (for example, pan-European public procurement online [PEPPOL]). This is the case in Denmark, Sweden, and most recently in Singapore. Another option is to use XML formats; in this model, taxpayers convert invoices to a government-defined XML format and transmit them through an online portal, such as in Italy and the Slovak Republic. Finally, tax authorities can use online cash register (OCR) initiatives that require retailers to use OCR software to instantly upload sales data to the tax administration portal (Korea, Russian Federation). In Russia, each receipt generated by the cash register contains a QR code that the consumer can use to verify the correctness of the transaction registration (with the transfer of information to the tax authority).

The online cashier system is based on the "Internet of Things" technology and Big Date analytics. Information on settlements is transmitted to the tax authorities in real time mode, thereby preventing fraud and allows assessing risks automatically, which, in turn, reduces the need for tax audits.

However, in some countries, the introduction of new digital technologies is proving problematic. Ambalangodage D., Kuruppu C., Timoshenko K. (2020) [5] investigated the implementation of the new revenue management information system (RAMIS) in Sri Lanka. They identified barriers and implementation problems such as poor infrastructure, resistance to change (especially among older workers), and a legal framework that prevents the system from fully functioning.

Switching to real-time systems can provide significant benefits for tax administrations, including greater control over taxpayer data and increased fraud prevention measures. The need to disclose information for tax purposes and ensure transparency of business processes in the global space strengthens the role of international financial reporting standards, exchange of tax and financial information in the development of the digital economy.

3 Tax Control: Terminology and Forms

Tax authorities organize a system of tax control measures within the tax administration framework. Tax control is one of the most effective strategies for protecting the interests of the state from tax evasion. In the international terminology, the term tax audit is used for control measures implemented by state tax authorities, whereas in the legislation of the Russian Federation the Tax Code doesn't define the term «tax audit», but establishes the concept of tax control and tax verification. However, the role of audit in modern tax administration goes beyond the framework of checking the taxpayer's reporting obligations and identifying discrepancies between the taxpayer's submitted returns and the payment confirming documentation.

Tax audits are conducted in order to verify that the taxpayer complies with the provisions of tax legislation (OECD, 2006) [6]. A tax inspection is an investigation conducted by a tax authority to verify the accuracy of tax returns and identify inappropriate behavior and activities (Kirchler, 2007) [7].

The tax authority's audit program performs a number of important functions which, if implemented effectively, can make a significant contribution to improve administration of the tax system. The effectiveness and resultiveness of audit activities by tax authorities depends to a decisive extent on the nature and scope of the powers enshrined in the current legal framework, including the provision of sufficient authority to access information, as well as the appropriate system of sanctions for tax offenses. A tax inspection is a random check of tax liabilities declared by taxpayers, which includes checking the accounting system for taxpayer transactions related to business activities (accounting data, primary documents, information about the movement of funds on a Bank account, information from third parties).

Tax audit is the most important and essential component of the activities of state tax authorities to ensure compliance with tax legislation through the proper application of the legislation on taxes and dues, that is, the audit officers conduct the appropriate verification of the correctness of the taxpayer's Declaration of tax obligations, including checking the taxpayer's systems, accounting books and other relevant information. It may include cross-checking the taxpayer's accounts with data from the taxpayer's suppliers or with other government departments and agencies. The source of the information and its effectiveness and efficiency must be guaranteed through appropriate procedures and the use of modern audit tools and methods [6]. The main preventive form of tax control is registration and listing of taxpayers in special registers.

The most effective tools for tax control are tax audits, which allow the most complete and comprehensive control over the correctness of calculation, timeliness and completeness of tax payments by comparing evidence obtained during control measures with data from tax declarations and other accounting documents submitted by taxpayers to the tax authorities. In accordance with the Tax code of the Russian Federation, there are cameral and on-site tax field inspections, as well as checks on transactions between affiliated entities.

According to the results of a study by D'agosto, Manzo, Modica and Pisani, cameral inspections are the most effective in increasing tax compliance [8]. In particular, tax payers who passed a full cameral desk inspection increased their declared regional business tax by about 19% in the year of the audit, declared VAT by about 18%, and personal income tax by 14.7%. Field inspections appear to have a weaker impact on compliance with tax laws.

Based on the provisions of article 89 of the Tax Code of the Russian Federation, inspectors have the right to request any primary documents, including accounting registers, when conducting an on-site inspection. The audit compares the accounting and taxation data obtained as a result of the audit with the data of tax declarations and accounting forms submitted to the Federal Tax Service (FTS). Accounting and taxation have the same documentary basis [9], so when conducting tax audits, the accounting registers are examined. The necessity to disclose information for tax purposes and for ensuring the transparency of business processes in the global space strengthens the role of international financial reporting standards [10].

4 The Role of Tax Control in the Tax Administration System

According to Chan C. W. (2000), tax audit is aimed at improving compliance of tax payers' behavior with tax legislation in the self-assessment system [11]. The main goal of tax authorities' compliance activities is to improve overall compliance with tax laws and, in the process, inspire public confidence that the tax system and its administration are fair. Violations of tax laws are unavoidable due to many reasons, including ignorance of taxpayers, negligence, recklessness and deliberate tax evasion, or immaturity of the tax administration system [6]. Audit remains one of the main tools for combating non-compliance with legislation, and in most OECD countries, tax authorities allocate the most resources to the administration of tax legislation.

In most OECD countries, the share of the tax authority's total human resources allocated to audit and control activities exceeds 30% [12]. In Russia, half of the employees of the Federal tax service perform functions related to audits, investigations and other checks.

At the conceptual level, control measures should include the following three elements:

1) a structured algorithm for selecting taxpayers who are candidates for in-depth tax audits;
2) implementation of control measures;
3) continuous monitoring of the quality and effectiveness of inspections.

Considering that it is not possible to audit all taxpayers, especially by the same mode, a critical aspect is that the first two elements of the audit function use some form of case selection. Types of inspections can range from simple cameral desk inspections to in-depth field inspections at the taxpayer's premises. The choice depends on the calculated risk based on compliance, the amount of turnover, the results of previous audits, the length of time since the last audit, reliability of declarations, etc. Many tax administrations use the Pareto "80/20" principle (20% of taxpayers pay 80% of the collected taxes). Although such a policy has many advantages, it should not be implemented at the expense of the exclusion of taxpayers with high risk and low income [13].

It should be considered reasonable to exclude the top 20% of taxpayers with the highest turnover from any risk selection plan or program and control them individually. In Russia, the structure of tax authorities includes 9 interregional inspections that focus on controlling the largest taxpayers. In addition, since 2016, the legislation has been supplemented with a new specific type of tax control in the form of tax monitoring for the largest companies in terms of assets, revenue and the amount of taxes paid. The essence of tax monitoring is that taxpayers who meet certain conditions voluntarily provide the tax authority with access to the organization's information systems, which increases the transparency of business processes for tax authorities and frees them from on-site inspections.

The rating of Paying Taxes (2020) indicates that the tax system of the Russian Federation demonstrates positive development trends, as it becomes easier for businesses to fulfill their tax obligations in Russia [14]. Due to the use of new information technologies and tax reforms, Russia managed to significantly improve its rating-to rise from 134 to 58 in the rating for the period 2009–2018. Despite the positive changes in tax administration, it is difficult for Russia to achieve the transparency, coherence and clarity of the mechanism of action of tax institutions that are typical for Western countries due to the significant differences between the Eurasian socio-cultural tradition and the western one [15]. In order to improve tax control, Russian tax authorities should continue to work in the field of digitalization of technologies, strengthening investigative powers, and encouraging professional development [16].

5   Principles of Building an Effective Tax Administration and Control System

As follows from official documents of international organizations (European Commission (2007). Fiscal Blueprints) the overall construction of an effective tax administration system must meet the following fundamental requirements [17]:

- The tax administration is responsible for all operations subject to control and evaluation.
- Management and evaluation of the tax administration is based on the performance management system.

To address the global problem of tax base erosion and profit shifting (BEPS), the organization for economic cooperation and development (OECD) and the G20 have joined forces on the basis of an equal partnership [18]. Their joint action Plan for BEPS enabled them to involve more than 100 countries, both developing and developed, in the development and implementation of rules, aimed at ensuring compliance of profit generation and taxation sites, as well as increasing the predictability, transparency and flexibility of the international business tax environment [19].

In order to ensure maximum efficiency of the audit in order to solve the global problem of tax base erosion and profit shifting (BEPS), the following areas are important:

1. Reliable legal framework for tax control, providing for access to information systems for taxpayers' accounting.
2. Developing a risk-based tax audit strategy.
3. Using the most appropriate and modern tools and methods of tax audit to control tax and accounting systems of taxpayers, including computerized ones.
4. Regular monitoring and evaluation of the results and effectiveness of tax audits.

The most important indicators of control function of tax authorities include [20]:

- Operating expenses of the audit Department compared to the revenue received.
- The total number of registered taxpayers compared to the total number of tax inspectors engaged in inspections.
- Number of auditors compared to the total number of employees.
- Degree of implementation of the annual audit plan.
- Average time spent on on-site/off-site inspection in comparison with additional charges.
- Comparison of the total number of on-site/off-site inspections performed and the total number of off-site adjustments.

The above list is only a basis that can be used for setting benchmarks and then monitoring and analyzing audit results. However, the actual percentage or set of indicators should depend on local conditions and the maturity of tax administration.

The number of employees of the tax administration can vary from 1 employee for 20 tax payers to 1 employee for more than 1000 taxpayers. Differences in staffing levels in different countries mean differences in government employment policies as well as audit policies. In developing tax administrations, where the government may well be the largest employer, the drive to improve efficiency will not mean cutting staff, but reallocating and reorganizing processes, so the ratio is likely to be much lower. The ratio of the number of employees of tax authorities per taxpayer will have a significant impact on the audit policy in terms of the number of inspections of each type and the time of their completion. All tax authorities should strive to reduce the burden of intervention and administration on the taxpayer. To do this, they must develop an audit policy that makes the best use of available resources, focusing on the most risky taxpayers, but not without limiting the time spent on individual cases or excluding other important risks.

The tax administration must have a justification for the percentage of established checks of taxpayers, the average time set for each check, the time spent on large taxpayers, and the percentage of cameral and field checks. The targets set in the plans must meet the "SMART" criteria (be specific, measurable, achievable, realistic, and time-based).

At a high level, the audit function should consist of three elements: 1) selecting taxpayers to conduct the audit during the audit planning process; 2) conducting the audit; and 3) continuously monitoring the quality of the audit.

According to Bird (2010), political readiness, a clear strategy, and adequate resources are the components necessary for effective tax administration [21].

6 The Risk-Based Approach and Digitalization of the Tax Control

Tax authorities in most countries have developed and implemented risk-based audit strategies. The risk-based approach takes into account various aspects of the business, such as historical compliance, industry and corporate characteristics, the debt-to-credit ratio for businesses registered with VAT, and the size of the business to better assess which businesses are most likely to avoid paying taxes. One study found that data mining technologies for auditing, regardless of the method, captured more dissenting taxpayers than random audits. Introducing sophisticated analytic models allows administrations to better identify revenues, claims, or transactions that may require further review or be fraudulent.

However, in a risk-based approach, the exact criteria used to audit case selection must be hidden so that taxpayers were unable to purposefully plan how to avoid detection, and allow a certain degree of uncertainty to encourage voluntary compliance. In most countries, there are risk assessment systems for selecting companies for tax audits, and the grounds on which these companies are selected are not disclosed. Although this is a post-filing procedure, audit strategies can have a fundamental impact on how companies file and pay taxes.

The risk-oriented approach is based on an analysis by the supervisory authority of various information about the company and determining the degree of risk of this firm, committing illegal actions. At the same time, scheduled inspections for companies with the lowest level of risk are rarely assigned, or are not assigned at all.

In Russia, the transition to a risk-based model of control and supervisory activities in the FTS began in 2007 with the adoption of the field tax audit planning system concept, aimed at creating a unified, open and understandable system for taxpayers and tax authorities for planning field tax audits. The introduction of a new risk management model was prompted by the understanding of the physical impossibility to carry out on-site inspections of the entire range of taxpayers. It was necessary to set priorities, with minimal labor and financial expenses, to achieve the maximum result for control measures. The document indicates 12 criteria for identifying risks and determining the number and direction of planned control measures. The Federal tax service not only identified the risks, but also made them public. Open data is one of the most significant, effective, but undervalued risk management tools.

The main principles of the risk-based model are:

- most-favored-nation treatment for bona fide taxpayers;
- timely response to signs of possible tax violations;
- inevitability of punishment of taxpayers in case of detecting violations of the legislation on taxes and dues;
- validity of the choice of objects of verification.

Thus, since 2016, the priorities of state regulation have changed. The efforts of state agencies are aimed at creating a system of "smart control", which is primarily focused on achieving socially significant results. First of all, the FTS turned to setting up a goal-setting system. The strategic management system in the Federal tax service is at the forefront. The use of a risk-based approach has reduced the period for carrying out mandatory inspections of declarations on excise duties on ethyl alcohol, alcohol and alcohol-containing products, and declarations on value-added tax from three months to two.

Russia is already ahead of most jurisdictions in terms of digitalization of tax services. The process of tax digitalization in Russia takes place in 3 stages:

1. The "digital maturity" model (websites, personal electronic services, electronic document management and reporting).
2. "Fully digital organization" of administration processes (mobile applications (My tax), individual proactive services).
3. The current stage is the "Adaptive Platform" (2019–2025), which combines the IT platforms of the Federal Tax Service and tax payers in real time mode, when tax obligations are performed automatically and "effortlessly" with built-in tax compliance.

In the field of tax control, new breakthrough information services and technologies have been put into operation, the main of which is an automated system for controlling value-added tax. The introduction of the ASK VAT-3 system allows to combine the results of the analysis of bank documents and information about purchase and sales books to identify gaps in the chains of "taxpayer-counterparty", the interruption of formal workflow and dubious financial transactions.

FTS of the Russian Federation is the leader among government agencies in terms of the volume of stored and used digital information. Each regional department has all the information about the economic activity of individuals and legal entities: from personal data, information about property, income, and ending with information about transactions, operations, and relationships, as well as the history of their changes and other data received from other government agencies.

This information base makes it possible to effectively carry out work to prevent evasion schemes. Active tax control actions are carried out in respect of taxpayers, who conduct aggressive tax evasion schemes, that cause significant damage to the budget of the Russian Federation. In this regard, the tax authorities are particularly concerned about a number of issues, including:

1. Investigation of tax evasion scheme signs - "business splitting", when legally separated legal entities actually conduct joint coordinated activities.
2. Deliberate control of pricing and distribution of expenses and revenues.
3. Creating artificial conditions for the formation of losses and their subsequent transfer to other periods.
4. Obtaining an unjustified tax benefit from artificially creating such conditions for financial and economic activities, that allow understating of tax liabilities.

Thus, the tax authorities are already using the opportunity to prevent damage, caused by tax evasion schemes in real time. Using all the control and analytical tools, the tax authorities carry out systematic work to encourage voluntary refusal of taxpayers from illegal tax optimization. When tax violations are detected, the first step is to encourage taxpayers to independently clarify their obligations, and tax planning focuses on those areas of tax risks that could not be neutralized by the results of the work carried out. The system of centralized analysis of reporting indicators, storage of tax audits results and control measures provides for an effective assessment of tax risks. Automated access to global data allows to calculate tax damage from a single inspector's workplace and determine measures to prevent tax evasion schemes. Thanks to the introduction of modern tools, it was possible to increase tax revenues and fees at all levels of the budget system. The implementation of a set of control measures provided for by the tax code of the Russian Federation is invariably initiated on the basis of data available to the tax authorities.

Those taxpayers who did not agree with the arguments of the tax authorities and did not consider it necessary to voluntarily clarify their tax obligations are included in the tax audit plan. Thanks to this approach, it was possible to reduce the number of tax audits while increasing their effectiveness. The introduction of sophisticated analytical models allows administrations to better identify revenues, pretensions, or transactions that may require further review or be fraudulent.

The future of tax administration involves the creation of a virtual transactional environment - a closed digital ecosystem in which all business entities will make transactions, and the Federal Tax Service of the Russian Federation will be able to automatically calculate and withhold taxes at the time of transaction. Tax services should in the future be fully integrated into the business environment. The boundary of the transition to the "new transparency" in the version of the Federal Tax Service is the practical implementation in the economy of the Internet of Things (IoT) in 2025–2035. In the future, the Federal Tax Service sees itself as a service department, whose IT infrastructure will closely interact in real time mode with "digital processes" within the taxpayer company, checking the correctness of tax payments in the same mode ("automatic fulfillment of tax obligations without effort").

7   Analysis of the Main Approaches to the Formation of Models for Evaluating the Effectiveness of Tax Administration and Control

The implementation of tax control provides for the economic and social impact. The economic effect is achieved by increasing the revenue side of the budget by the amounts that are additionally accrued based on the results of tax control. When determining the economic effect, it is necessary to take into account the costs of tax control, since budget revenues increase by the difference between the amounts received and the costs incurred to obtain them.

The social effect is manifested in creating conditions for an even distribution of the tax burden by reducing tax offenses, which makes it possible to fully implement the social regulatory functions of taxes. In addition, the application of penalties to violators provides certain competitive advantages to law-abiding taxpayers, stimulating an increase in tax culture.

An urgent task is to evaluate the effectiveness of the control work performed - post-inspection analysis, aimed at diagnosing the factors, that affect the effectiveness of the control work, as well as developing an optimal strategy for selecting taxpayers for the field audit programs, ensuring the rational use of labor costs.

To assess the budget efficiency of tax authorities, one can use the classic cost effectiveness analysis (CEA). It is aimful to use the ratio of the achieved result and the amount of resources spent, as the basis for evaluating the effectiveness of control activities of tax authorities. In case the costs of labor and material resources are not taken into consideration it is necessary to provide for a possible distorted perception of the control work results.

We can formulate the main thesis: the higher the degree of control efficiency, the lower the cost of achieving the set target. The main performance indicators of the tax authority are: a) the amount of tax revenues; b) the difference between the amount of tax revenue and the costs of the tax authority; c) the ratio of tax revenue to the costs of the tax authority. These indicators are influenced by factors such as the quantitative and qualitative composition of taxpayers, the quality of control work of the tax authority, including the organization of business processes in the tax authority, the legislative framework, etc.

However, the correlation of tax revenues to the budget with the funds spent on the operation of the tax authority is not sufficient to make an objective comparison of the effectiveness of tax inspections. This approach is more applicable to the entire tax system functioning during a certain period of time than to a single territorial tax authority.

Tax control quality management involves the use of a scientifically based methodology for evaluating the effectiveness of tax authorities in the field of control and verification activities with clear algorithms and a system of indicators. We will analyze differentiated approaches to assessing the quality of tax authorities control work. All methods can be structured in two groups: normative and author's.

7.1   Normative Methods for Assessing the Effectiveness of Tax Control

Intradepartmental methods have a normative basis. They are developed and applied by authorized tax authorities for planning and analysis, as well as in the distribution of material incentives for public servants. In Russia, the first official methodology for evaluating the effectiveness of control and verification activities was approved in 2004.

Since 2007, the performance of Russian tax authorities has been regulated by the internal regulations, according to which the assessment is based on the calculation of an integral indicator. The calculations are based on determining the values of quantitative and qualitative parameters of the tax authorities' activities, taking into account the set value for each criterion (Table 1), while the methodological base for establishing these value coefficients requires a critical revision.

Table 1. Criteria for evaluating the effectiveness of tax authorities

Criteria	Criterion weight
1. Collection of taxes and dues, %	18%
2. Percentage of the amounts of claims considered by the courts in favor of the tax authorities, relative to the total amounts of legal disputes with taxpayers, %	16%
3. The share of the number of decisions of tax authorities declared invalid by the court in the number of decisions of tax authorities issued by the results of tax control, %	16%
4. Reduction of tax arrears to the budget system of the Russian Federation, %	12%
5. Increase in the share of taxpayers who satisfactorily assess the quality of tax authorities work, %	6%
6. Number of citizens and organizations that receive information from the Unified state register of legal entities and the Unified state register of individual entrepreneurs using Internet technologies	6%
7. Share of taxpayers who have access to personalized information about the state of settlement with the budget via communication channels and the Internet, %	11%
8. The ratio of the number of complaints on tax disputes considered in pre-trial procedure by higher tax authorities to the total number of claims on tax disputes submitted to the tax authorities and considered by the courts, %	15%

Based on the positive experience of OECD [22] countries, the tax control activities are considered satisfactory if the following specific quantitative guidelines are met:

- The audit coverage of taxpayers is clearly related to the volume of the tax administration resources. In OECD countries, tax audit coverage ranges from 1% of tax payers (for example, Austria) to 5–10% of taxpayers (for example, Australia, the Netherlands).

- 100% of large taxpayers must pass an audit.
- About 30–50% of the entire tax service working time should be devoted to audit and disclosure of the facts of obtaining unjustified tax benefits.
- With regard to tax control conducted on a risk-oriented basis, at least 70% of inspections must be "positive", that is, to identify violations and present additional charges for taxes. In the Russian Federation, due to deep pre-verification analyses using digital information techniques, 98% of cases of field tax audits resulted in 2019 with additional tax charges.

7.2   Author's Methods of Assessing the Effectiveness of Tax Control

In the Tax Administration 2019 report the OECD provides a fairly narrow list of indicators for tax control and tax audits when conducting a comparative assessment of the OECD and other countries with developed and emerging economies:

Number of completed checks

- including the number of audits during which tax adjustments were made. Additional charges based on the results of audits for the financial year, including fines and interest
- including the total amount collected in the financial year as a result of audit evaluations.

A review of monographs and scientific studies has shown that the authors justify various sets of quantitative and qualitative, absolute and relative (coefficient) indicators, which can be difficult to apply in the practice of tax authorities.

Among the scientific studies that affect the definition issues, it should be noted a subgroup of works, the authors of which suggest using a single universal coefficient, as an indicator of the effectiveness of tax control (Table 2).

Table 2. Universal performance indicators of control activities

Indicator	Method of calculation
The ratio of tax assessment	Share of taxes independently accrued by the taxpayer in the total amount of accruals
Tax collection rate	The ratio of total tax revenues for the region to the tax potential of the region
Coefficient of fiscal efficiency of tax administration	The ratio of the absolute amount of tax revenue to tax administration expenses
Assessment of the «tax potential reserve»	The difference between the expected and actual amount of tax payments transferred to the budget
The ratio of tax assessment	Share of taxes independently accrued by the taxpayer in the total amount of accruals
Tax collection rate	The ratio of total tax revenues for the region to the tax potential of the region
Coefficient of fiscal efficiency of tax administration	The ratio of the absolute amount of tax revenue to tax administration expenses

Not all researchers share the idea of using a single universal coefficient and develop systems of relative indicators to assess the level of effectiveness of tax authorities control work. At the same time, judgments based on the results of such studies are usually based on horizontal and vertical analysis of the relative indicators included in the model. The presence of numerous absolute and relative (coefficient) indicators that characterize the effectiveness of tax control measures does not always make it possible to uniquely determine its effectiveness as a whole.

The method of analyzing the effectiveness of field tax audits (A. A. Sofyin) [23] allows ranking tax inspections based on 14 performance evaluation coefficients and a Matrix of pairwise comparison criteria to build their rating depending on the effectiveness of the control work. The method is quite complex in practical application due to the use of the hierarchy analysis method.

The method of comprehensive assessment of the effectiveness of tax audits (V. G. Gruzdeva) [24] includes a multi-level set of tax control performance indicators based on a system of private indicators structured by the main functional areas of tax authorities' activities "personnel – quality of inspections – relations with taxpayers – financial results". The use of expert assessment methods limits and complicates the methodology.

The methodology for evaluating the effectiveness of tax administration (E. V. Ivanova, 2011) [25] is based on the use of indicators of 3 areas (effectiveness, efficiency and intensity) that are subject to a point assessment. The author uses a methodology for comparative analysis of the work of regional tax authorities.

It is quite justified, that a number of authors underline the need to develop integrated indicators when developing a methodology for evaluating the effectiveness of the tax authority. Among the known methods of determining the value of the integral indicator the following methods are used: the method of sums, method of works, method, arithmetic average, geometric average method, distance method, the method of standardization of the scale of the indicators, matrix, logical (semantic) convolution method, arithmetic weighted average method, methods of expert estimates, as well as the construction of a synthetic index.

For example, Yashina N. I. and Alexandrov E. E. [26] we justified the system of private indicators of tax control efficiency, which is a fairly systematic approach due to the grouping of indicators by areas. They solve the question of choosing the optimal way to integrate private criteria into a single generalizing indicator by constructing an integral indicator based on a linear transformation of private indicators.

Taking into account that Russian Federation includes 85 subjects (regions), and as a result the complex multi-level structure of the Federal tax service and its regional inspections, many scientists (Yashina N. I., Alexandrov E. E., Gruzdeva V. G., Sofyin A. A.) attempts have been made to develop a methodology for ranking tax authorities based on their internal efficiency.

8 Results

Most of the existing methods are aimed at evaluating the activities of tax authorities in general and do not allow to clearly determine the effectiveness of a particular area of work. In particular, the development of methods for assessing the quality of administration of specific types of taxes has not been widely used.

Criteria-based assessment of the consequences of tax control involves monitoring its impact on the behavior of taxpayers. Thus, tax control should be considered effective when it entails a real reduction in the number of tax offenses committed. The effectiveness of tax control can be characterized by the ratio of the amounts of taxes independently accrued by taxpayers and the tax potential of the territory. At the same time, the tax potential should be understood as the maximum amount of taxes and dues, that can be accrued under the current legislation.

The use of this indicator most objectively reflects the effectiveness of the tax control. However, we have to state the complexity of calculating the tax potential of the territory, which makes it difficult to apply this indicator to assess the effectiveness of a particular tax administration.

A comprehensive system for evaluating the performance of tax authorities can be structured by levels:

1. Evaluating the effectiveness of tax administration.
2. Evaluating the effectiveness of tax control.
   2.1 Assessment of the effectiveness of registration and accounting of taxpayers.
   2.2 Evaluating the effectiveness of tax audits.
      2.2.1 Evaluating the effectiveness of desk (cameral) tax audits.
      2.2.2 Evaluating the effectiveness of field tax audits.
   2.3 Evaluating the effectiveness of tax monitoring.

At the same time, as part of a broader assessment of the tax administration system, it is necessary to consider such an important aspect as the level of expenses for tax compliance by taxpayers (time for documenting, calculating and making payments to meet tax obligations and compliance with procedures after filing documents) [27].

For this purpose, "customer satisfaction surveys (audited taxpayers)" sent by tax inspectors after each tax audit project or report can be used. We should understand the complexity of interpreting such surveys. The indicator of an increase in the share of taxpayers who satisfactorily assess the quality of work of tax authorities does not allow assessing the effectiveness of performing functional duties by tax inspectors. Ultimately, the tax authority represents the interests of the state in tax relations and controls the timeliness and completeness of the taxpayer's debt repayment. Therefore, when conducting such surveys, the majority of payers negatively assess the actions of tax authorities, even if the rights and legitimate interests of the payer are not violated. In addition, such studies need to involve a third party, which entails an increase in expenditure on the formation of the estimates and therefore reduced effectiveness in the use of funds allocated for implementation of activities of the tax authorities.

A comprehensive model for evaluating the effectiveness of control work of tax authorities should include the aspect of preventive effectiveness of control measures, which can be implemented on the basis of taking into account the organizational and technological and preventive and disciplining effectiveness of control checks. It should be noted that this parameter of tax control is difficult to measure directly. The generalizing indicator of preventive and disciplining effectiveness can be calculated taking into account the following indicators:

- quantitative measurement of the disciplining effect of tax control can be achieved through an increase in voluntarily paid taxes in the post-verification period;
- preventive impact of tax audits (rank);
- intensity of interaction between tax authorities and taxpayers (rank).

Thus, it is proposed to supplement the assessment of tax control measures from the point of view of tax payers, compliance with their rights, and the adequacy of tax inspectors' actions. In the current realities of tax administration, for analytical purposes, it is recommended to introduce into the system of indicators for assessing the effectiveness of tax control a relative indicator of the "tax gap" for VAT, reflecting the level of discrepancy between the declared VAT deductions and the actual amounts of tax paid. Calculations show that by using the ASK VAT-3 and increasing the transparency of economic operations, the tax authorities of the Russian Federation have reduced the VAT tax gap to a record low of 0.6% in 2019. Moreover, in the world practice the value of the "tax gap" for VAT is much higher, in European countries it reaches 10%, which demonstrates the limited fiscal and control functions of the tax.

9 Conclusion

Tax administrations are becoming more and more like leading digital companies in working with data arrays, attention to interaction with customers, flexibility of the structure, as well as in the acquisition and development of new competencies. Thanks to the transition to work with large volumes of information, constant interaction with taxpayers, and increased transparency of operations, tax administrations are now turning into digital business platforms. In the near future, all relationships in the economy will be reflected in the virtual network of fiscally transparent taxpayers. In an atmosphere of trust, the latter will be ready to share personal data with the inspection in exchange for complete transparency in regard to their position. Tax systems will switch from automatic filling of declarations based on available information to the complete abolition of these reporting forms as unnecessary. For bona fide taxpayers, paying taxes will become a routine that requires no effort and takes place almost without their participation. Inspection control activities will focus only on potential violators.

Creating digital tax ecosystems based on replacing traditional tax control technologies with information and digital ones requires new approaches to evaluating the effectiveness of tax administration, the effectiveness of tax control, and the efficiency of tax authorities' employees. As a result of the study, a comprehensive 3-tier system for assessing the activities of tax authorities with regard to the implementation of the control function is justified.

When assessing the results of tax control, the authors substantiated the feasibility of taking into account sociological surveys of taxpayers, aspects of the preventive impact of tax control and the dynamics of taxpayer compliance costs. In relation to the effectiveness of control over the completeness of VAT payment, a dynamic assessment of the level of the "tax gap" is relevant - the difference between the amount of VAT that theoretically should be paid by taxpayers and the amount of tax actually paid. The proposed expansion of the set of indicators for assessing the effectiveness of tax control will facilitate ongoing monitoring of the achievement of the strategic goals of tax administrations and effective tax communication in the conditions of a digital economy.

Most of the indicators included in the proposed methodology do not require to organize an additional information collection, since they are mainly based on official tax and budget reporting data. As a result, the presented set of indicators is characterized by minimal labor intensity of the control process.

References

1. Heng, L.L., Alifiah, M.N., Chen, L.E.: A proposed GST compliance model of GST registered entity in Malaysia. Int. J. Recent Technol. Eng. (IJRTE) 8(1S), 200–206 (2019)
2. Giriuniene, G., Giriūnas, L.: Identification of the complex tax system evaluation criteria: theoretical aspect. J. Secur. Sustain. Issues 7(2), 258–265 (2017)
3. Porollo, E.V.: Tax control: its essence and place in the system of state financial control. TERRA ECONOMICUS 11(3.3) 84–88 (2013). https://te.sfedu.ru/arkhiv-nomerov/2013/83-nomer-3-3/1068-nalogovyj-kontrol-sushchnost-i-mesto-v-sisteme-gosudarstvennogo-finanso vogo-kontrolya.html. Accessed 28 Jan 2020
4. Fleischacker, S.: On Adam Smith's "Wealth of Nations": A Philosophical Companion. Princeton University Press, Princeton (2004). https://jstor.org/stable/j.ctt7ss85. Accessed 29 Jan 2020
5. Ambalangodage, D., Kuruppu, C., Timoshenko, K.: Digitalising tax system in sri lanka: evidence from inland revenue department. In: Antipova, T. (ed.) Integrated Science in Digital Age. ICIS 2019. Lecture Notes in Networks and Systems, vol. 78. Springer, Cham (2020)
6. OECD: Strengthening tax audit capabilities, general principles and approaches, Paris, France (2006). https://www.oecd.org/tax/administration/37589900.pdf. Accessed 28 Jan 2020
7. Kirchler, E.: The Economic Psychology of Tax Behavior. Cambridge University Press, Cambridge (2007)
8. D'Agosto, E., Manzo, M., Modica, A., Pisani, S.: Tax audits and tax compliance – evidence from Italy (2017). https://www.irs.gov/pub/irs-soi/17resconmodica.pdf. Accessed 03 Feb 2020
9. Lugovsku, D., Kuter, M.: Accounting policies, accounting estimates and its role in the preparation of fair financial statements in digital economy. In: Antipova, T. (eds.) Integrated Science in Digital Age. ICIS 2019. Lecture Notes in Networks and Systems, vol. 78. Springer, Cham (2020)
10. Chumakova, I., Oleinikova, L., Tsevukh, S.: Formation of the information space for audit and taxation as a factor for the improvement of investment attractiveness of the ukrainian economy. In: Antipova, T. (ed.) Integrated Science in Digital Age. ICIS 2019. Lecture Notes in Networks and Systems, vol. 78. Springer, Cham (2020)
11. Chan, C.W., Troutman, C.S., O'Bryan, D.: An expanded model of tax payers compliance: empirical evidence from the United States and Hong Kong. J. Int. Acc. Audit. Taxat. 9(2), 83–103 (2000)

12. Tax Administration 2019: Comparative Information on OECD and other advanced and emerging economies. OECD Publishing, Paris (2019). https://doi.org/10.1787/74d162b6-en. Accessed 01 Feb 2020
13. Jacobs, A., Crawford, D., Murdoch, Y., Hodges, Y., Lethbridge, C.: Detailed guidelines for improved tax administration in Latin America and the Caribbean. USAID, W. (2013). https://www.usaid.gov/sites/default/files/LAC_TaxBook_Entire%20Book%20-%20ENGLIS H.pdf. Accessed 04 Feb 2020
14. Paying Taxes 2020: 14th edition. PWC, World Bank Group (2020). https://www.pwc.com/ gx/en/services/tax/publications/paying-taxes-2020.html. Accessed 01 Feb 2020
15. Vishnevsky, V.P., Goncharenko, L.I., Gurnak, A.V.: Evolution of tax institutions and the problems of transition to economic growth. Space Econ. **4** (2016), https://cyberleninka.ru/ article/n/evolyutsiya-nalogovyh-institutov-i-problemy-perehoda-k-ekonomicheskomu-rostu. Accessed 29 Jan 2020
16. Antipova, T.: Governmental auditing systems in Indonesia and Russia. In: Antipova, T., Rocha, Á. (eds.) Information Technology Science. MOSITS 2017. Advances in Intelligent Systems and Computing, vol. 724. Springer, Cham (2018)
17. Fiscal Blueprint: A path to robust, modern and efficient tax administration. European Communities (2007). https://ec.europa.eu/taxation_customs/sites/taxation/files/docs/body/ fiscal_blueprint_en.pdf. Accessed 30 Jan 2020
18. OECD/G20: Base erosion and profit shifting project. Explanatory Statement. Final Reports. OECD (2015). https://www.oecd.org/ctp/beps-explanatory-statement-2015.pdf. Accessed 30 Jan 2020
19. Shelepov, A.: The BEPS project: global cooperation in the field of taxation. Bull. Int. Organ.: Educ. Sci. New Econ. **11**(4), 36–59 (2016). https://cyberleninka.ru/article/n/proekt-beps-globalnoe-sotrudnichestvo-v-sfere-nalogooblozheniya. Accessed 29 Jan 2020
20. Detailed Guidelines for Improved Tax Administration in Latin America and the Caribbean An Overview (2013). https://www.usaid.gov/sites/default/files/LAC_TaxBook_Entire% 20Book%20-%20ENGLISH.pdf. Accessed 30 Jan 2020
21. Bird, R.: Improving tax administration in developing countries. J. Tax Adm. **1**, 23–45 (2015)
22. Tax Administration in OECD and Selected Non-OECD Countries: Comparative Information Series (2010). Forum on Tax Administration. OECD (2011). https://www.oecd. org/ctp/administration/CIS-2010.pdf. Accessed 30 Jan 2020
23. Sofin, N.A.: Analysis of the effectiveness of accounting and control procedures of tax authorities. Econ. Anal.: Theory Pract. **32**(335) (2013). https://cyberleninka.ru/article/n/ analiz-rezultativnosti-uchetnyh-i-kontrolnyh-protsedur-nalogovyh-organov. Accessed 28 Jan 2020
24. Gruzdeva, V.G.: Model for assessing the effectiveness of the main forms of tax control. Entrepreneurship **4**, 97–100 (2009)
25. Ivanova, E.: Tax administration efficiency assessment methodology. Financ. J. **4**, 109–118 (2011)
26. Yashina, N.I., Alexandrov, E.E.: Methodological tools for determining the effectiveness of tax audits. Financ. Credit **17**(689), 28–39 (2016)
27. Dabla-Norris, E., Misch, F., Cleary D., Khwaja, M.: IMF working paper. Tax administration and firm performance: new data and evidence for emerging market and developing economies. International Monetary Fund (2017). https://www.imf.org/en/Publications/WP/ Issues/2017/04/14/Tax-Administration-and-Firm-Performance-New-Data-and-Evidence-for-Emerging-Market-and-44838. Accessed 29 Jan 2020

Digital Technologies Development in Industry Sectors and Areas of Activity

Elena Alexandrova$^{(\boxtimes)}$ ⓘ and Marina Poddubnaya ⓘ

Economy Department, Kuban State University, Stavropolskaya Street, 149,
350040 Krasnodar, Russia
al-helen@mail.ru

Abstract. The paper considers the directions and problems of digital technologies development in certain industries and areas of activity. Currently, the
industries demonstrate different levels of digital maturity. In some industries,
there are so-called "superstar" firms that have, unlike traditional companies, a
higher level of productivity and derive maximum benefits from digital technologies. "Superstars" set trends in the development of digitalization in their
industries. Other industries are far from their potential to use digital technologies. The speed of implementation of innovative solutions in the business
processes of companies depends both on the external environment, and on the
problems and risks that are generated by such solutions. Some of such problems
are discussed in this article. On an example of tourism and network retail, the
author describes the industry specifics of the advantages and restrictions of the
development of some digital technologies, including digital platforms, the
Internet of Things, blockchain technology, and mobile applications. For many
companies, the introduction of digital technologies is a complex process not
only from the technical, but also from the management side, requiring a digital
strategy, implementation plan, and a comprehensive assessment of possible risks
and benefits for firms. The paper highlights the necessary components of a
digital strategy that meets the goals of the digital transformation of companies.
Special attention is paid to the industry features and specifics of the market in
which the company operates.

Keywords: Digital technology · Industries · The challenges of digital
transformation

1 Introduction

Digital technologies are now affecting many industries and areas of activity, leading to
processes of digital transformation. As part of the Fourth Industrial Revolution,
countries, industries and companies are faced with the need to change their business
models under the influence of modern technologies, including artificial intelligence,
machine learning, chat rooms, block bots, cloud technology, Internet of Things, etc.
The preconditions for digital transformation in certain industries are determined by "the
desire of new digital generation clients for timeliness, availability, quality and personalization" [1]. Their implementation, according to The World Economic Forum
(WEF) estimates, can increase productivity in companies by 40% [2]. Active

© Springer Nature Switzerland AG 2021
T. Antipova (Ed.): ICIS 2020, LNNS 136, pp. 112–124, 2021.
https://doi.org/10.1007/978-3-030-49264-9_10

introduction of digital tools occurs in all of the fields and scopes of activities: marketing; production, sales; finances; accounting [3, 4].

In modern scientific literature various problems associated with the introduction and development of digital technologies in industries and sectors of the economy are presented. Don Tapscott identifies some "key areas in the economy", including retailing and distribution, manufacturing, tourism and other, in which "the new technology and business strategies are transforming... business processes" [5].

The effectiveness of digital technologies is mainly related to cardinal changes in the business processes of companies in order to take advantage of the information provided by these technologies [6]. Companies must use different digital technologies that are integrated across staff, processes and functions to achieve an important business advantage. The digital maturity of companies and areas in the economy determines the effectiveness of their digital strategies. In other words, "simply implementing or using digital technologies is not enough" for successful development.

Digital technologies open up not only new and unknown opportunities for their application in industries but also have disruptive social and economic consequences [7]. According to Berkhout and Hertin [8], there are several effects of the information and communication technologies (ICTs) and the Internet's impact on the environment, including "direct impacts – resource use and pollution related to the production of infrastructure and devices, electricity consumption of hardware, etc., and indirect impacts related to the effect of ICTs on production processes, products and distribution systems". Hirsch-Kreinsen considers the consequences of the progressive use of digital technologies in industrial work [9]. Particular attention is paid to the problems of "data migration and the integration of the new systems into the existing production structures and databases", "skeptical attitude towards the automation and the efficiency promised by the smart systems", as well as security and safety of the complex databases. The author points out that the widespread diffusion of digital technologies, "with their structure-changing character, is confronted by technical, economic and social barriers that are hard to overcome".

The key objective of introducing digital technology in companies of different sectors and industries is to collect, process and provide the information needed by the user [10]. The needs of a particular industry play an important role. Thus, for network retailers, to meet the needs of customers at any time and in any place, omni-channel business models that synchronize data and information in all digital and physical channels of interaction with it becomes important. The tourism industry is characterized by a large number and variety of information flows, which are accompanied by constant updating and high speed of exchange transactions. This leads to the need to use digital technologies (social networks, mobile phone apps, voice assistants, Internet of Things, augmented reality technologies (AR), etc.), which allow delivering information to target consumers, including on a global scale, in almost real-time. Such technologies allow to reduce costs, increase operational efficiency, and improve service and quality of customer service in tourism.

In the Bain & Company Digital Insights survey of more than 1,200 senior executives, 54% of respondents indicate that digital technology will have a significant impact on their industry over the next five years [11]. Almost all industries are undergoing digital transformation and are starting to use digital services to some extent.

At the same time, some companies are rapidly increasing the volume of digital services provided, while others are only switching to digital technologies in order to increase their competitiveness, speed up decision-making, improve the effectiveness of risk analysis and accuracy of their forecasts. The most digitally sensitive high-tech industries are those related to software production and distribution. The banking sector, the service sector, the chemical industry, FMCG networks, and mechanical engineering are modernizing relatively quickly within the framework of the digital economy. As for traditional industries such as agriculture, digital transformation is also present in them, although at a relatively slow pace.

The widespread digitalization gradually affects traditional sectors of the economy, as well as leads to new markets and niches opening up. For example, big data analysis and artificial intelligence technologies help find new sources of value creation by studying digital consumer profiles and patterns of their economic behaviour. For instance, the BASF chemical company provides its customers with detailed recommendations on what fertilizers to use, in what amount and on what plants at a given time, based on monitoring and analysis of data on soil, plant health, weather conditions and other parameters [12]. Applications of the Internet of Things make it possible to evaluate the parameters of product use and the effects achieved. On this principle is built a popular model of carsharing, payment for car insurance depending on the kilometres covered [13].

Directions and possibility of mass introduction of digital technologies in business processes of the companies are connected both with industry-specific issues, and a number of the general conditions predetermining their development in concrete branch and the country as a whole. Among the main drivers of the digital economy and the digitalization process is undoubtedly the attention paid by the government to the topic of digital economy. But the main driving force is business. First and foremost, it is about the readiness of companies themselves for digital transformation through the development of strategies and new business models based on the intensive use of digital technologies. One of the important aspects of digital transformation in the industry of a particular country is its hardware and software, the lack of which in many cases determines the dependence of national industry companies on foreign suppliers. It is obvious, that a mature sector of technological supply at the country level should meet the sectoral demand for the necessary technologies both at the expense of its own resources (development of national technological solutions) and the possibility of rapid transfer and adaptation of foreign digital technologies and business solutions based on them. Finally, it is important to take into account not only the technologies themselves, which change the business of companies in the sector, but also the reaction and attitude of consumers to such technologies [14]. It is "consumers' needs and opportunities that ultimately determine the demand for digital technology from organizations, primarily in the B2C sector, that is adequate for them" [13].

The first part of the work analysed the development of digital technologies in various spheres of the world economy, on the example of individual sectors considered the directions of their use, taking into account the industry specifics, characterized the main advantages and problems of their implementation. In the second part of the study, the main conclusions and results are presented, and areas for further work in this area are identified.

2 Digital Technologies Development Sector-Wise: Trends and Challenges

Digitalization covers an increasing number of industries and sectors, from agriculture to construction. It covers a wide range of new applications of information technology in products and business models. Industries that are leaders in digital transformation processes are typically services or sectors associated with the manufacture of products that are less tangible and more intangible than physical products. Other sectors that demonstrate faster digitalization include those with direct consumer connections, faster capital turnover and are more global than local. Sectors with significant digitalization gains are the media and finance; those lagging include pharmaceuticals and large industrial firms [15].

Today digital technologies are actively implemented in companies of different industries and sectors - Nasdaq, Facebook, Google, IBM, VISA, Master Card, Bank of America, HSBC, AT&T, Coca Cola, Starbucks, Netflix, etc. Superstar firms, including Google, Amazon, Facebook, Apple, Alibaba, Baidu and Tencent, are emerging in certain industries based on global digital platforms. These companies, in contrast to traditional ones, have a higher level of productivity, and their efficiency and global scale make them part of a small and increasingly concentrated group of firms that create value for shareholders above the cost of capital [16]. Superstars have been able to take advantage of digital technology because of their accumulated and constantly evolving technical expertise, IT infrastructure development and significant investments. These and other factors allow them to compete for rare and expensive data analysis specialists, create powerful data centres and special processors, and enter new markets for them. For example, China's Baidu, Alibaba and Tencent (BAT) are actively investing in smart technologies to enter markets where U.S. companies traditionally hold leading positions: chip development, virtual assistants and unmanned vehicles, etc. Tencent has access to over 1B users on its platform, Baidu is the country's largest search provider, and Alibaba is its biggest e-commerce platform [17]. Now, these companies are actively promoting intellectual technology development platforms and individual applications in international markets, based on their own experience with these tools.

At present, no country, industry or areas has achieved full digital transformation on a global scale. Even when conditions are created to accelerate the introduction of digital technologies (through special public policy measures, development of appropriate infrastructure, etc.), many companies in different industries face the fact that the different types of technologies used in practice are quite complex and their implementation ultimately remains slow. Leveraging McKinsey Global Institute (MGI) most recent 2018 digital survey, they find «that digital maturity across ICT-using sectors by traditional incumbent firms is still relatively low, growing thus far to only around 25 percent of the potential» (Table 1) [18].

Table 1. Digital maturity level of selected industries (the application of digital technologies by companies).

N	Industries	Digital frontier gap (100% = ICT frontier), %
1	Travel	51,0
2	Retail	46,0
3	Automotive and telecom assembly	31,0
4	Financial services	29,7
5	Consumer packaged goods	28,5
6	Media	25,0
7	Healthcare system	24,3
8	Business and professional services	17,0
9	Pharmaceuticals/medical products	13,4

Source: McKinsey Digital Survey 2018; McKinsey Global Institute analysis.

Table 1 presents digital maturity levels in selected sectors of global economy for 2018. In particular, the tourism industry is most susceptible to the process of digital transformation: its digital maturity level is 51%. According to the data provided, many sectors are far from their potential to use information technology. Only 26% of global sales were made through digital channels, 31% of transactions were digitally automated and 25% of supply chain interactions were digitized [19]. In companies, the largest share of digital technology is automated internal operations - 30% [18].

Integrating technology into business processes, companies of various industries increasingly benefit in terms of improved labour efficiency. According to the estimations by MGI in 2018, by 2030 global GDP will grow by 13 trillion dollars due to digitization, automation, and artificial intelligence, as these technologies create significant new business opportunities and productivity growth is reinvested in the economy [20].

Modern digital technologies create new industry problems and risks, which also affects the speed of their widespread implementation in companies' business processes:

- data protection threats;
- cyber threats;
- the need for radical changes in the structure of companies and their business processes;
- the area of companies' activities is too difficult to understand and control;
- the lack of specialists.

The priority of the above-mentioned threats differs with industry. Thus, in the field of consumer goods, the most important threats are those related to data protection, and in the field of high technologies, media, and telecommunications – cyber threats.

Among the limitations of using digital technologies, it should be noted that many companies lack the experience and resources to fully take advantage of them. Working with artificial intelligence requires teams of experts in the field, large amounts of data, specialized infrastructure, and computing power. Also, companies with such resources should identify suitable ways to use smart technologies, develop specialized solutions and implement them in business processes. The development of firms' competencies based on digital solutions requires investment and time, and this in itself is a problem for many of them.

Cloud computing, mobile technologies, sensors, Internet of things (IoT), big data, cognitive technologies(AI), augmented reality (AR), robotics, additive manufacturing (3D printing), drones and other technologies can extract information from physical devices (sensor data on the state of a physical device), quickly distribute it (using mobile technologies), store it in the cloud, instantly analyse it (using Big Data and advanced analytics), thereby integrating products, services and processes, and often having a devastating impact on the established business models of traditional companies [21].

The manifestation of threats to information security leads to industry and corporate losses as a result of data privacy violations, disclosure of trade secrets, the possibility of industrial espionage, unforeseen problems in business processes, intellectual piracy, and reduced quality of products and services. From the information security, the least controlled areas among the many digital technologies are Big Data, the Internet of things and artificial intelligence technologies. For example, the number of cyberattacks on IoT devices is growing, as more companies acquire "smart" devices, such as routers or video cameras, but not all of them provide an appropriate level of protection. In the context of digital transformation, it is clear that companies must implement intelligent, innovative and effective controls to detect and prevent various cyberattacks. Most cyberattacks over the past few years have targeted the financial industry (primarily banking) and the industrial sector. For example, for retail, the need to ensure information security leads to an increase in legal services, a trend towards outsourcing of it security expertise, as well as growing threats associated with the use of new technologies in trade and the development of the e-commerce network. The lack of effective protection and counteraction to threats to information security ultimately leads to a decrease in the growth rate of digitalization in individual companies, industries and countries. At the same time, more than 79% of companies indicate that they are learning new technologies faster than they can solve the security problems associated with them. According to experts, dependence on the Internet is growing, while confidence in Internet security remains low and is projected to fall from 30% in 2018 to 25% in 2023 [22].

The introduction and application of digital technologies, despite some common problems and trends, has industry-specific characteristics. Let's consider some areas and problems of digital technology development using the example of tourism and retail networks.

The introduction and application of digital technologies, despite some common problems and trends, has industry-specific characteristics. Let's consider some areas and problems of digital technology development using the example of tourism and retail networks.

The use of digital technologies in the tourism sector is determined by the need for large flows, turnover and updating of information to inform target markets of new opportunities and products, the development of innovative communications with consumers, and the expansion of the list of traditional marketing tools. The consumer of tourist services needs accurate and complete information about the product, on the one hand, and available technological capabilities for receiving, processing, combining and transmitting information with automatic feedback, on the other hand. There are various technological solutions available to optimize business processes of tourism organizations:

- artificial intelligence can significantly improve the quality of customer service with less human resources involved. Research Booking.com shows that a third of the world's travel audience is interested in helping artificial intelligence plan trips;
- augmented reality (AR) technologies, such as the use of three- dimensional views of hotel rooms in mobile applications, are an important marketing and personalization tool in the tourism industry;
- voice-activated digital assistants are aimed at simplified and accelerated information search and data communication from the company to the consumer;
- the Internet of things provides a "seamless" journey – flight, transfer, hotel, car rental. By exchanging data, devices can minimize any waiting times, preventing various problems-from lack of Parking space to loss of orientation in an unknown city;
- blockchain technology has a significant potential in the field of tourism, which allows you to make secure transactions. For example, Lufthansa Group cooperates with the blockchain start-up Winding Tree, which allows you to get direct access to the offers of travel market service providers using a decentralized B2B platform for conducting booking transactions based on blockchain technologies. As a result, there is no need to interact with intermediaries like travel agencies, aggregators, distribution systems, etc. By integrating the application programming interfaces (APIs) that the Lufthansa Group participated in developing with the public blockchain platform Winding Tree, Lufthansa provides its partners with access to a decentralized platform with offers from travel providers, which excludes intermediaries.

Booking.com, Airbnb, Uber, Gett, etc. are among the most trending digital applications in the tourism sector. They are based on certain digital platforms that allow planning travels individually and provide a wide range of tourist services. The success of such platforms ' business models depends directly on the growth of the number of services they offer. Various blocks, services and mobile applications aimed at the tourism development are implemented through digital platforms. Digital platforms are the most important channel for companies to diversify their activities in the field of tourism beyond industry (transport, communications, accommodation, etc.) and country borders. Eventually, such diversification provides increased company profit due to:

- providing high-quality, convenient and diverse information content for end users, developing their consumer experience and personalization of services – the more services that customers start using, the more dependent they become on the platform and it becomes more difficult for them to switch to other services outside the platform. For example, global databases of travel services that provide tourists with solutions for the main elements of their trip according to specified criteria, while simultaneously selling and collecting selected trip elements;
- reducing online payment costs;

- new income sources for digital platform members;
- reducing the cost of promoting a tourist product;
- responding timely to new (changing) customer needs;
- developing modern types of communications with stakeholders (government organizations, clients, competitors) regardless of the country location.

The development of digital technologies leads to the growth of e-commerce in the tourism sector. Online sales of tourism services in the world is growing faster than the sale of books, music and computers. Numerous online stores, payment systems, auctions, travel companies and other tourism organizations are engaged in e-commerce. For example, in 2019, the volume of online sales of travel services in Russia exceeded 1 trillion. RUB and continues to grow at a rate of more than 20% per year [23].

However, digitalization of the tourism industry is associated with many challenges related to both general digital development constraints in the country (e.g. lack of legislation, broadband Internet access, digital inequality in the regions of the country, limitations on the infrastructure of fibre-optic lines, etc.) and industry-specific features. So, major challenges in digital tourism platforms development include the mechanisms for their design and implementation, as well as the procedure for connecting professional participants of the tourism industry (global accommodation facilities, associations, tour operators, payment systems, etc.) to these platforms. It is necessary to take into account the cost of investment projects in IT solutions, which help to increase the level of satisfaction of tourists. Today not all organizations are ready (both organizationally and financially) to connect services that allow registering in a hotel using a smartphone, to implement loyalty programs based on digital technologies. There is also a need for high-quality and effective software that allows automating the internal activities of travel companies, hotels, restaurants and other tourism businesses.

Digital technologies bring opportunities and challenges for chain retail companies. Digital technologies introduction and management in chain retail encourages digitalization of all business processes of a company, from procurements to sales, including:

- creating a brand image in social networks;
- simplifying payment;
- personalization of the offer and development of the consumer experience (for example, Amazon GO with its fully automated stores) based on CRM, Big Data;
- improving the efficiency and flexibility of operating activities;
- digital technologies in the development of employee competencies.

Digitalization is facilitated by the widespread Internet and e-commerce development. Also, companies that previously operated exclusively on the Internet (Alibaba, Tencent) are entering the field of offline retail and expanding the use of their knowledge in the field of digital technologies to improve the quality of the shopping process in the physical world. JD.com has invested 4.5 billion dollars in the development of an AI-enabled mall. Its goal is seamless integration of online and offline platforms and creation of virtual fitting rooms and automated stores. At the same time, physical retail retains a significant development potential due to the use of digital technologies. For example, self-service kiosks, smart carts, etc. thus, "digital" chain retail companies are actively using digital technologies to improve management efficiency and develop the consumer experience in both online and offline environments (an opportunity to provide a useful innovative digital experience). It is these companies that are expected to determine the specifics of the industry's development in the future.

Such digital technologies as smart contracts, VR and AR, Big Data, the Internet of things, etc. are being applied in retail. Intelligent technologies (for example, robots or self-service checkout counters) allow automating routine operations, and useful information is extracted from a huge array of data as much as possible. For example, the Internet of things (IoT) technology in retail studies customer preferences. In advertising communications, IoT is based on RFID tag technology, which quickly provides and analyses information about the movement of goods from the moment they are produced and delivered to stores until they are purchased. The French supermarket chain Carrefour uses iBeacon beacons to collect consumer data. Customers can use mobile phones or tablets attached to shopping carts to get in-store routes and personalized promotions.

Digital technologies in network retail allow companies to significantly reduce costs, automate monitoring of competitors' prices and significantly optimize the pricing process, predict demand and sales in conditions of uncertainty, optimize store management, develop consumer experience through personalized offers, etc. For network retail companies, the risks of implementing and developing digital technologies are related to both the specifics of network trading and the limitations of a particular technology. For example, in the case of the Internet of things, there are high risks associated with the constant accumulation and growth of huge amounts of information, protection of competition and non-discrimination, and threats of information loss through hacker attacks. Among the obstacles to the development of IoT, we note the lack of unified standards, which makes it difficult to combine wireless networks of objects into a single network.

3 Results

Companies that adopt digital technologies are well versed in the application of these technologies, and as a result, they demonstrate higher productivity and profit growth. In many industries, especially in professional services and retail, the lack of investment in digital technology leads to the fact that the cash flow of such companies is on average 30% lower than that of their colleagues using digital solutions. At the same time, if we are talking about the high-tech sector, this gap can be quite significant-80% [15].

Regardless of digital transformation goals in different industries and areas of activity, companies face problems that result from the nature of digital technologies (for example, legislative regulation, threats to information security, etc.) and the challenges of managing digital technologies in practice. Naturally, the process of digitalization is associated with high risks of failure, especially in customer experience digital transformations. Being digital natives, consumers are used to placing and receiving orders promptly and conveniently (everything can be done through a mobile phone app). So, they dictate the conditions, which manufacturers have to accept, as well as they have to make goods and services more available and user-friendly, and develop consumer experience, taking into account their individual characteristics and interests.

Many companies face problems precisely at the stage of practical implementation of digital technology. In other words, even having a digital development strategy, companies are not incapable of failure. As a rule, projects on the development of a digital solution are costly and require appropriate qualification of specialists. In this case, companies should invest in staff development, especially their skills in data analysis and digital technology. It is also important that digital solutions being developed can be scaled up. For example, a successfully implemented project on the implementation of digital technology in business processes of the company in one market could be implemented with the same success in another market.

In terms of the performance of digital transformation management, it is important to ensure company resources mobilization and consider the industry-specific peculiarities of implementing a particular technology. Many companies do not implement digital technologies in most business processes. There are objective reasons for this (e.g. traditional industry, where digital transformation takes place rather slowly and at the level of point projects, limitations in the development of digital infrastructure at the industry and/or national level, availability of technologies, etc.) and managerial reasons - lack of a clear understanding of the ways and goals of the application of such technologies to the benefit of a particular business. Management problems can be solved within the framework of a digital strategy that best meets the objectives of a company in a particular industry. Such digital strategy should include the following components:

- goals and objectives of the application of digital technologies;
- expected results of the implementation of a digital solution (new models of interaction with consumers, increased security, informed decision-making, increased average income per client, development of new sources of income, expansion of channels of interaction with consumers in online and offline space, etc.);
- directions and tools of organizational changes in the company.

Not only the industry characteristics are taken into account, but also the specifics of the market in which the firm operates. Thus, consumers in mass markets (B2C) are traditionally receptive to new innovative technologies and products based on these technologies. They encourage companies to introduce and master the most advanced technologies not only in business processes that remain "invisible" to end customers (for example, logistics, warehouse automation, which are "forced" to undergo digital transformation to improve the quality of customer service) but above all in processes that provide direct interaction with customers. In the field of tourism, for example, this includes the development of digital platforms that allow data to be processed and used to implement effective solutions of the company in various country markets, contactless payment system, "seamless" travel. In network retailing - collection and integration of marketing data, self-service cash registers, mobile applications, personalized loyalty programs, innovative channels of interaction with customers.

Thus, the consequence of digitalization projects in companies of different industries within the framework of digital strategies is an increase in the level of their competitiveness and increase in shareholder value, when a "traditional" company undergoing digital transformation reaches efficiency and productivity comparable to that of "technological" companies.

4   Conclusion

The aim of the study was to identify the main directions and problems of the development of digital economy technologies, taking into account the industry context. According to the results of the conducted study, the application of digital technologies in companies depends on the industry sector and area of activity. In some industries today there are large companies - "superstars", deriving maximum benefits from digital technology through accumulated and constantly evolving technical expertise, the development of digital infrastructure and significant investment costs. In industries with a high level of digital maturity, "traditional" companies focus on their colleagues who demonstrate high performance as a result of implementing digital solutions, which in some cases makes them follow the strategy of digitalization. Another reason is the inevitability of digitalization processes in some industries. Thus, in the tourism industry, mobile applications (allowing users of mobile devices to book flights, hotels, cars from anywhere) are becoming increasingly important, which in the next few years will become the basis for the development of competitive advantages of a number of organizations in the industry. An increasing number of companies and countries now have greater opportunities to take advantage of the growing popularity of online tourism offerings, platforms, information distribution and marketing opportunities. Companies that ignore the inevitability of digital transformation in the industry and miss opportunities to introduce information technologies will gradually disappear from the market. Consider intermediaries in tourism as an example: they have turned into online intermediaries, while traditional ones have disappeared from the market.

The study determined that digital transformation processes are penetrating an increasing number of industries and segments, including traditional sectors of the economy. New markets and niches are emerging as a result of digital technologies. As companies and industries undergo digital transformation, they benefit increasingly from productivity gains. However, to date, virtually no country, industry, or field of activity has achieved complete digital transformation on a global scale. In various industries, firms are faced with managerial and organizational challenges in implementing digital technologies into practice. In addition, modern digital technologies are creating new industry problems and risks, including data protection threats, cyber threats, specialist shortages and legislative regulation. Some companies have financial and time constraints that prevent them from taking full the benefits of digitalization.

The specific areas and challenges of digitalization of companies in the context of the industry have been identified through the example of tourism and network retailing. It has been concluded that digitalization in these areas is facilitated by the widespread development of the Internet and related e-commerce. Implementation should take into account the technologies and problems that arise when using a specific digital solution.

While the characteristics of the industry undoubtedly influence the successful implementation of innovative solutions, there are also important common challenges related to effective device management, development of global standards for data collection and subsequent synthesis, identification of digitalization ideas and actions based on them on an ongoing basis across the industry, and ensuring security when using the corporate and personal information of consumers. From the point of view of the effectiveness of the company's digital transformation management, the digital strategy should focus on goals and objectives of applying specific digital technologies, taking into account the industry specifics and market characteristics of the company's operation. Another area of digital strategy should be a comprehensive approach to assessing the expected results of the implementation of digital solutions, as well as related areas and tools for organizational change in the company.

The study does not claim to comprehensively characterize the promising areas and key challenges of the impact that digital transformations have on company development, taking into account the level of digital maturity of the corresponding industry. Monitoring and reviewing the opportunities and existing limitations of digital transformation in individual industries will help us get closer to solving this problem. The vector of further research on the chosen topic is seen in the field of evaluating the contribution of digitalization to the development of industries and spheres of activity, measuring the contribution of digital technologies to labour productivity growth in various industries.

References

1. Akatkin, Y.M., Karpov, O.E., Konyavsky, V.A., Yasinovskaya, E.D.: Digital economy: conceptual architecture of the digital industry ecosystem. Bus. Inf. 4(42), 17–28 (2017)
2. Digital Transformation Initiative. Unlocking $100 Trillion for Business and Society from Digital Transformation. World Economic Forum. http://reports.weforum.org/digital-transf ormation/wp-content/blogs.dir/94/mp/files/pages/files/dti-executive-summary-20180510.pdf. Accessed 15 Jan 2020
3. Kuter, M.I.: Introduction to Accounting: A Textbook. Prosveshchenie-Yug, Krasnodar (2012)
4. Lugovsky, D., Kuter, M.: Accounting policies, accounting estimates and its role in the preparation of fair financial statements in digital economy. In: Antipova, T. (ed.) Integrated Science in Digital Age, ICIS 2019. Lecture Notes in Networks and Systems, vol. 78, pp. 165–176. Springer, Cham (2020). https://doi.org/10.1007/978-3-030-22493-6_15
5. Tapscott, D.: The Digital Economy: Promise and Peril in The Age of Networked Intelligence: A Textbook, p. 368. McGraw-Hill, New York (1994)
6. Kane, G.C., Palmer, D., Phillips, A.N., Kiron, D.: Is your business ready for a digital future? MIT Sloan Manage. Rev. 56(4), 37–44 (2015)
7. Avant, R.: The third great wave. In: The Economist, 4 October 2014, Special Report (2014)
8. Berkhout, F., Hertin, J.: De-materialising and re-materialising: digital technologies and the environment. Futures 36(8), 903–920 (2004)
9. Hirsch-Kreinsen, H.: Digitization of industrial work: development paths and prospects. J. Labour Mark. Res. 49(1), 1–14 (2016)

10. Alexandrova, E.: Digital economy in competitiveness of modern companies. In: Antipova, T., Rocha, Á. (eds.) Digital Science 2019, DSIC 2019. Advances in Intelligent Systems and Computing, vol. 1114, pp. 114–125. Springer, Cham (2020). https://doi.org/10.1007/978-3-030-37737-3_11

11. Anderson, N., O'Keeffe, D., Lancry, O.: What's Hot and What's Not. Bain & Company, https://www.bain.com/insights/whats-hot-and-whats-not-snap-chart/. Accessed 10 Jan 2020

12. What is Digital Farming? Discussion with Tobias Menne, Head Digital Farming Unit, BASF. Capgemini. https://www.capgemini.com/2018/04/what-is-digital-farming-discussion-with-tobias-menne-head-of-digital-farming-basf/. Accessed 10 Jan 2020

13. What is the digital economy? Trends, competencies, measurement. HSE Report (2019). https://www.hse.ru/data/2019/04/12/1178004671/2%20Цифровая_экономика.pdf. Accessed 02 Feb 2020

14. The Digital Enterprise: Moving from experimentation to transformation. Insight Report. World Economic Forum. https://www.bain.com/contentassets/7279619637c1423d9603f6b87518e13e/digital_enterprise_moving_experimentation_transformation_report_2018.pdf. Accessed 10 Jan 2020

15. Bughin, J., Manyika, J., Catlin, T.: Twenty-five years of digitization: ten insights into how to play it right. McKinsey Global Institute (MGI). https://www.mckinsey.com/business-functions/mckinsey-digital/our-insights/twenty-five-years-of-digitization-ten-insights-into-how-to-play-it-right. Accessed 12 Jan 2020

16. Superstars: The dynamics of firms, sectors, and cities leading the global economy. McKinsey Global Institute. https://www.mckinsey.com/ ~ /media/McKinsey/Featured%20Insights/Innovation/Superstars%20The%20dynamics%20of%20firms%20sectors%20and%20cities%20leading%20the%20global%20economy/MGI-Superstars-Discussion-paper-Oct-2018.ashx. Accessed 23 Jan 2020

17. Rise of China's Big Tech In AI: What Baidu, Alibaba, And Tencent Are Working On. Cbinsights. https://www.cbinsights.com/research/china-baidu-alibaba-tencent-artificial-intelligence-dominance/. Accessed 23 Jan 2020

18. A winning operating model for digital strategy. McKinsey Digital. https://www.mckinsey.com/ ~ /media/McKinsey/Business%20Functions/McKinsey%20Digital/Our%20Insights/A%20winning%20operating%20model%20for%20digital%20strategy/A-winning-operating-model-for-digital-strategy.ashx. Accessed 23 Jan 2020

19. Bughin, J., Catlin, T.: 3 digital strategies for companies that have fallen behind. Harvard Business Review, 12 February 2019. https://hbr.org/2019/02/3-digital-strategies-for-companies-that-have-fallen-behind. Accessed 02 Feb 2020

20. Assessing the economic impact of artificial intelligence. ITUTrends. https://www.itu.int/dms_pub/itu-s/opb/gen/S-GEN-ISSUEPAPER-2018-1-PDF-E.pdf. Accessed 02 Feb 2020

21. Spremić, M.: Governing digital technology – how mature it governance can help in digital transformation? Int. J. Econ. Manage. Syst. 2, 214–223 (2017)

22. Securing the Digital Economy. Reinventing the Internet for Trust. Accenture. https://www.accenture.com/_acnmedia/thought-leadership-assets/pdf/accenture-securing-the-digital-economy-reinventing-the-internet-for-trust.pdf. Accessed 23 Jan 2020

23. eTravel in Russia-2019: Statistics and trends. Data Insight. https://ict.moscow/research/etravel-v-rossii-2019-statistika-i-tendentsii/. Accessed 23 Jan 2020

IPO – The Pattern of Hierarchy with a Variety of Alternatives upon Criteria

Igor Shevchenko$^{(\boxtimes)}$ (iD), Svetlana Tretyakova (iD), Natalya Avedisyan,
and Natalya Khubutiya (iD)

Economy Department, Kuban State University,
Stavropol'skaya Street, 149, 350040 Krasnodar, Russia
exclusi@list.ru

Abstract. An article is dedicated to the discussion about the ways to find
solutions for the questions and closely related questions of IPO's strategy as the
pattern of hierarchy with a variety of alternatives upon criteria. The type of
public offering in which shares of a company are sold to institutional investors
and usually also retail (individual) investors. An IPO is underwritten by one or
more investment banks, who also arrange for the shares to be listed on one or
more stock exchanges. Fundamentally, this article consists of a systematic
introducing to the theory of relative value of criteria, and this theory had been
developing the author during the period of a couple of decades. In the article,
there is the axiomatic methodology of presenting is utilized. This methodology
is the type of method where previously the set of requirements, named axiom, is
formed which then are associated with the group of tasks being under consideration. The most important resources for this are the information about the
relative values of the criteria.

Keywords: IPO · Decision-making strategy · Hierarchy analysis

1 Introduction

The transformation of the economic system of the Russian Federation into the market
economy caused escalating difficulties in raising capital within the industry sector in
order to strengthen the capability and to maximize profit. Therefore, the need for
determining the strategies of augmentation of the funds by attracting third-party
investors appeared. The strategy of initial public offering (IPO) is a strategy of raising
funds through the enterprise entry into the world financial market; in other words, a
company turns into a public enterprise, which gains additional funding by the means of
publishing shares on the stock exchange or several of them [1, 3]. The key aspect of
successfully implementing this strategy is the legitimacy of the open financial and
economic activity of a company. The domestic enterprises experience difficulties in
finding a decision about realizing the strategy of IPO [2, 4].

The general approach for the implementing this method could be considered as a
model including further steps: firstly, the preliminary analysis of implementation the
strategy of the initial public offering; secondly, defining the alternative options of the
strategy and the key criteria of the selection; thirdly, collecting the quantitative data and

© Springer Nature Switzerland AG 2021
T. Antipova (Ed.): ICIS 2020, LNNS 136, pp. 125–134, 2021.
https://doi.org/10.1007/978-3-030-49264-9_11

generating the model of mathematical modeling; then mathematical process of the gathering data [5, 6].

The implementing of the IPO as a decision-making approach could be considered as the appearing approach. This kind of approach is included into the dynamic category, due to the fact that it requires constant repeating of the decision making in various time intervals, consequently, the character of its integration could be considered as regular. There are several options exist which could be used for finding the way to make a decision about realizing the initial public offering strategy [7, 8, 10]. In order to achieve success, the indispensable condition is to define the most suitable alternative. The target of IPO is considered as the non-trivial method due to the fact that it consists of several criteria of decisions making process [9, 11]; thus, to decide the method is based on type of gathered information, particularly, the information which illustrates the preferences founded on multi-criteria and multiple-choice in the condition of uncertainty (Fig. 1) [12–14].

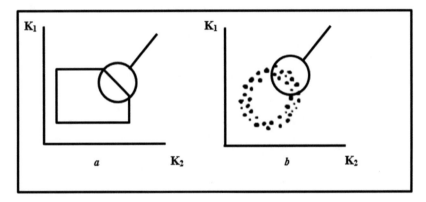

Fig. 1. Area of non-dominate effect. Choice of the alternatives relying on two criteria: a) The case of the sustained field of alternative; b) The case of discrete alternative.

2   Establishing the Modification of IPO Decision Making

The literature in this field shows several approaches, which are subsidiary in decision-making within the conditions of multi-criteria. The further explanation explores this issue more extensively. The method of hierarchy analysis (MHA) is a method of separating the problems or targets into the elements of the lower level and assessing a person, which is officially assigned for the responsibility of decision-making. The method of hierarchy analysis (MHA) is graded by the feature of links among criteria and alternatives [15, 16]. There are two types of hierarchy, which are defined by the nature of the connectivity among the criteria and alternatives.

The first type of hierarchy constitutes the pattern where each criterion has a connection with the particular alternatives and at the same time have the connection with every considering set of alternatives [11, 12]. In other words, this is the type of hierarchy which has a particular number of alternatives and the similar functioning of these alternatives within the group which is determined by criteria. Another type of hierarchy contains the criteria, which are linked to the alternatives, are not in connection with every other considering alternative. This is a type of hierarchy which includes the alternatives with the differences in the number and the diversity in functions of alternatives which are settled by certain criteria (1, 2). In accordance with this statement, the modification of the multicriteria choice upon hierarchy with a variety of number and structure of alternatives formed by criteria appears for practitioners in practice as the extremely beneficial for getting closer to the main target, particularly to the finding the most beneficial strategy in the initial public offering [16–25]. Considering the practical experience, there is the approach of the ranking of numerous criteria and alternatives for finding the right solution, and these criteria and alternatives could be assessed by the expert review and evaluation as well as using the established standards and norms. An example of the hierarchy of decision-making strategy within the initial public offering field could be illustrated by the method where each criterion Kj out of the variation {K1, K2, … Kp} includes the different number of alternatives out of the variation {A1, A2, … Ai}. Alternatives "Ar" are estimated accordingly the criteria Kp. The pattern of hierarchy with a variety of alternatives upon criteria is illustrated in Fig. 2, where part "a" is synthesis and part "b" is decomposition.

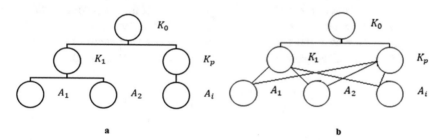

Fig. 2. The pattern of hierarchy with a variety of alternatives upon criteria

In the case when the hierarchy has one level, there could be emphasized the methodology of identifying the vector of prioritized alternatives combined by the focus relying on the value of the criteria and the number of alternatives of the criteria. This methodology implies the completing of the series of particular procedures in order to frame the information for conducting the estimation [3, 33, 36]. The implementation of this methodology for meeting the challenge in IPO consists of several stages. It is essential to make the initial step which is to structure the task in the hierarchy which defines the interconnection among the multiplicity of comparing alternatives {A1, A2, … Ai} and multiplicity of criteria {K1, K2, … Kp}. Criteria and alternatives for the compilation of the hierarchy of initial public offers are taxons which are illustrated (1).

The further step includes the formation of a binary matrix based on the hierarchical structure [B] establishing the accordance among the alternatives and criteria. The matrix [B] consists of the components bij = {0, 1}. With the conditions of the conduction of measurement in a way where the estimation of alternative Ai is made upon criteria Kj, the element bij = 1; otherwise the element bij = 0. The following stage is to conduct the expert estimation of alternatives according to the suitable set of criteria. In order to achieve this aim, the method of doubled-paired comparison, the method of comparison of relative standards, or the method of replication. Taking as a basis the expert assessing results, which relying on matrix [B], the matrix [A] is built reflecting the model what is illustrated in number of mathematical quantities expressed by conventional signs (1) [33, 34].

$$
[A] = \begin{array}{c} \\ A_1 \\ A_1 \\ \dots \\ A_i \end{array} \begin{array}{cccc} K_1 & K_2 & \dots & K_p \\ \left| a_{11} & a_{12} & \dots & a_{1p} \right. \\ a_{21} & a_{22} & \dots & a_{2p} \\ \dots & \dots & \dots & \dots \\ a_{i1} & a_{i2} & \dots & a_{ip} \end{array} \tag{1}
$$

The matrix [A] comprises the components of expert judgment {aip} are represented by vectors of alternatives priority regarding the criteria Kp. If the alternative Ai is not evaluated according to the criteria Kp, the matrix [A] has the meaning of aip = 0. The vectors of the considering matrix have various AIP meanings, therefore, they could be classified as nominalized or non-nominalized depending on the applied method of alternatives comparison. The determination of the category of the vector of prioritized criteria $\overline{Y}$ is the result of the doubled-paired comparison of the Kp criteria. In this case, the criteria G and S are generated named structural criteria, which are reflected in the relevant diagonal matrix [G] and [S]. The matrix [G] has the pattern illustrated in the number of mathematical quantities expressed by conventional signs (2).

$$
[G] = \begin{bmatrix} E_1 & E_2 & \dots & E_p \\ \left(\sum_{i=1}^{r} a_{i_1}\right)^{-1} & 0 & \dots & 0 \\ 0 & \left(\sum_{i=1}^{r} a_{i_2}\right) & \dots & 0 \\ \dots & \dots & \dots & \dots \\ 0 & 0 & \dots & \left(\sum_{i=1}^{r} a_{i_p}\right) \end{bmatrix} \tag{2}
$$

The normalizing of the vectors of priority alternatives forming the matrix [A] is ensured by means of matrix [G] if the matrix [A] is completed by the method of comparison regarding standards or coping regardless the preliminary normalizing. The matrix [S] has the pattern shown in the number of mathematical quantities expressed by conventional signs (3).

$$
[S] = \begin{bmatrix} E_1 & E_2 & \dots & E_p \\ Q_1 / N & 0 & \dots & 0 \\ 0 & Q_2 / N & \dots & 0 \\ 0 & 0 & \dots & 0 \\ 0 & 0 & \dots & Q_p / N \end{bmatrix} \tag{3}
$$

The element Qp is the number of alternatives Ai which is placed regarding the criteria Kp, also the cumulative number of the alternatives referred to all criteria (16–33).

$$N = \sum_{j=1}^{P} Q_j \tag{4}$$

The implementation of the structural criteria S allows practitioners to alter the weight of alternatives if there is a necessity for it. The changing of the alternatives could be made if they are linked to the relevant criteria according to the proportionate ratio: Qp/N. This proportion guarantees the increase of the value of the priority alternatives represented by the large groups, as well as ensures the decreasing of the priority of the group which has a relatively small number [24–33]. It could be explained in another way where the group is determined by the set of alternatives constituted by the "projection" of the criteria Kp. The necessity of the provided pattern of calculation is caused by the fact that the criteria appearing as "founders", which have the high priority in the hierarchy, can contain a large number of alternatives classified as "followers". However, the category of the criteria "founders" with a low level of priority has significantly less number of alternatives of the "follower" category [27–33]. The algorithm of determination the vectors of prioritized alternatives within hierarchy consisting of several directions. The example of the process of determining the vector of prioritized alternatives in the hierarchical structure of the initial public offering is illustrated in detail further. This includes several directions formed according to Fig. 3.

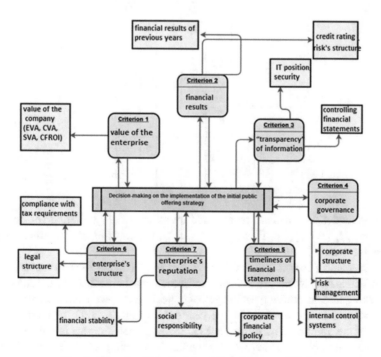

Fig. 3. The algorithm for hierarchy with several limbs

The further scheme is the computations of the vectors (5) of prioritized alternatives depending on criteria K_{ip}:

$$\left\{ \begin{matrix} W^A_{K_{11}} & W^A_{K_{12}} & \cdots & W^A_{K_{1m}} \\ W^{A'}_{K_{21}} & W^{A'}_{K_{21}} & \cdots & W^{A'}_{K_{2n}} \\ W^{A^n}_{K_{n1}} & W^{A^n}_{K_{n2}} & \cdots & W^{A^n}_{K_{np}} \end{matrix} \right\} \tag{5}$$

The next step is the construction of the matrix [Ai] where the rows are named regarding the alternatives and the columns are the criteria K_{ip}. The element "i" of the vectors of prioritized alternatives $W^A_i (i = 1, r)$ is calculated according to criteria Ki. This is illustrated in the next model (6):

$$\begin{aligned} W^A_1 &= [A_1][G_1][S_1]\bar{x}_1[B_1] \\ W^A_2 &= [A_2][G_2][S_1]\bar{x}_2[B_2] \\ &\cdots \cdots \cdots \cdots \cdots \\ W^A_r &= [A_r][G_r][S_r]\bar{x}_r[B_r] \end{aligned} \tag{6}$$

The element [Gi] is the matrix for rationing the matrix [Ai]; the element [Si] is a structured matrix for modification of the value of alternatives. The process of building the resulting matrix [Ao] is depicted in the number of mathematical quantities expressed by conventional signs (7).

		K_1	K_2	...	K_r
	$\begin{matrix}A_1\\A_2\\\cdots\\A_m\end{matrix}$	W^A_1	0	...	0
$[A_0] =$	$\begin{matrix}A'_1\\A'_2\\\cdots\\A'_s\end{matrix}$	0	W^A_2	...	0
	$\begin{matrix}A''_1\\A''_2\\\cdots\\A''_t\end{matrix}$	0	0	...	W^A_r

$$\tag{7}$$

In this model the rows are represented by all considering alternatives on the way ($\{Ai\}, i = 1, t; \{A^{\wedge'} i\}, i = 1, s; \{A''i\}, i = 1, t$); the columns are the criteria Kp. The result is the matrix [Ao].

This research paper is based on the information relating to the companies of the Russian Federation which are intended to implement the strategy of the initial public offer. Taking into account the challenges of collecting the data from earlier periods, the period included in this research is one year before applying IPO for preliminary IPO measure of the traditional performance [35, 36, 38]. Therefore, the initial data gathered for twelve Russian enterprises which include the meaning of vectors of alternatives during the estimation of them regarding the explained criteria of the one-year period before starting IPO and the one-year period after implementation the strategy of IPO (5).

The further calculation needs to be continued with computing the right site vector of the doubled-paired comparison matrix. Within this process, the experts answered the questions related to defining the alternatives, which have higher impact, and to what extent is this impact when the strategy of the company's entry into the global market is realized (6).

The subsequent step after assessing is to conduct hierarchical synthesis resulting in the determination of the integral vector of alternative priorities regarding the relative focus of hierarchy [34, 37, 38]. For considered matrix the integral vector is:

$$W = \{\, 0,5 \quad 0,6 \quad 0,5 \quad 0,5 \quad 0,2 \quad 8,2 \,\}$$

The analysis of this vector allows providing the conclusion where there most suitable alternative is A7 only if it has the index 8,2 conventional units, what, thereby, specifies the most valuable alternatives, which is the cost of the enterprise, where the indicators are: EVA, CVA, SVA, CFROI [12, 36]. There are several issues need to be solved during the process of forming the matrix of the alternative priorities significations within the estimation procedure according to their criteria. The first question is relates to the selection of prioritized criteria for comparison; the anther question is what category of functioning could influence the company's strategy for making the decision about entering into the stock market.

3   Conclusion

The success of the selected solution within multi-criteria conditions relating to applying the IPO requires utilizing the diversity of information about the cost of the company. In this case, the most valuable criteria for the assessment of the efficiency of the company's cost are EVA, CVA, SVA, CFROI. The element EVA means the short-term perspective, specifically the period of the cycle reporting, where the profit of accounting, consisting of more than 150 variants, serves as the basis for calculation. The accounting profit is characterized by an overstatement of the indicators, therefore, the alternative of decision-making regarding this characteristic appears as the mistaken for practitioner form the beginning. The next element named CVA is the short-term and long-term perspective where the basis for calculation is the time of the payback period for investments. The other two components, which are SVA and CFROI, similarly to CVA are connected to the money flood, nevertheless, they have a difference in the duration of the forecasting (or estimation) and the area of implementation. There are several advantages, which are needed to be taken into consideration, appearing due to usage of the main criteria for estimating the cost of a company. These advantages are: it allows to determinate the type of funding and the amount of capital for reaching the necessary outcome; it allows closely monitor the spending on procedures for fundraising; it is could be implemented for the estimation of investment attractiveness in general and also for the variety of separated directions of the company' activity.

The results of the model do not depend on the accounting policy of standards and norms established in a company. It is suitable for public companies as well as for the private sector of business. The expenditure on vocational education and training the employees during the implementation of this approach is extremely low. There is more accuracy of capitalization calculation of the net profit NOPAT based on the market information about the initial value of invested capital. The method provides certainty to some extent about the effectiveness of committed investments considering the uneven distribution of the amount of value added by years.

The computation of the money flows is conducted with regard to inflation; and the accountant reserve funds are eliminated from the computation, due to this, the manipulations with the indicators are infeasible; the amount of shareholders refunds is measured in percent rather than in monetary terms, what improves the process of comparison of the profitability among the companies.

One of the most important resources for this is the information about the relative values of the criteria. However, before starting learning how to identify and to utilize this information, there are several issues which are needed to be solved. The first one is highly important to understand what does mean the information about the relative values of the criteria. Also, the question, which could appear for specialists, which are recently involved in this area, is what the meaning of criteria, or a group of criteria, is more valuable than other criteria, or a group of other criteria. Furthermore, being in the state of awareness of information about the relative value of the criteria, how this information can make a contribution to the process of making a decision. An additional question is: whether there are the principles of building the border for using the set of similar random questions and the questions about decision-making strategy. A article is dedicated to the discussion about the ways to find solutions for these questions and closely related questions. Fundamentally, this article consists of a systematic introducing to the theory of relative value of criteria, and this theory had been developing the author during the period of a couple of decades. In the article, there is the axiomatic methodology of presenting is utilized. This methodology is the type of method where previously the set of requirements, named axiom, is formed which then are associated with the group of tasks being under consideration. Moreover, the core concepts are rigorously determined and the results are formed in the theorem, which is substantiated by applying the applicable mathematics tools.

References

1. Penzin, K.: On the way to Russian IPOs, Money and Credit, 6, (2015)
2. Marttunen, M., Belton, V., Lienert, J.: Are objectives hierarchy related biases observed in practice: a meta-analysis of environmental and energy applications of multi-criteria decision analysis. Eur. J. Oper. Res. 265(1), 178–194 (2018)
3. Nihat, A., Kathleen, A., Ettore, C., Ozdakak, A.: Stock market development and the financing role of IPOs in acquisitions. J. Bank. Finance 98, 25–38 (2019)
4. Bomas, V.V.: Support for making multi-criteria decisions according to user preferences, DSS DSB, UTES, p. 172. Ed.: Publishing House of the Moscow Aviation Institute (2006)

5. Corner, T.: Prerequisites and factors for the development of business activity in IPO markets. T.V. Corner, Actual problems of international relations: political, economic, legal aspects: AI materials Int. (Correspondence.) Nauch. practical. conf. (Lviv, 19 September 2012), Council of Young Scientists of the Faculty of International Affairs. rel. Lviv. nat Un to them I. Franko. Lviv. nat Un to them I. Franko, 175 (2012)

6. Federal Law of December 26, 1995 No. 2008-ФЗ On Joint-Stock Companies (harm of Federal Laws of June 13, 1996 No. 65-ФЗ, as of May 24, 1999, No. 101-FL), Rossiyskaya gazeta. M., 29 December 1995, no. 248, p. 3; Meeting of the legislation of the Russian Federation. - M., №1 Article 1. (with subsequent changes), (1996)

7. Hafiz, H., Shaolong, M.: Partial private sector oversight in China's A-share IPO market: an empirical study of the sponsorship system. J. Corp. Financ. **56**, 15–37 (2019)

8. Shevchenko, I., Puchkina, E., Tolstov, N.: Shareholders' and managers' interests: collisions in russian corporations. High. School Econ. **23**(1), 118–142 (2019)

9. Database on IPOs and mergers and acquisitions. http://dealogic.com. Accessed 21 Dec 2019

10. Instruction of the Central Bank of the Russian Federation of 10.03.2006 No. 128-I: On the rules for issuing and registering securities by credit organizations in the territory of the Russian Federation (2019)

11. Lotov, A.V., Pospelova, I.I: Multicriteria decision-making problems. Achievable goals method, p. 203. UNPress, Moscow (2018)

12. Rife, G.: Analysis of solutions (introduction to the problem of choice in the condition of uncertainty). Per. Sangl., p. 408. Nauka, Moscow (1977)

13. Bebchuk, L.A., Brav, A., Jiang, W.: The long-term effects of hedge fund activism. Columbia Law Rev. 83–115 (2015)

14. Kim, M.: Valuing IPOs. J. Financ. Econ. **53**, 409–437 (1999)

15. Kini, R., Rife, X: Decision making under many criteria: preferences and substitutions: Per. Sangl. I.R. Shakhova, p. 560. Radio communication, Moscow (1981)

16. Kostogryzov, A.I., Nistratov, G.A.: Standardization, mathematical modeling, rational management and certification in the field of systems and applied engineering, p. 396. Publishing House of the military-industrial complex and 3 Central Research Institute of the Ministry of Defense of the Russian Federation, Moscow (2004)

17. Napolov, A.V.: Structuring of initial public offerings of shares of Russian companies, Dis. Cand. econ Sciences: 08.00.10, p. 247 (2019)

18. Global IPO Trends 2010 Ernst & Young Homepage. https://drivkraft.ey.se/wp-content/uploads/2011/06/EY-Global-IPO-trends-2010. Accessed 21 Dec 2019

19. Global IPO Trends 2011 Ernst & Young Homepage. https://drivkraft.ey.se/wp-content/uploads/2012/08/EY-Global-IPO-trends-2011. Accessed 21 Dec 2019

20. Global IPO Trends 2012 Ernst & Young Homepage. https://drivkraft.ey.se/wp-content/uploads/2013/10/EY-Global-IPO-trends-2012. Accessed 21 Dec 2019

21. Global IPO Trends 2013 Ernst & Young Homepage 2013. https://drivkraft.ey.se/wp-content/uploads/2014/12/EY-Global-IPO-trends-2013. Accessed 21 Dec 2019

22. Global IPO trends 2014 Ernst & Young Homepage. https://drivkraft.ey.se/wp-content/uploads/2015/14/EY-Global-IPO-trends-2014. Accessed 21 Dec 2019

23. Global IPO trends 2015 1Q Ernst & Young Homepage. https://www.ey.com/Publication/vwLUAssets/EY-global-ipo-trends-2015-q1. Accessed 21 Dec 2019

24. Global IPO trends 2016 2Q Ernst & Young Homepage. https://www.ey.com/Publication/vwLUAssets/EY-global-ipo-trends-2016-q2. Accessed 21 Dec 2019

25. Global IPO trends 2017 3Q Ernst & Young Homepage. https://www.ey.com/Publication/vwLUAssets/EY-global-ipo-trends-2017-q3. Accessed 21 Dec 2019

26. Global IPO trends 2018 4Q Ernst & Young Homepage. https://www.ey.com/Publication/vwLUAssets/EY-global-ipo-trends-2018-q4. Accessed 21 Dec 2019

27. Hearn, B., Filatotchev, I.: Founder retention as CEO at IPO in emerging economies: the role of private equity owners and national institutions. J. Bus. Ventur. **34**(3), 418–438 (2019)
28. Bellman, R., Zade, L.: Decision-making in vague conditions.Issues of analysis and decision-making procedures. Trans. Eng., pp. 172–175. Mir, Moscow (1976)
29. IPO Watch Europe Q1 2016 Electronic resource Review compiled by PricewaterhouseCoopers (2016). www.pwc.ru/ru/capital-markets. Accessed 21 Dec 2019
30. IPO Watch Europe Q1 2017 Electronic resource, Review compiled by PricewaterhouseCoopers (2017). www.pwc.ru/ru/capital-markets. Accessed 21 Dec 2019
31. IPO Watch Europe Q2 2018 Electronic resource, Review compiled by PricewaterhouseCoopers (2018). www.pwc.ru/ru/capital-markets. Accessed 21 Dec 2019
32. IPO Watch Europe Q3 2016 Electronic resource, Review compiled by PricewaterhouseCoopers (2016). www.pwc.ru/ru/capital-markets. Accessed 21 Dec 2019
33. IPO Watch Europe, Report for the third quarter of 2017 Electronic resource, the report is published by the World Federation of Exchanges World Federation of Exchanges, December 2017. https://www.pwc.ru/ru/publications/ipo-watch-europe.html. Accessed 21 Dec 2019
34. Pinelli, M.: Global IPO trends, Electronic resource, Report prepared by Ernst & Young (E & Y). http://www.ey.com/Publication/vwLUAssets/ey-q4-14-global-ipo-trends-report/$FILE/ey-q4-14-global-ipo-trends-report. Accessed 27 Dec 2019
35. Russian IPO Pioners-2: a review of Russian and CIS companies, PNB. http://www.pbnco.com. Accessed 25 Dec 2019
36. Sieradzki, R.: Does he pay to invest in IPO's: evidence from the Warsaw Stock Exchange. National Bank of Poland Working Paper, p. 139 (2018)
37. Tailor, M.A: The role of IPO scales in Russia and the socio-economic consequences of these processes. Electronic resource, Russia and America in the XXI century, p. 1 (2019)
38. Xunjie, G., Zeshui, X., Huchang, L., Francisco, H.: Multiple criteria decision making based on distance and similarity measures under double hierarchy hesitant fuzzy linguistic environment. Comput. Ind. Eng. **126**, 516–530 (2018)

The Problem of Agency Conflicts in Russian Corporations and Ways to Overcome It

Nikolai Tolstov$^{(\boxtimes)}$ and Igor Shevchenko

Kuban State University, 149, Stavropolskaya Street,
Krasnodar 350040, Russian Federation
ntolstov@yahoo.com

Abstract. This article is devoted to the problem of agency conflicts between shareholders and the hired managers in large Russian corporations. The senior management decisions that don't meet the interests of the owners can significantly weaken the company business model, which will lead to a decrease in the business efficiency. This may cause a significant drop in the company's market capitalization within a short period of time in a digital economy, which means a decrease in the value of the business for the shareholders. The models and indicators used in Western practices of developed markets in most cases are based on a limited set of financial information and don't make it possible to fully assess the quality of the financial management in economic realities of the corporate sector in Russia. In this regard, a new approach involving three stages of assessment is introduced, where interrelated indicators are selected and calculated according to the established procedure, all significant information environments are identified; a qualitative assessment of the indicators which allows identifying the strengths and weaknesses of the corporation taking into account its industry affiliation is carried out and the results of the second stage are consolidated in the final assessments of the financial management quality, which allows not only to determine the quality of the financial management, but also to identify weaknesses of the current business model. This approach has been tested in 20 largest corporations in Russia and the obtained results prove its effectiveness in Russian economic realities.

Keywords: Agency conflicts · Value based management · Management quality evaluation

1 Significance of Agency Conflicts in the VBM Concept

The beginning of the 21st century is characterized by a complete change in the established economic relations, rapid development and introduction of market mechanisms new for the economy of the country, active development of which continues in the situation of a widespread digitization of the economy as a whole, as well as the individual actors.

A natural development stage in forming the market economy in Russia was predominating majority of capital concentrated in the corporate sector, on the effectiveness of which almost all spheres of the society activities became directly or indirectly dependent. At the same time, problems related to the corporate management especially

T. Antipova (Ed.): ICIS 2020, LNNS 136, pp. 135–143, 2021.
https://doi.org/10.1007/978-3-030-49264-9_12

those which concerned conflicts of interest and understanding between the owners and senior management of the company goals and development strategy, became increasingly important and widespread.

It should be noted that in the context of a digital economy, many large corporations undergo IPO procedure, which allows estimating their value on the basis of the functioning market. At the same time, depending on the events taking place in a corporation, the cost of the business can change significantly within a short period of time, thus, the importance of the agency conflicts problem, expressed in inefficient management, has increased.

According to one of the key principles of the modern financial theory, the ultimate goal of any commercial organization including a corporation, is to maximize the wealth of its owners, that is shareholders, whose "welfare" directly depends not only on the dividends, but also on the share price [1].

Speaking of the welfare of shareholders and their interests in a corporation, it should be taken into account that corporate management in Russia has switched to the value categories [2], expressed in the concept of value based management (VBM). This is due to the fact that the usual accounting evaluation criteria often do not show the actual position of a company on the market, do not allow assessing the value of the company for its owners, while the retrospective nature of the indicators does not provide all the necessary information for making management decisions and executing value based management. The main task of VBM is to maximize the company value and it implies that the senior management will take only those decisions which will strengthen the current business model and promote its development in the future.

On the strength of this it is concluded that for the majority shareholders the business value and prospects of its activity are often more important than current financial results which may not coincide with the interests of the senior executives in such aspects as, for example, implementation of key performance indicators and compliance with the targets set out in the business plan, which the annual bonuses for senior executives [3] often depend on and in large Russian corporations may reach the size of annual salary, their interest in stable wage increase, use by the senior executives unnecessarily conservative low-risk investment strategies to minimize the probability of mistakes and their consequent dismissal, which may lead to a gradual deterioration of the company position in the market due to its lower rates of return and profitability compared to the market average values. There are often cases when senior executives have their personal interest in certain aspects of the financial and economic activity, and make management decisions that contradict the interests of the owners, and sometimes even cause direct damage to the financial well-being of the companies which to some extent has negative impact on the fundamental value of the company, its market capitalization, as well as causes a number of other financial costs and lost profits [4]. Thus, one of the most urgent today's problems of the corporate management theory arising in modern corporations is agency conflicts between shareholders and the senior management.

Topicality of the agency relations problem has increased in the course of financial market development, investment processes and digitalization of the economic relations. Detailed research of the agency problem influence on business capitalization were conducted and introduced in works of Macpherson [5], ideas of the national corporate systems and their features were formulated by R. La Porta, A. Schleifer, F. Lopez de Silaneso. and R. Vishny, [6], researches in the field of protecting interests of the minority shareholders [7, 8] were also conducted. Studies comparing developed national corporate management systems with the emerging ones, including those covering the corporate sector in Russia, have become widespread [9].

However, in scientific works the nature and manifestation of the problem of agency relations together with the financial and economic models widespread in Western financial science are discussed and analyzed. At the same time, due attention to peculiarities of the national model of corporate management in Russia as well as advice on minimizing the financial consequences of agency conflicts by strengthening monitoring of the quality of the financial management and timely prevention of agency conflicts under conditions of the Russian model are not given attention to.

In our opinion, one of the most effective measures to settle the main corporate conflict of interests may be by changing the system of assessing the quality of financial management based on practices that can monitor not only individual economic indicators of a company, but also integrate them into a single indicator that allows increasing objective character of assessing quality of the financial management. This will increase the level of control over the senior management from the owners, assess the company efficiency for the medium- and long- term perspective under the current system of corporate management, strengthen the current business model and promote its development in the future.

2 Problems of Using Common Approaches in Assessing the Quality of Financial Management

The company's focus on maximizing value has led to the need to develop new indicators and models that can correctly assess changes in business value for shareholders. At the same time, the approaches to evaluate the corporation performance differ not only in the calculation methods and key indicators taken as a basis, but also in the categories of information sources upon which the calculations are made in conformity with a particular indicator. In this regard, all most popular indicators can be divided into 3 main groups depending on the information sources: the accounting group – is calculated basing on the accounting approach to evaluation using the profit and loss statement and balance sheet data, the cash group – based on the assessment of cash flows and data from cash flow statements, and the market group – where the company is evaluated basing on the market institutions, while some indicators may have several calculation bases, as shown in Fig. 1.

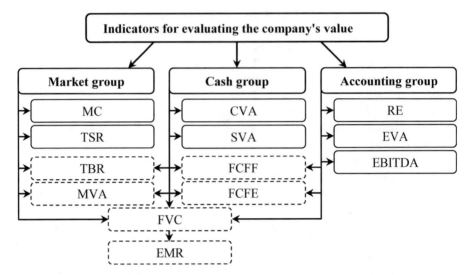

Fig. 1. Classification of indicators evaluating the company's value depending on the information sources.

Among the accounting indicators group the most common in scientific circles and financial management practices are residual earnings (RE) [10], economic value added (EVA) [11] and earnings before interest, taxes, depreciation and amortization (EBITDA). The indicators of this group are based on the accounting estimates and are mainly reduced to adjusting one of the profit indicators in order to assess the newly created value.

In our opinion, the fundamental advantage of this group of indicators is their simplicity of calculation with relative efficiency, which allows us to make the conclusions about the company performance, however this group of indicators is under the influence of the corporate accounting policy, and there is also possibility for senior the executives to have an impact on it, which can distort the results to some extent and affect the management decision. The analysis is based on retrospective data, which cannot be used to correctly assess the company development prospects and the current market sentiments, which does not allow giving comprehensive assessment of operation and manage the business value. Thus, indicators based on the accounting estimates can give the shareholders an idea of the business value dynamics from the perspective of the internal environment of the company, however non-financial factors of the value formation which can influence the future financial results are not taken into account.

Cash Value Added (CVA) [12] and Shareholder Value Added (SVA) have became the most common in the group of monetary indicators. In general, the CVA indicator is similar to the residual earnings indicators, however it is based on the adjusted net operating cash flow minus interest and economic depreciation, which, in contrast to the accounting indicators, allows you to take into account reinvesting possibility of the depreciation charges. Thus, the CVA indicator provides information about whether the available cash flow is sufficient to maintain the value of the invested capital and to what extent.

In its turn, the SVA indicator is focused solely on changing the business value, which is determined as the difference between the shareholder values at the end and at the beginning of the period under consideration. Calculation of the shareholder value involves the use of the discounted cash flow and residual value of the business, which puts the size of the shareholder value of the business in strict dependence on the current values of the cash flow and weighted average cost of the capital.

This group, like the previous one, allows you to take into account only financial factors of the value formation, regardless of such important aspects as transparency of the company activities, attitude to the investors, its reputation in the market, etc.

The third group includes indicators based on the market estimates, where the most common are total shareholder return (TSR) [13] and market capitalization (MC). Unlike the previous groups, this set of indicators does not involve the analysis of the corporation performance, but the direct assessment of the total benefits of the shareholders, which are expressed in the amount of dividends paid and changes in the market value of the financial instruments of the corporation traded on the exchange. However, in practice, the market is not always able to reliably assess the value of the company or take into account all the processes going on in it. Market capitalization and the TSR indicator based on it, in our opinion, are extremely volatile and not suitable for managing business value due to economic and political factors that can disrupt the usual work of market institutions and not only determine the dynamics of the market value of individual companies, but affect national markets at certain periods at the macro level as a whole.

Among the indicators based on several calculation bases, we consider it appropriate to point out free cash flow to the firm (FCFF) and free cash flow to equity (FCFE) derived from it, together with the market value added (MVA) [14], total business return (TBR) and the fundamental value of a company (FVC), calculated on the basis of the Federal Valuation Standards of the Business Value. FCF indicators allow you to estimate the actual amount of funds attributable to the capital suppliers in the reporting period, which, in its turn, affects the value of the company and allows using these indicators as the basis for calculating the amount of the dividends to be paid, debt repayment, repurchases of shares from the stock exchange, mergers and acquisitions.

The MVA indicator characterizes the increase in the market value of the company compared to the total amount of the invested capital, thereby determining the total amount of the value created over the entire period the corporation existed. It should be noted that this indicator does not take into account actually paid dividends as the way to measuring the value, and includes many factors that do not depend on the company performance, therefore we believe that this indicator is of exclusively analytical use. More complete indication of the created value basing on the indicator of the market capitalization for the period under consideration is achieved when using the TBR indicator, which, beside the dynamics of the market capitalization involves the amount of free cash flow, one part of which is to be used as payment of dividends and the other part is capitalized.

The last of the indicators we consider is FVC and its derivative EMR (Economic Market Recognition) [15]. The fundamental value itself does not allow determining the increment of the company value, however, the analysis of FVC in dynamics makes it possible for us to get not only most complete evaluation of the cost but also to determine factors influencing the change of value to some extent. Change in the value of the income approach indicates the change in the structure of cash flows and weighted average cost of the capital, change of cost approach shows change in the net asset value, their possible deterioration, change of market indicates that there are non-financial market factors of the value formation. It should be noted that the calculation of this indicator is extremely time-consuming, and the result may be variable depending on the methods used in the assessment and the evaluator's judgments, it is especially the case in the process of compiling the results into the final cost, and therefore the calculation procedure should be strictly regulated.

Most of the considered indicators were formed in the conditions of the Anglo-US model of corporate management, where they proved to be effective and where spread to all developed national markets. As the result, these approaches have become widespread in Russia among the major national corporations. At the same time, the conditions of functioning of the national model of corporate management in Russia differ significantly from the conditions of the model in which these indicators were formed, which reduces their effectiveness in the domestic economic environment.

3 Formation of Comprehensive Assessment of Quality of the Financial Management

In our opinion, quality of the financial management assessment tool should not be based on a single model, but a set of interdependent indicators resulting in the final number with the help of priority coefficients. On the one hand, it will be a stimulus for top management to improve efficiency of the most vulnerable areas of the corporation activities, while on the other hand, it will make ineffective the attempts to influence key performance indicators due to the relatively low impact of each individual criterion on the final assessment.

In order to create a model that can reflect the limitations of the existing financial management system that negatively affect the market capitalization of a business, in addition to EMR, we will need to take into account a number of indicators that assess the effectiveness of individual tasks and business processes. At the same time, we believe that in order to improve the objective character of the model, the selected indicators must meet a number of specific requirements.

Firstly, the selected indicators should provide qualitative assessment of the performance of a particular task, which will allow determining the effectiveness of a business in comparison with other companies in the industry. Such comparison is necessary to neutralizing the effect of macroeconomic factors that do not depend on the effectiveness of the management.

Secondly, the calculation of indicators should be simple and unambiguous. We believe that this criterion to select indicators will limit the ability of the management to manipulate the evaluation results. When calculating complex economic indicators based on the market depreciation, discount rates, cost of the capital, etc. the calculation methods may often be variable and not fully regulated, which on the one hand allows the management to influence the evaluation result to some extent, and on the other hand makes it difficult to verify these results.

Thirdly, the indicators must be balanced. It means that the inclusion of a certain indicator in the calculation should not lead to the management's efforts to maximize it at the expense of other performance indicators. In this regard, if maximizing a single indicator can lead to weakening a company's current business model, then the evaluation model should include related indicators that characterize the reverse side of the evaluated process.

In our study, we used 7 indicators, namely: EMR, TSR, SVA, Sales Growth Rate, Equity to Total Assets, Return on Sales and Current Ratio. Consequently, basing on the selection of 20 major Russian corporations, we obtained correlation between the received estimates and the future financial results of the companies under study, which was discussed in detail in the previously published researches [15].

We evaluated the quality of the corporate financial management in 3 stages. At the first stage, the selected indicators corresponding to the specified principles were calculated. At the second stage, the indicators were evaluated, and each indicator was assigned a score basing on its value. At this stage we developed a rating scale in accordance with which all allowed values of the selected indicators were split into 4 ranges, each of them having strictly assigned score. At the third stage, using weight coefficients based on the points received, we formed the final assessment of the financial management quality for each corporation.

It should be noted that the received scores estimate effectiveness of the financial management according to the actual results of the reported period and characterise a company at a certain point of time. However, large corporations are known for low mobility structure level of their assets and liabilities, and in many cases are not able to significantly change business activities within a short period of time. This allowed us to verify efficiency of the proposed model under the conditions of the naturally functioning Russian market.

To this effect evaluation of the financial management quality was conducted using actual results of 2016, while TSR indicators and rates of MC increase were evaluated on the basis of the results received in 2018. Basing on correlation analysis of the data received in the course of score evaluation and business value indicators, a steady relation which made 70% was found. This shows that the higher the performance of a company was in 2016, the higher growth in value it provided to the shareholders by the results of 2018. In their turn the companies with low performance demonstrated rather modest value indicators, which testifies to the problems in the system of their financial management and presence of agency conflicts affecting the business value which confirms efficiency of the proposed model under the conditions of the Russian national corporate management practice.

4 Conclusion

The conducted analysis demonstrated that the given methodology allows not only assessing the level of financial management of a company but also marking main limitations of the current business model with the purpose of its further development and strengthening, which in the long-term prospects will not only improve financial and economic activity but promote increase of its value as well. It should also be mentioned that use of the given methodology will promote better control of shareholders over the financial and economic activities of a company, which will make it possible to identify and prevent agency conflicts thereby minimizing their potential financial consequences.

However, a number of conditions providing credibility of the obtained results are worth mentioning. Firstly, the companies should be of public ownership and possess free-trading shares. It should be also noted that equity financial instruments of the compared companies should be traded on one stock market and in a single currency. Secondly, financial reporting of the appraised corporations should be organized according to the unified standards, at the same time preference should be given to reports, made according to the International Financial Reporting Standards (IFRS). Thirdly, the appraised companies should have history of trading on the exchange and comparable financial reporting for the period of at least 3 recent years. The fewer number of periods will significantly reduce credibility of the obtained values of the greater number of indicators and distort the final evaluations. In the fourth place, appraised and compared companies should be representatives of the same sector of industry. Comparing companies from different sectors will considerably reduce credibility of the obtained results which is explained through differences in sectoral liquidity requirements, profitability of sales, sectoral growth rates, as well as specific character and situation in the industry.

However, evaluation of the quality of the corporation financial management conducted in accordance with the above-mentioned conditions may eventually turn out to be beneficial both for its shareholders and managers. On the one hand, by the results of the second stage of evaluation the senior management will define the weakest areas of the company business model which require close attention and possible correction of the financial policy. On the other hand, the shareholders will be able to independently conduct financial monitoring, hereby strengthening their control over the management.

Thus, digitalization of economic relations in Russia provides the researchers with a wide range of directions to improve efficiency in the corporate sector. We are of the opinion that using this approach in solving the problem of agency conflicts in Russian corporations, it will possible to strengthen control over the company by the owners, will help to detect weaknesses of the financial and economic activities of a company, and will lead to formation of a more objective criterion as the basis for stimulating financial management. It will not only help to settle the conflict between shareholders and managers, but also improve the efficiency of the corporate sector of the Russian economy as a whole.

At the same time we suppose that further research in this area should be conducted in respect of increasing the number of analysed sectors and estimated periods. However, many Russian companies began to make reports in accordance with the International Financial Reporting Standards not long time ago, which does not often make it possible to evaluate companies before the factual data of the year 2016, testing the results on the basis of future exchange quotations at the same time.

The area of research may be expanded due to changing the set and number of indicators used in the model, analysing the possibility of adapting the model to the needs of investors, who do not have inside information about the operation of the company, justifying the choice of the weighting coefficients when forming the final evaluation. We also think that issues of methodological support of the processes, such as calculation of fundamental value, shareholder value added, together with determining market and sector average characteristics are of no small importance.

References

1. Jensen, M.: Value maximization, stakeholder theory, and the corporate objective function. J. Appl. Corp. Financ. **14**(3), 8–21 (2001)
2. Aleksandrova, A.V., Gorohova, A.E., Sekerin, S.V.: The evolution of views on corporate governance. Izvestiya MGTU «MAMI» **4**(18), 141–147 (2013)
3. Murphy, K., Jensen, M.: CEO bonus plans: and how to fix them. Harvard Business School NOM Unit Working Paper, no 12-022 (2011)
4. Jensen, M., Meckling, W.: Theory of the firm. Managerial behavior, agency costs and ownership structure. J. Financ. Econ. **3**(4), 305–360 (1976)
5. Macpherson, C.B.: Property, Mainstream and Critical Positions. University of Toronto Press, Toronto (1978)
6. La Porta, R., Lopes-De-Silanes, F., Shleifer, A., Vishny, R.W.: Legal determinants of external finance. J. Financ. **LII**(3), 1131–1150 (1997)
7. Djankov, S., La Porta, R., Lopez-de-Silanes, F., Shleifer, A.: The law and economics of Self-dealing. J. Financ. Econ. **88**(3), 430–465 (2008)
8. La Porta, R., Lopez-de-Silanes, F., Shleifer, F.: Law and finance after a decade of research. Handb. Econ. Financ. **2**, 425–491 (2013)
9. Fox, M.B., Heller, M.A.: Lessons from fiascos in Russian corporate governance. Working paper No. 99-012. University of Michigan Law School (1999)
10. Bernard, V.: Accounting based valuation methods, determinants of market to book ratios, and implications for financial statement analysis. Working paper University of Michigan. Business School. Faculty Research, no. 9401 (1994)
11. Stewart, B.: The Quest for Value: A Guide for Senior Managers. Harper Business, New York (1999)
12. Ottosson, E., Weissenrieder, F.: Cash value added – a new method for measuring financial performance, Gothenburg studies in financial economics. Social Science Research Network, no. 1996/1. Social Science Electronic Publishing, Inc., New York (1996)
13. Bressan, Bocardo A., Weijermars, R.: Total shareholder returns from petroleum companies and oilfield services (2004–2014): capital gains and speculation dissected to aid corporate strategy and investor decisions. J. Financ. Acc. **4**(6), 351–356 (2016)
14. Black, A.: Questions of Value: Master the Latest Developments in Value-Based Management, Investment and Regulation. Bell and Bain Ltd., Glasgow (2004)
15. Shevchenko, I., Puchkina, E., Tolstov, N.: Shareholders' and managers' interests: collisions in Russian corporations. HSE Econ. J. **1**, 118–142 (2019)

Implementation of Digitization and Blockchain Methods in the Oil and Gas Sector

Zhanna Mingaleva[1,2(✉)] ⓘ, Elena Shironina[1] ⓘ,
and Dmitriy Buzmakov[1] ⓘ

[1] Perm National Research Polytechnic University,
Perm 614990, Russian Federation
mingal1@pstu.ru
[2] Perm State University, Perm 614990, Russian Federation

Abstract. Blockchain technologies are now as familiar in many areas of society as television, mobile communications or the Internet. However Blockchain technologies are not only unused, but they are not even seriously thought about in the industrial sectors of the economy and at many industrial enterprises. At the same time, Blockchain technologies have great potential for application in industry, and their implementation is not something too complicated and expensive. Currently, as part of the modernization of ordinary business processes, there is a total digitalization and the transformation of each company and each business process into "data-driven processes", creating data streams, including data on the registration of the state/origin of physical objects, which is the place of application of blockchain protocols. The use of Blockchain technology is already on the agenda for many Russian companies, if they want to maintain their competitiveness not only in world markets, but even in national and local ones.

Keywords: Digitization · Network · Blockchain technologies · Oil and gas sector

1 Introduction

Blockchain - (a chain of blocks) is a continuous sequential chain of blocks (linked list) built up according to certain rules, containing information. Most often, copies of block chains are stored on many different computers independently of each other [1]. The blockchain is often presented as an electronic registration journal or file cabinet containing a list of records, for example, information about financial transactions. Its peculiarity is that this journal is accessible from any computer connected to a certain network. This is a decentralized database (distributed registry), where all records (or blocks) are encrypted using reliable cryptographic methods and are interconnected, updated and synchronized on computers throughout the network [2]. In theory, it provides complete protection against hacking and unauthorized changes, since hackers will not only have to decrypt the information, but also make changes on all computers in the system [3]. At the same time, participants in transactions and other operations

© Springer Nature Switzerland AG 2021
T. Antipova (Ed.): ICIS 2020, LNNS 136, pp. 144–153, 2021.
https://doi.org/10.1007/978-3-030-49264-9_13

themselves do not need to be programmers, users only make text entries about operations and transactions.

As a result, the blockchain enables the exchange of assets without central intermediaries. Therefore, as the main advantage of the blockchain, a complete rejection of intermediaries and maximum protection against fraud is declared [4]. Decentralization of trust allows entities to exchange everything that they own, without the participation of third parties that hold access keys to assets, and without verifying their authenticity. This is one of the most important features and advantage of the Blockchain that distinguishes it from many other data storage databases.

The potential of the blockchain is not limited to the monetary and financial spheres. Any information that is entered and stored in the registration journal is perceived by the blockchain as a sequence of entries. Blockchain technology does not disclose personal information. The only transaction data that is accessible to outsiders is the sum of the transaction and the hash (the cipher obtained by passing the details of the transaction through a cryptographic function).

At the moment, the main consumers of Blockchain technology are financial institutions, since cryptocurrency is most closely associated with this technology. In addition, thanks to the advent of the blockchain, it has become possible to implement the so-called Smart-contract [5], with the help of which it became possible to simplify work in many areas of business and life, including the public sphere (elections, referenda, voting on various issues), logistics, management, land registry [6], education [7], and other. Also, this tool is increasingly used by personal users and small enterprises in order to protect copyright and intellectual property rights [8–10], to store any important documents or information about transactions. Today, blockchain is one of the elements of the formation and development of the digital economy [11].

In the last few years, Blockchain solutions for the oil and gas industry have been actively developed. It is determined by the peculiarity of the financial and economic situation that has developed in the industry to date and as well as the features of the blockchain technologies themselves and the digitization capabilities of various technological and business processes in the oil and gas industry [12].

2   Theoretical Background

The industry situation in which companies from various segments of the oil and gas industry are located has posed a number of challenges for companies over the past decades - the long-term period of low oil prices (E&P/production), the revision of contracting relations and relations with the supplier ecosystem (OFSE/oilfield services), price reduction and integration of the international market (gas companies and LNG producers), regulatory enforcement of liberalization (gas midstream, wholesale gas and gas distribution). This fully applies to leading Russian oil companies [13].

A feature of oil and gas companies is the presence of a large volume of physical assets, the high cost of investment projects, the high risks of making operational decisions, the significant size of production and sales of products, and the critical importance of supply chain management. The composition of the physical assets of oil and gas companies, the features of production and marketing technologies, the importance of the physical processes of control over all stages of the business, determine the high need for the use of new technologies, including Internet and Blockchain.

An important consequence of the deterioration of the external and intra-industry competitive environment in the global oil and gas sector is the need to save investment and increase operational efficiency [14]. According to PwC research, the global oil and gas industry annually provides over $ 700 billion in investment [15]. Within the framework of these amounts, the issue of saving on investment to maintain the competitiveness of industry enterprises is estimated at about $ 140 billion per year. The need to optimize the operating costs of the global oil and gas industry, the value of which is estimated at $ 1 trillion annually, has shown that achieving the necessary savings can be obtained on the basis of technology for digitizing business processes. PwC estimate "that use of digital technologies in the upstream sector could result in cumulative savings in capital expenditures and operating expenditures of US$100 billion to $1 trillion by 2025" [15]. The calculations made by PWC showed that the digitization and application of Blockchain technologies in the hydrocarbon production sector will provide operational savings of $ 30–50 billion, and in processing and marketing - in the amount of $ 15–20 billion per year (savings forecasts for the main types of operations of oil and gas companies are shown in Fig. 1 [16]).

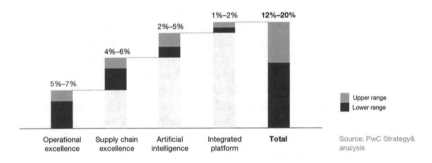

Fig. 1. Efficiency increase from digitization (% saving on total operating expenditure)

Finally, the use of Blockchain technologies will also optimize the costs of the information technologies themselves used in the industry. So, according to a study oil and gas companies in the world spend more than $ 50 billion on information technology, counting on a return on investment of at least 40%, and compared with these costs, the costs of implementing a blockchain are small, and the expected ROI, respectively, is higher [17]. The relevance of this area of application of the blockchain is determined by the fact that in the oil and gas industry, as in any other, there are intermediaries, manual operations and the associated costs. Automation of processes within the company and between contractors is the main focus of all services and departments. Using a distributed chain of registry blocks can help create a single trusted space between contractors, and the use of smart contracts automates some functions. It should be noted that specialists single out one important area in the activities of oil and gas companies where blockchain is not allowed - this is an investment discipline related to the digitalization of production.

Nevertheless, in all other areas of activity and spheres of cost reduction, they may contain the use of blockchain in a natural way. Among them there are the optimization of asset management, the reduction of inventory and material inventories, the reduction in operating costs of sales, the optimization of the logistics of materials and products, the optimization of procurement and sales.

The most advanced companies, oil and gas companies are already starting to use blockchain, make prototypes of solutions, invest in developers of relevant platforms and are the first to get access to technology [18]. In 2018, BP, Shell, Repsol, Gazprom and other large oil and gas companies in the world as a whole launched pilot block-chain projects to solve logistics, trading and financial problems [12]. Let us consider in more detail the possibilities and expediency of using Blockchain technologies in the oil and gas sector of Russia.

3 Research and Results

Currently, the most progressive oil and gas companies are already starting to use blockchain technology in the most affordable and efficient areas for blockchain use. They include:

1) workflow management;
2) improving the efficiency of equipment usage (eliminating duplicate and redundant equipment);
3) logistics and IoT technology;
4) hydrocarbons market;
5) information security.

Let us analyze the experience of using Blockchain in each of these areas.

1. *Workflow*
   Since most of the data in the document flow of individual market participants is shared information, it creates optimal conditions for the implementation of the Blockchain. Based on the technology, you can create a single network for digitizing all the processes of interaction and their automation.

Thanks to such a network, the following advantages can be achieved:

- reducing the number of errors associated with data mismatch, fraud, reconciliation and settlements between counterparties;
- reduction of efforts and time for checking expenses;
- transparent distribution ledger.

In practice, it means a significant reduction in the number of disputes between counterparties due to inconsistencies in invoices, bills of lading and payrolls, a reduction in accounting and audit costs, and an acceleration in the time for tracking goods and services in the supply chain.

In this area of implementation of Blockchain technologies, two projects are currently launched in the world.

The first project is the platform of IBM and the National Oil Company of Abu Dhabi (ADNOC) for launching an automated accounting and transaction management service for trading in petroleum products (and other extracted raw materials) - from wells to end consumers [19, 20].

The second project is related to the development of OOC Oil & Gas Blockchain Consoryium, which included ConocoPhillips, ExxonMobil Corp, Chevron Corp and 4 other major oil and gas companies [21]. They check all possible options for implementing blockchain in the oil and gas industry, including automation of accounting and audit. According to experts, the reduction in administrative and commercial costs in vertically integrated oil companies throughout the decision-making chain due to the blockchain will be 5–10%, which will allow large oil companies to save 0.4–0.7 billion dollars a year [22].

2. *The task of improving the efficiency of equipment usage by eliminating duplicate and redundant equipment*
   When equipment is connected to the blockchain, all parties will be able to obtain accurate data on the volume, density and impurities in the raw materials using one set of control and measuring equipment. Therefore, there is no need for duplication of this equipment.

In this area of implementation of Blockchain technologies, the implementation of one large project has now begun in the world. This is a project introduced in May 2018 by Quisitive (a subsidiary of Fusion Agiletech Partners) called Blockchain Oil Pipeline [23]. The project is being implemented in the field of pipeline monitoring and data collection, which oil producers and pipeline operators use to maintain crude oil purity standards. In its development, Quisitive used Blockchain Workbench, Microsoft Azure and Internet of Things (IoT) to create a platform that can serve as a single, secure and reliable data source for all interested parties [24]. Blockchain Oil Pipeline works according to the following algorithm:

1. A single cloud environment is created in Microsoft Azure, access to which is available to all participants in the events.
2. Provers connect to Azure using the Internet of Things (IoT) technology. Testing equipment records important events, measurement results and certified calibration measurements.
3. Data is entered into the blockchain and is provided to all participants in the events. If necessary, regulatory authorities, environmental organizations and structures that must respond to safety problems (excessive pressure, high temperature, etc.) can be connected to the system.
4. During the transportation of raw materials, IoT devices located along pipelines send pressure, flow and temperature data to the blockchain. As a result, all interested parties can see the measurement data collected in one common medium.

Quisitive demonstrated the final results of its project at Microsoft Technology Center industry events in Toronto, Denver and Dallas [24].

3. *Logistics and IoT technology*

Unlike the financial sector, blockchain technology in the oil and gas industry does not remove intermediaries in supply chains. Instead, the distribution register, smart contracts and IoT tags allow you to track the movement of products and raw materials, as well as other assets in real time and collect this information into a single pool of reliable data [25, 26].

In practice, this provides the following benefits:

1. The ability to track the source of raw materials at any stage of the product life cycle. This will increase the purity and legitimacy of the market, as well as reduce the costs associated with product recalls.
2. The ability to analyze big data on supply chains and understand how effective they are and whether there is an opportunity to improve them.
3. The ability to more accurately predict processes, as market chaos decreases: the number of errors, downtime, time to fill out documents and financial calculations will decrease.
4. In case of failure of certain equipment or its component, the operator can turn to the blockchain to find out when, where and by which company the component was produced.
5. Manufacturers of technological equipment can monitor how often equipment is serviced, how recommendations for current and major repairs of equipment, and its condition are followed.

A practical example of the application of Blockchain technologies in this area is the Accenture project, which is designed to accompany the processes of purchase, maintenance and monitoring of equipment for wells [27]. The second significant project in this area is the process of delivering valves to the Prirazlomnaya MLSP implemented by Gazprom based on the blockchain platform [28].

4. *Hydrocarbons market*

Trading operations can be carried out through smart contracts that automate most of the bureaucratic procedures associated with inspections and guarantees. This solves the problem with the speed of operations and significantly reduces the size of fees. It was this area that was first tested on the possibility of using Blockchain technologies [29, 30].

In 2017, two hydrocarbon trading projects were implemented.

First, Natixis, IBM, and Trafigura launched an IBM Bluemix pilot project to buy and sell raw materials [31]. Testing took place in the US crude oil market. The site allowed in real time to share relevant data on payment, shipping and delivery of goods. Usually this requires complex "paper" procedures, couriers for sending paper documents and fax/email for digital ones.

The second example is a project launched by the Swiss trading company Mercuria Energy Group to organize the supply of crude oil from Africa to China. Trading was carried out using the Easy Trading Connect blockchain trading platform, created in collaboration with ING and Societe Generale [32]. The experiment showed that blockchain reduces the time of operations from several hours to 25 min. It was especially liked by traders who increased their efficiency by 33% due to a reduction in commission.

The success of Easy Trading Connect has led to the creation of an international oil consortium aimed at using blockchain in trading operations to increase their efficiency and traceability. The consortium included oil giants Shell, British Petroleum and Statoil, trading houses Koch Supply & Trading, Mercuria Energy Group and Guvnor Group, as well as banks ING, ABN Amro and Societe Generale. The result of cooperation within the consortium was the creation of the VAKT Global trading platform, which is currently being tested in closed mode, that is, only members of the consortium have access [33, 34].

5. *Information security*
   In the oil and gas industry, so far many problems have been identified related to the safety and reliability of data [35–37]. At the same time, some of them not only lead to monetary costs, but also create threats of theft and data leakage, cyber espionage and unauthorized interventions in the system, which is fraught with a loss of reputation.

The Indian multinational company Tata Consultancy Services, which provides information business and IT consulting services, conducted an oil and gas market research regarding the adoption of the blockchain in the context of data security and revealed the following positive effects from the implementation of the distribution registry and smart contracts [38].

1. Data leakage will become almost impossible, since hacking the blockchain will require very large capacities. True, there remains the possibility of breaking into individual nodes that supply this or that information to the blockchain.
2. The integrity and security of data will rise to a new level due to the immutability of information in the blockchain and the inability to fake it.
3. Data processing will become faster and more reliable through asset tokenization and the use of smart contracts.
4. The integration of participants will be quick, easy and transparent.

At the same time, it is noted that any blockchain solutions contribute to increasing information security, since they use blockchain and smart contracts, which in themselves are protection from hackers, fraud and errors [39–41]. Nevertheless, specialized blockchain solutions in the field of information security have been created and are already being implemented. It is the use of blockchain to consolidate back-office processes, protect against cyber threats, as well as to reduce long-term costs, which is implemented by BP in collaboration with O&G and Eni and Wien Energie [42].

4 Conclusions

The possibilities and expediency of using Blockchain technologies in the oil and gas sector of Russia are largely determined by the successful examples of blockchain technology implementation in the world. The study showed that currently in the world there are already examples of the successful implementation of such projects in all areas most significant for the activities of companies. In general, the results of the creation and testing of blockchain prototypes, test projects, various specialized studies have shown a rather high efficiency of using blockchain in the oil and gas industry [43]. So, at the World Economic Forum in Davos in 2018, it was noted that the transition to Blockchain technology in the oil industry will increase market liquidity by $ 1.6 trillion, market volume by $2.5 trillion, and bring companies an additional $ 1 trillion and society in a total of another $640 billion [44]. Also, BP, Shell and Statoil experts calculated that the implementation of Blockchain in the trading operations of oil companies will reduce transaction time by 30%, and it will become equal to the delivery time of the goods from seller to buyer. Vygon Consulting believes that the introduction of blockchain in markets with high tax burden will save companies up to 10% of costs. As for the overall performance of companies, Gazprom Neft calculated that digitizing the oil and gas business increases the company's efficiency by 10–15%.

Acknowledgment. The work is carried out based on the task on fulfilment of government contractual work in the field of scientific activities as a part of base portion of the state task of the Ministry of Education and Science of the Russian Federation to Perm National Research Polytechnic University (topic # 0751-2020-0026 "Development of innovation management methods for industrial enterprises in the context of digitization and networking").

References

1. Iansiti, M., Lakhani, K.R.: The truth about blockchain. Harvard Bus. Rev. **95**(1), 118–127 (2017)
2. Allison, I.: Blockchain Pioneer Lisk on the philosophy of decentralization. International Business Times UK. https://www.ibtimes.co.uk/blockchain-pioneer-lisk-philosophy-decentralisation-1665678. Accessed 14 Jan 2020
3. Joshi, A.P., Han, M., Wang, Y.: A survey on security and privacy issues of blockchain technology. Found. Comput. Math. **1**(2), 121–147 (2018)
4. Gervais, A., Karame, G.O., Wüst, K., Glykantzis, V., Ritzdorf, H., Capkun, S.: On the security and performance of proof of work blockchains. In: Proceedings of ACM SIGSAC Conference on Computer Communication Security (CCS), Vienna, Austria, pp. 3–16 (2016)
5. Christidis, K., Devetsikiotis, M.: Blockchains and smart contracts for the internet of things. IEEE Access **4**, 2292–2303 (2016)
6. Klimenkov, G.V.: Digital economy program - implementation paths. Bull. PNIPU. Soc.-Econ. Sci. **2**, 127–136 (2018)
7. Sweden tests blockchain technology for land registry, Reuters. https://www.reuters.com/article/us-sweden-blockchain/sweden-tests-blockchain-technology-for-land-registry-idUSK CN0Z22KV. Accessed 21 Jan 2020
8. Mingaleva, Zh., Mirskikh, I.: Globalization in education in Russia. Proc.-Soc. Behav. Sci. **47**, 1702–1706 (2012)
9. Roberts, J.J.: Microsoft and EY launch blockchain tool for copyright, Fortune, 20 June 2018
10. Mingaleva, Zh., Mirskikh, I.: The problems of legal regulation and protection of intellectual property. Proc. Soc. Behav. Sci. **8**, 329–333 (2013)
11. Mingaleva, Z., Mirskikh, I.: Small innovative enterprise: the problems of protection of commercial confidential information and know-how. Middle East J. Sci. Res. **13** (SPLISSUE), 97–101 (2013)

12. Andoni, M., et al.: Blockchain technology in the energy sector: a systematic review of challenges and opportunities. Renew. Sustain. Energy Rev. **100**, 143–174 (2019)
13. Antipova, T.: Streamline management of arctic shelf industry. In: Antipova, T., Rocha, Á. (eds.) Information Technology Science. MOSITS 2017. Advances in Intelligent Systems and Computing, vol. 724, pp. 114–121 (2018)
14. Postnikov, V.P., Timirova, K.A.: Investigation of the dependence of the dynamics of indicators of investment activity of an oil producing company on market conditions. Vestnik PNIPU. Socio-Econ. Sci. **1**, 260–278 (2019)
15. BlockChain—An Opportunity for Energy Producers and Consumers? PwC Global Power Utilities, London, U.K. (2017). https://www.pwc.com/gx/en/industries/assets/pwc-blockchainopportunity-for-energy-producers-and-consumers.pdf. Accessed 13 Nov 2019
16. Drilling for data: Digitizing upstream oil and gas. https://www.strategyand.pwc.com/gx/en/insights/2018/drilling-for-data.html. Accessed 21 Jan 2020
17. Oil and Gas Transformation with Blockchain Technology. https://www.oocblockchain.com. Accessed 01 Feb 2020
18. Lakhanpal, V., Samuel, R.: Implementing blockchain technology in oil and gas industry: a review. In: Procedinmng SPE Annual Technical Conference and Exhibition, Dallas, TX, USA, pp. 1–12 (2018)
19. World Oil. ADNOC Has Implemented IBM Blockchain Technology to Streamline Daily Transactions. https://www.worldoil.com/news/2018/12/10/adnochas-implemented-ibm-blockchain-technology-to-streamline-dailytransactions. Accessed 11 Jan 2020
20. Zaher, A.: ADNOC team up with IBM to pilot a blockchain application for its full value chain. https://www.forbesmiddleeast.com/adnoc-team-upwith-ibm-to-pilot-a-blockchain-application-for-its-full-value-chain. Accessed 11 Jan 2020
21. OOC Oil & Gas Blockchain Consortium Successfully Tests Digitizing the AFE Balloting Process Leveraging Blockchain Technology. https://www.businesswire.com/news/home/20191218005033/en/OOC-Oil-Gas-Blockchain-Consortium-Successfully-Tests. Accessed 01 Feb 2020
22. Petroteq Energy: Announce Blockchain-based Initiative to Optimize Oil & Gas Supply Chain Management. http://www.marketwired.com/pressrelease/petroteq-energy-inc-first-bitcoin-capital-corp-announceblockchain-based-initiative-2239517.htm. Accessed 05 Dec 2019
23. Fusion Agiletech Partners has acquired Quisitive. https://www.channele2e.com/news/blockchain-as-a-service-acquisition-fusion-agiletech-buys-quisitive. Accessed 10 Dec 2019
24. Blockchain Could Bridge Data Divide Between Oil Producers, Pipeline Firms. https://www.rigzone.com/news/blockchain_could_bridge_data_divide_between_oil_producers_pipeline_firms-30-may-2018-155769-article. Accessed 05 Dec 2019
25. Makhdoom, I., Abolhasan, M., Abbas, H., Ni, W.: Blockchain's adoption in IoT: the challenges, and a way forward. J. Netw. Comput. Appl. **125**, 251–279 (2019)
26. Samaniego, M., Deters, R.: Blockchain as a service for IoT. In: Proceedings of IEEE International Conference on Internet Things (iThings) IEEE Green Computing Communication (GreenCom) IEEE Cyber, Physical and Social Computing (CPSCom) IEEE Smart Data (SmartData), Chengdu, China, pp. 433–436 (2016)
27. Energy consulting capabilities. https://www.accenture.com/us-en/industries/energy-index. Accessed 06 Dec 2019
28. Blockchain for Prirazlomnaya. Gazprom Neft applied blockchain technology in logistics. https://www.gazprom-neft.ru/press-center/sibneft-online/archive/2018-april/1533012. Accessed 20 Dec 2019

29. Oil trade is put on a "chain." How blockchain will help oil industry. https://www.forbes.ru/biznes/355179-torgovlyu-neftyu-sazhayut-na-cep-kak-blokcheyn-pomozhet-neftyanikam. Accessed 15 Dec 2019

30. Lee, W.H., Miou, C.-S., Kuan, Y.-F., Hsieh, T.-L., Chou, C.-M.: A peer-to-peer transaction authentication platform for mobile commerce with semi-offline architecture. Electron. Commer. Res. **18**(2), 413–431 (2018)

31. Natixis, IBM and Trafigura introduce first-ever Blockchain solution for U.S. crude oil market. https://www.natixis.com/natixis/jcms/rpaz5_59312/en/natixis-ibm-and-trafigura-introduce-first-ever-blockchain-solution-for-u-s-crude-oil-market. Accessed 05 Dec 2019

32. A soybean shipment to China became the first commodity deal to use blockchain tech. https://www.insider.com/energy-and-commodity-companies-use-blockchain-tech-for-trading-2018-1. Accessed 30 Nov 2019

33. Post Trade Management Platform. https://www.vakt.com/technology. Accessed 08 Dec 2019

34. From PoC to Production: Implementing an Enterprise Blockchain Solution. https://www.vakt.com/from-poc-to-production-implementing-an-enterprise-blockchain-solution. Accessed 08 Dec 2019

35. Felin, T., Lakhani, K.: What problems will you solve with blockchain? MIT Sloan Manage. Rev. **60**(1), 32–38 (2018)

36. Dorri, A., Roulin, C., Jurdak, R., Kanhere, S.: On the activity privacy of blockchain for IoT. https://arxiv.org/abs/1812.08970. Accessed 18 Dec 2019

37. He, Q.S., Xu, Y., Liu, Z., He, J., Sun, Y., Zhang, R.: A privacy-preserving internet of things device management scheme based on blockchain. Int. J. Distrib. Sens. Netw. **14**(11), 1–12 (2018)

38. Blockchain Adoption in Oil & Gas: A Framework to Assess Your Company's Readiness, Tata Consultancy Services, India. https://www.tcs.com/blockchain-oil-gas. Accessed 20 Dec 2019

39. Mendling, J., et al.: Blockchains for business process management challenges and opportunities. ACM Trans. Manage. Inf. Syst. **9**(1), 1–16 (2018)

40. Blockchain in Energy and Sustainability. https://consensys.net/blockchain-use-cases/energy-and-sustainability. Accessed 04 Feb 2020

41. How does blockchain impact the oil and gas industry? https://consensys.net/blockchain-use-cases/energy-and-sustainability. Accessed 04 Feb 2020

42. BP Energy Outlook, Brit. Petroleum, London, U.K. (2019). https://www.bp.com/content/dam/bp/businesssites/en/global/corporate/pdfs/energy-economics/energy-outlook/bpenergy-outlook-2019.pdf. Accessed 05 Dec 2019

43. Oil and Gas Industry—Blockchain, the Disruptive Force of the 21st Century, Infosys, Bengaluru, India (2018). https://www.infosys.com/industries/oil-and-gas/features-opinions/Documents/blockchain-disruptive-force.pdf. Accessed 05 Dec 2019

44. Digital Transformation Initiative Oil and Gas Industry, World Economic Forum, Cham, Switzerland (2017). http://reports.weforum.org/digital-transformation/wp-content/blogs.dir/94/mp/files/pages/files/dti-oil-and-gas-industry-whitepaper.pdf. Accessed 05 Dec 2019

Crowdfunding as a Source of Financing in Russia: PEST Analysis

Svetlana Tsvirko[✉] iD

Financial University under the Government of the Russian Federation,
Leningradskiy prospekt 49, Moscow 125993, Russian Federation
s_ts@mail.ru

Abstract. The article focuses on the features of crowdfunding in Russia. The method of research is PEST analysis. The objective of the study was to analyze opportunities as well as problems and unsolved issues that are connected with crowdfunding in Russia. The main drivers of the development of crowdfunding in Russia, revealed in the paper, are as follows: new legislation, devoted to the crowdfunding; real need of small and medium enterprises as well non-governmental organizations in additional financial resources; demand from individual and institutional investors for new investment opportunities, especially under the conditions of lowing of the interest rates; active use of financial technologies. The most serious factors, that constrain crowdfunding in Russia, identified in the paper, are as follows: some unsolved regulatory issues; poor investors' protection and possibility of fraud; low levels of real income and savings; low profitability of many businesses. The analysis provides a basis for different participants (projects' founders, investors, policy makers) to exploit opportunities and minimize threats of this new approach in the sphere of finance.

Keywords: Crowdfunding · Crowdinvesting · Crowdlending · Platform · PEST analysis

1 Introduction

There is a serious problem of the lack of financial resources for the development of breakthrough investment projects in the Russian Federation. Demand from various market participants for alternative sources of financing is significant.

In general, the main sources of financing are as follows: bank loans, issuance of securities, venture financing, money from business angels, government financing; but under some circumstances they are unavailable for the borrowers or can be accompanied by risks and disadvantages for them. One of the promising approaches to the search for funding is the mechanism of "new money", which is understood as "a system of alternative mechanisms for the involvement of private and currently inactive capital in technology projects" [1, p. 7]. Among the mechanisms of "new money" we can name crowdfunding, the features of which will be discussed in this paper.

There is significant amount of publications with the research of crowdfunding aspects – both in common and devoted to the experience of particular countries. Among pioneers of the research in this sphere we can name such authors, as Gerber

© Springer Nature Switzerland AG 2021
T. Antipova (Ed.): ICIS 2020, LNNS 136, pp. 154–163, 2021.
https://doi.org/10.1007/978-3-030-49264-9_14

[2, 3] Bretschneider, Leimeister [4, 5], Wieck [5] and others. Adamo, Federico, Intonti, Mele and Notte [6] revealed the specific features of crowdfunding in Italy. Karadogan concentrated on the experience of China and the United Kingdom in the sphere of crowdfunding [7].

Ilenkov and Kapustina [8], Genkin and Mikheev [9], Kuznetsov [10], Chulanova [11], Zvonova [12] and others pay attention to the peculiarities of the Russian financial market.

As far as we know there are no papers with PEST analysis of crowdfunding for the Russian Federation, that can be our contribution to more integrated assessment of the drivers and barriers of this relatively new approach to attract funds.

The main goal of the research is to evaluate factors influencing the attractiveness of crowdfunding as a financial source in Russia. The paper is organized in the following way: the first part is devoted to the essence of crowdfunding and its practice in Russia and abroad, then methodology of research is revealed and results of PEST analysis are presented. The analysis will result in evaluation of factors that have impact on crowdfunding in Russia and in formulation of general recommendations for the development of this source of finance. The findings can be useful for projects' founders, investors, policy makers (financial regulators).

2 Essence of Crowdfunding and Its Practice in Russia and Abroad

Crowdfunding implies attracting financing from individuals ("public funding") in order to sell a product or service. The development of crowdfunding is a current trend abroad; it is advisable to consider its advantages and disadvantages for use in Russia.

There is no consensus in the expert community on the issue of crowdfunding definitions. Michael Sullivan, the founder of Fundavlog, used the word «crowdfunding» for the first time to explain the essence of the idea of the platform to raise funds in 2006. Overview of the definitions made by Ilenkov and Kapustina led them to the conclusion that «generally crowdfunding is associated with collecting money from a large number of people for specific purpose primarily via Internet-based platforms» [8, p. 403]. From the point of view of Genkin and Mikheev, crowdfunding is «the collective cooperation of people (donors, contributors) who voluntarily pool their money or other resources, usually via the Internet, to support the efforts of other people (owners, creators of a startup or project) or organizations (recipients)» [9, p. 584].

Three types of participants in the process of crowdfunding are as follows:

1) project founder who wants to attract capital,
2) project's sponsors ("backers") who pledge money into a project,
3) crowdfunding platform which connects them together [13].

Kuznetsov gave the following author's definition to crowdfunding companies: «economic entities operating in the Russian Federation as financial intermediaries within the existing legal framework of the Russian Federation, which are not subject of legislation connected with the national payment system, anti-money laundering, consumer credit (loan) and do not fall under regulation and supervision (in the context of the existing legal field) of the Central Bank of the Russian Federation - the financial market regulator» [10, p. 68]. Thus, this definition emphasized the presence of gaps in the regulation of crowdfunding in Russia as of 2017.

There is no unity among experts on the issue of crowdfunding classifications. So, Genkin A.S. and Mikheev A.A. believe that crowdfunding and crowdinvesting are synonymous. Kuznetsov V.A. reveals crowdfunding subcategories: crowdinvesting and crowdlending, charitable crowdfunding and conditional-return crowdfunding. In world practice crowdlending currently prevails among all types of crowdfunding.

An important characteristic of crowdfunding is that it is a new instrument for financing projects that can be used if traditional sources of financing are not available. This is especially important for small companies (business start-ups), which need the opportunity to develop interesting innovative projects. As it was revealed in [13] the advantages of crowdfunding include:

- reduction of costs associated with raising funds;
- attraction of investments from a wide range of investors, diversification of investors;
- acceleration and simplification of operations on raising funds due to the electronic type of transactions.

The positive consequences of the use of crowdfunding mechanisms are as follows: support for the start-ups and innovative solutions, creation of new jobs, increase in budget revenues and, ultimately, economic growth [13].

There are more then 600 crowdfunding platforms in the world and the total amount of funds collected with them is more than 35 billion US dollars. In the report «Global Crowdfunding Market 2018–2022» the analysts have predicted that the crowdfunding market will register a CAGR of more than 17% by 2022 [14]. The main players in the global crowdfunding market are as follows: GoFundMe, Indiegogo, Kickstarter, Patreon, and Teespring. As an example, at Kickstarter since its launch on April 28, 2009 17 mln people have backed projects; 4,8 bln US dollars have been pledged and 177937 projects have been successfully funded [15]. The most popular categories in terms of number of launched projects at Kickstarter are Film&Video, Publishing, Games, Technology and Design. In terms of total amount of funds gathered the leading categories are Music, followed by Video, Society, Publishing and Technology. The leaders among the countries in the development of crowdfunding include the UK and the USA.

As for the Russian Federation, at present the crowdfunding market is at the initial stage of development; but there are a number of successfully functioning crowdfunding platforms, including StartTrack, VentureClub, Planeta.ru, Boomstarter, etc.

So, at the StartTrack site 122 companies raised funding totaling 2949.5 mln rubles from 12224 investors. According to data presented on the StartTrack site, the volume of transactions made using crowdfunding sites in Russia in 2017 exceeded 11 billion rubles. It should be noted that not only individuals participate in the crowdfunding process; thus, investment of VEB-Innovations in StartTrack for 200 million rubles was announced [16].

Among other platforms for the investment into small business we can name Sbercredo (launched at the end of 2019), Potok (that belongs to Alfabank and FinTechCapital fund), Ozon.Invest, Gorod Deneg, Penenza.

«Planeta» positions itself as a social service ecosystem that helps monetize ideas, projects, creativity and author's content in various ways. Within this platform, as of February 2020, 1189.5 mln rubles were raised in over 5400 projects; every third project is successful [17].

At Boomstarter during the period from June 2012 until June 2019 375 mln rubles were raised by 1900 projects [18].

Similar to foreign practice, in Russia crowdfunding is popular in creative industries. There are also some projects connected with charity and social initiatives that get support at crowdfunding platforms in Russia.

3   Methodology of Research

It is necessary to understand factors influencing crowdfunding in Russia. We will use PEST analysis for this research. PEST analysis is a method for analyzing the macro environment (external environment). The PEST analysis technique is often used to assess key market trends in the industry, and the results of the PEST analysis can be used to determine the list of threats and opportunities. PEST analysis is a tool for long-term strategic planning and is compiled 3–5 years in advance, with annual data updates. It can be made in the form of a matrix of 4 quadrants or in tabular form.

PEST analysis is an abbreviation of the following industry indicators:

P (Political) - factors of the political and legal environment.
E (Economical) - factors of the economic state of the market.
S (Socio - cultural) - factors of the social and cultural state of the market.
T (Technological) - factors characterizing technological progress in the industry [19].

4   Results of PEST Analysis

As for the political factors, they create a system, in which the companies perform their activities. The system is given by power interests of different political parties, and development of political situation in the country. Factors connected with the laws and regulations are of great importance for the development of crowdfunding.

Among important factors of increasing the volume of crowdfunding activity we can name legislation. In the Russian Federation there has been created a comprehensive legal regulation of crowdfunding in 2019. Federal Law dated 02.08.2019 № 259 regulates crowdfunding activities [20]. Now, for persons attracting investments and investors, restrictions are envisaged. For example, one person will be able to attract no more than 1 billion rubles in year. Individuals without the status of a professional investor will be entitled to invest no more than 600,000 rubles. Investment platforms must have at least 5 million rubles of equity; they should also be included in the register, which will be maintained by the Bank of Russia. In addition, the new law establishes what are the utilitarian digital rights and how they are taken into account.

The concept of crowdfunding's regulation developed by the Bank of Russia establishes consumer rights' protection as a priority. For example, the "regulatory sandbox" is used - a mechanism for piloting new financial services and technologies that require changes in legislation [21, p. 12]. As part of the "sandbox", modeling of the processes of application of innovative financial services, products and technologies was carried out to test hypotheses about the positive effects of their implementation. This can be regarded as a balanced approach and a reflection of the basic principle applied to the innovation economy, which consists of proportional regulation and supervision in accordance with the perceived risks, as well as taking into account the needs of the participants, whose access to financial services is limited [22]. In spite of significant improvements in the legislation, connected with crowdfunding, unsolved issues and risks remain. So, in particular, Chulanova O.L. draws attention to the fact that one of the most discussed and often realized risks, connected with crowdfunding, in practice is the risk of fraud [11].

Some of unsolved regulatory issues are as follows:

1. Tax consequences of participation in crowdfunding projects for investors.
2. Fulfillment of duties connected with responsibilities to conduct internal control, countering laundering of proceeds from crime and the financing of terrorism.
3. Legislative mechanisms to combat fraud that is possible in the crowdfunding industry [13].

As for the factors hindering development of business in the country, we can name level of bureaucracy and corruption. Potential problems also include conflicts of interest during crowdinvesting. In general, the investors' rights protection remains poor.

Economic factors, influencing crowdfunding, come out of the economic situation of the country and the economic politics of the country. Among these factors we can name macroeconomic indicators, such as pace of the economic growth, unemployment, phases of the business cycle, inflation, changes in the interest rates, exchange rates, etc. As for the Russia's macroeconomic indicators, they are relatively stable. We had budget surplus of 500 billion rubles, or $8 billion, last year; significant foreign currency reserves were accumulated; inflation is low (in January 2020 annual inflation declined to 2.4%.). At the same time economic growth has slowed to well below the global average. Russia's gross domestic product grew only 1.3% in 2019, down from 2.5% the previous year. Economic activity continues to be constrained by a number of factors [23]. Under such conditions its necessary to increase investments.

The need for additional finance sources for small and medium enterprises and non-government organizations can be an important factor for the development of crowd-funding. There are different methods to attract funds, but they are not entirely available for business in Russia. Existing traditional financial instruments are usually oriented towards financing of large and stable companies and, therefore, they are not applicable to new projects. In recent years, there has been a decline in available financing for innovation from banks. Venture financing has long been a traditional source of financing small and medium-sized innovative projects. However, for several reasons venture financing does not satisfy all the needs for financing innovation. Venture financing currently can be characterized by concentration on certain industries, as well as popularity of projects in the late stages of development with a low level of risk. Under such conditions it is very difficult to attract resources for projects at their early stages and with higher levels of risk [13].

Interest rates on bank credits are relatively high in Russia. At the same time individual and institutional investors are in search of investment opportunities. Interest rates in the deposit and credit market continue to go down. It can increase interest in crowdinvesting.

As for the economic challenges constraining development of crowdfunding we should name low levels of real incomes and savings as well as low profitability of many businesses.

The social factors are mainly predetermined by the society, its structure, social structure of the population and by social and cultural habits by the citizens of the country. Currently we observe such trend as increasing citizens' social activity. One of the major drivers for the crowdfunding market is the social media as a source of free of cost promotion [15].

As for the factor that restrains and complicates the growth of this market, we can state that crowdfunding is the time-consuming process. Not all projects are successfully funded on crowdfunding platforms. According to statistics of different platforms, only small portion of projects is successfully funded (collected at least 100% of initial goal amount). Experts say that crowdfunding project success depends not only on the idea behind, but also on the way this idea is introduced to investors and on the level of the crowdfunding campaign advertising.

Lifestyle trends and the demographics effect investment decisions. People can be interested in investing in something new to them, explore new trends and new ways how to get money for a project or starting own business. As for the factor influencing contrary, we can name aging of the population: the older generations are less likely to invest in crowdfunding because it is an innovation of investing and they keen to stick with ways of investing that are already verified and working for a longer period.

We should also mention negative experience of the significant part of the population in Russia with investing, connected with financial reforms and financial pyramids.

Technological factors represent the innovational potential, the speed of development and changes in the technologies in different spheres of the country. From the point of view of Ilenkov and Kapustina, «prospects for the development of the Russian crowdfunding market are supported by high rates of financial technologies development» [8, p. 402].

Investing through crowdfunding is mainly done via web pages, so there is a need for computer device and the access to the internet or mobile connection. In Russia, the number of Internet users, according to Digital 2020, amounted 118 million [24]. This means that 81% of Russians use the Internet. The number of Internet users among Russians over 16 years old (which is more relevant for studying of the crowdfunding potential) during year 2019 has increased from 91 million to 94.4 million people. Now 80% of the adult population of the country have access to Internet [25].

The audience of social networks in Russia at the beginning of 2020 amounted to 70,000 million users, that is, 48% of the total population of the country [24]. Usage of Internet and social networks can be the important factor for crowdfunding.

Crowdfunding is carried out on the basis of crowding platforms. Some features of such platforms include the following: availability of a credit rating of borrowers, compiled by experts of the platform or external experts; possibility of automatic diversification, when the amount of invested funds is distributed to several investments based on the parameters determined by the investor; minimum and maximum levels of investment for both the borrower and the lender and others [13]. Such opportunities rely on the development of different technologies. At the same time, usage of the internet, social media and online marketing brings additional risks, including cyber risks.

Results of PEST analysis for the implementation of crowdfunding in Russia are presented in Table 1.

Table 1. PEST analysis of crowdfunding in Russia.

Political factors	Economic factors
1. new legislation devoted to crowdfunding introduced in 2019 (federal law);	1. relatively stable macroeconomic indicators;
2. unsolved regulatory issues;	2. very slight, insufficient economic growth;
3. generally low level of investors' protection and high risks of fraud;	3. low levels of real incomes and savings;
4. high level of bureaucracy and corruption;	4. low profitability of many businesses and high probability of default;
5. possibility of conflicts of interest	5. low inflation rate;
	6. high credit rates and generally low availability of loans for business;
	7. search for additional finance sources for small and medium enterprises and non-government organizations;
	8. decreasing interest rates on deposits and search for investment opportunities from the side of individual and institutional investors;
	9. absence of tax incentives to invest and some unsolved questions, connected with taxation of crowdfunding activities
Social factors	**Technological factors**
1. increasing citizens' social activity;	1. high rates of financial technologies development;
2.ability to influence societal issues;	2. digitalization of the economy;
3. social media as a source of free of cost promotion;	3. increase in quantity of the Internet-users and potential for further growth;
4. absence of information constraints;	4. growth of cyber risks and necessity to increase spending on cyber security
5. demographic changes (new generations of investors, aging of the population, etc.);	
6. negative experience of investing, connected with financial reforms and financial pyramids	

5 Recommendations for the Development of Crowdfunding in Russia

Initiators of the projects should know the benefits of crowdfunding. It can be recommended for the projects founders to pay attention to the key success factors of achievement a target amount of money on crowdfunding platform. Then it should act prudently in connection with attracted financial resources. Under the conditions of weak economic growth and low demand for different goods and services experts consider only selected business areas to be relatively safe (for example, small IT companies and Fintech companies that offer services to large enterprises and export software).

For investors it can be recommended to pay close attention to the quality of scoring, otherwise the risks increase significantly. Investors should understand that crowdfunding is usually used by exotic, innovation businesses, start-ups, for which credits of the banks are not available. Sometimes it is very difficult to evaluate the sustainability of the project even for the professionals. That is why it is necessary to diversify investments and limit the share of money invested into crowdfunding projects within the structure of the whole investment portfolio.

As for necessary actions from the side of the policy makers, complex approach to improvement of regulation in connection with new mechanisms for attracting funding is needed. The development of crowdfunding mechanisms should be linked to the promotion of electronic interaction mechanisms in the financial market; ensuring the protection of the rights of consumers of financial services and improving the financial literacy of the population; discouraging unfair behavior in the financial market. In general, an acceptable level of stability in the financial market should be ensured [13].

6 Conclusion

PEST analysis of crowdfunding in Russia can produce relevant data and provide some understanding of the necessity and possibility to implement this approach as a source of financing. As for further development of the method of research, PEST analysis can be expanded to PESTEL (adding environmental/ecological and legal aspects), PESTELI (adding industry analysis), STEEP (that includes PEST analysis +ethical factors) and LONGPEST (PEST analysis with local, national and global factors). As for limitations of research we should understand that within the method applied we need to have high quality data, access to which can be time consuming and expensive. Regular updates of the evaluation of the changing external environment are required. For example, some new factors for further investigation are on the way: the effects of pandemic and coronovirus, erupted in 2020, a significant decrease in economic activity worldwide and the consequences for financial markets, including crowdfunding. It can be the suggestion for future research.

Summing up, crowdfunding in Russia has both drivers and obstacles. Among the factors that support the development of crowdfunding in Russia we have figured out new legislation, devoted to the crowdfunding; significant need of different participants of the market in additional financial resources; demand from individual and institutional investors for new investment opportunities; active use of financial technologies. The most serious factors, that restrain crowdfunding in Russia, are as follows: some unsolved regulatory issues; poor investors' protection and possibility of fraud; low levels of real income and savings; low profitability of many businesses. The above mentioned results of PEST analysis of crowdfunding can be used by initiators of the projects, investors and policy makers (financial regulators).

Thorough study of foreign experience with crowdfunding is required with its subsequent introduction into Russian practice while implementing the necessary regulation and its further improvement.

References

1. New tools to attract funding for the development of technology companies: practice and development prospects in Russia. Analytical report, April 2018 (2018). (in Russian). https://www.csr.ru/upload/iblock/d43/d43abe96c5e5a9cc5dea8c673f5028e1.pdf
2. Gerber, E., Churchill, E., Muller, M., Irani, L., Wash, R., Williams, A.: Crowdfunding: an emerging field of research. Paper Presented at the Conference on Human Factors in Computing Systems Proceedings, pp. 1093–1098, April 2014. https://doi.org/10.1145/2559206.2579406
3. Gerber, E., Hui, J.: Crowdfunding: motivations and deterrents for participation. ACM Trans. Comput. Hum. Interact. 20(6), 1–12 (2013). https://doi.org/10.1145/2530540
4. Kunz, M., Bretschneider, U., Erler, M., Leimeister, J.: An empirical investigation of signaling in reward-based crowdfunding. Electron. Commer. Res. 1–37 (2016). https://doi.org/10.1007/s1066001692490
5. Wieck, E., Bretschneider, U., Leimeister, J.M.: Funding from the crowd: an internet based crowdfunding platform to support business setups from universities. Int. J. Cooper. Inf. Syst. 22(3), 1340007 (2013). https://doi.org/10.1142/s0218843013400078
6. Adamo, R., Federico, D., Intonti, M., Mele, S., Notte, A.: Crowdfunding: the case of Italy. http://dx.doi.org/10.5772/intechopen.90940
7. Karadogan, B.: CrowdFunding, October 2019. https://www.researchgate.net/publication/336318793_CrowdFunding. Accessed 20 Feb 2020
8. Ilenkov, D., Kapustina, V.: Crowdfunding in Russia: an empirical study. Eur. Res. Stud. J. XXI(2), 401–410 (2018). (in Russian)
9. Genkin, A., Mikheev, A.: Blockchain: How It Works and What Awaits Us Tomorrow. Alpina Publisher, Moscow (2018). (in Russian)
10. Kuznetsov, V.: Crowdfunding: current regulatory issues. Money Credit 1, 65–73 (2017). (in Russian)
11. Chulanova, O.: Risks and barriers when using modern crowd technologies. Afanasyev Read. 1(18), 49–63 (2017). (in Russian)
12. Balyuk, I.A., Bich, M.G., Zvonova, E.A., et al.: The impact of globalization on the formation of the Russian financial market. In: E.A. Zvonova (ed.) Knorus, Moscow (2018). (in Russian)
13. Tsvirko, S.: Crowdfunding as an approach to finance entrepreneurial initiatives in a digital economy. In: «New in Entrepreneurship Development: Innovation, Technology, Investment». Materials of the VII International Scientific Congress. Financial University under the Government of the Russian Federation, pp. 208–214. Dashkov and Co, Moscow (2019)

14. Global Crowdfunding Market 2018–2022. https://www.researchandmarkets.com/research/mmbr2m/global?w=4. Accessed 20 Feb 2020
15. Kickstarter. https://www.kickstarter.com/about?ref=global-footer. Accessed 20 Feb 2020
16. StartTrack. Investment site. https://starttrack.ru/. Accessed 20 Feb 2020
17. Planeta. Investment site. https://planeta.ru. Accessed 20 Feb 2020
18. Boomstarter. Investment site. https://boomstarter.ru/. Accessed 20 Feb 2020
19. PEST analysis in detail, http://powerbranding.ru/biznes-analiz/pest/. Accessed 20 Feb 2020
20. Federal Law of August 2, 2019, №259 "On attracting investments using investment platforms and on amending certain legislative acts of the Russian Federation" (in Russian). http://ivo.garant.ru/#/document/72362156/paragraph/1/highlight/259-%D0%A4%D0%97:1
21. The main directions of development of the financial market of the Russian Federation for the period 2019–2021 (in Russian). https://www.cbr.ru/Content/Document/File/71220/main_directions.pdf. Accessed 20 Feb 2020
22. The concept of crowdfunding regulation in Russia has been developed (in Russian). https://www.cbr.ru/press/event/?id=712. Accessed 20 Feb 2020
23. The Central Bank of the Russian Federation. The Bank of Russia cuts the key rate by 25 bp to 6.00% p.a, 7 February 2020. https://www.cbr.ru/eng/press/keypr/
24. Digital 2020: The Russian Federation, 18 February 2020. https://datareportal.com/reports/digital-2020-russian-federation?rq=Russia
25. https://www.gfk.com/ru/insaity/press-release/issledovanie-kazhdyi-pjatyi-vzroslyi-rossijanin-ne-polzuetsja-internetom/. Accessed 20 Feb 2020

Features of Accounting on Household Goods Accounts in Early Sole Proprietorships

Mikhail Kuter[✉] and Marina Gurskaya

Economy Department, Kuban State University,
Stavropol'skaya Street, 149, 350040 Krasnodar, Russia
{prof.kuter,marinagurskaya}@mail.ru

Abstract. The paper deals with early examples of the use of an account of household goods in sole proprietorship of Francesco Datini in Pisa. The receipt of household goods was reflected in the debit of the household goods account. Entrata e Uscita accounts or personal accounts of suppliers or buyers of household goods were used as entries. Particular attention is paid to the presence of unbalanced balance in the household goods account. Household goods accounts as well as personal accounts were not closed as there was no trial balance. It is indicative that depreciation procedures were not applied to household items in individual households, despite the existence of a "profit from sale of goods" account.

The study was conducted using logical-analytical modelling. Application of this element of the digital system of research in the field of history of accounting contributes to obtaining more accurate information due to the possibility of linking the indicators of various accounting registers in a single information system.

Keywords: Accounting history · Medieval accounting · Household goods account

1 Introduction

Investigations of household goods accounts are devoted to Alvaro Martinelli [1], Zerbi [2], Melis [3], de Roover [4], Sokolov [5], Chatfield and Vangermeersch [6], Rihll [7], Lee [8], Macve [9], Smith [10], Sangster [11], and Woodward [12]). As a rule, the mentioned researchers described the account from the book G. Farolfi (1299–1300), or Francesco del Bene and Partners (1318–1324), or the household goods account from the book J. Mellis [13]. And it seems that no one paid attention to the numerous household goods accounts of the companies of Francesco di Marco Datini in Pisa (1383–1400) and Barcelona (1393–1400). F. Melis himself [3, 14, 15] did not mention such studies either.

The authors have collected significant material on the organization of accounting on household goods accounts in the companies Francesco Datini in Pisa and Barcelona. Special attention was paid to asset impairment and depreciation procedures.

The aim of this paper attempts to investigate the early organization of accounting on house-hold goods accounts sole proprietorships of Datini in Pisa.

© Springer Nature Switzerland AG 2021
T. Antipova (Ed.): ICIS 2020, LNNS 136, pp. 164–173, 2021.
https://doi.org/10.1007/978-3-030-49264-9_15

2   Research Method

The principal research method adopted in this study is archival. It uses material found in the State Archive of Prato. This research team has been working with the material in this archive for over 13 years. Many of the records have been photographed and linked together using logical-analytical reconstruction. This is an approach that we developed for the purpose of enabling entries in the account books to be traced visually between accounts and books and from page to page. By adopting this approach, we are able to see the entire accounting system electronically, making entries and their sources clear in a way that is not possible if one only has the original set of account books. This allows us to review each transaction in detail, trace its classification and thus explain the accounting and reporting methods used, while avoiding misinterpretations. This approach represents a new method in how to analyze and interpret accounting practices for periods when there was no concept of either a standard method or a unified approach to either financial recording or financial reporting. In this paper, we present the bookkeeping method adopted by the accountant in Pisa. As far as we are aware, the entries in the account books included in this study have never been previously analyzed.

3   Organization of Activity in the Company in Prato

The first proprietorship of Francesco Datini in Pisa was founded at the end of 1382. The official introduction took place in February 1383. The sole proprietorship was created on an entirely new production base and undeveloped territory. All this required the attraction of a significant amount of household goods.

The peculiarity of the first solo proprietorship is that it was officially closed in 1386, and in fact, it continued to operate until 1406 until all merchandise acquired before the official closing date was sold and all Merchandise accounts were closed. And, only then was the "Profits on Merchandise" account prepared. Another peculiarity was that it seemed that using the same household goods, the second proprietorship of Datini, opened in 1387, worked in parallel.

While studying the cash register (Entrata e Uscita - Prato, AS, D. №403), the authors noticed that on pages 403, c. 109r and 403, c. 109v, the largest number of expense records refer to the purchase of household goods.

It should be noted that all entries in Entrata e Uscita related to household goods have cross-entries to Prato, AS, D. no. 357, c. 332r in Ledger (photocopy to Fig. 1, translation to Table 1). Thus, a Household goods account was found. In 1383 Household goods accounts were generally formed in paragraph form or "mingled account".

Fig. 1. The household goods account (Prato, AS, D. №357, c. 332r)

As we can see, the goods account was opened on April 21, 1383. On the account transferred from Memorial Prato, AS, D. No. 367, c. 3v, you can consider a set of household goods Nicolò and Lodovico di Bono purchased in Florence. The amount of the purchase was f. 55 s. 6 d. 1.

On the same day, reflecting the household goods receiving from Lorenzo Bianciardi and Co and previously accounted for in the Memorial of the Merchandises A at c. 178 of the old *ragione*[1]. The amount of household goods previously accounted for f. 32 s. 18.

Most entries in the account Prato, AS, D. №357, c. 332r, as mentioned above, have contra entries to accounts in the Entrata e Uscita (Prato, AS, D. №403). All entries are correct, subject to the principle of double entry.

With the exception of two entries only:

- The entry "For 1 pair of average scissors for the house as in the Uscita B at c. 109" in Ledger (Prato, AS, D. №357, c. 332r) has a cost estimate s. 2 d. 10, while the corresponding entry in Entrata e Uscita is equal to s. 2 d. 8;
- The entry "To Expenses for household goods on this day f. 1 s. 15 d. 8 *d'oro* we took from the old Memorial A at c. 5 and put in the Book of merchandises B at c. 332" in Entrata e Uscita wasn't reflected in Ledger.

---

[1] *QUADERNI DEI SALDI DI RAGIONE*. Similar to the Extracts of debtors and creditors, the writings in these Notebooks can also be considered preparatory to Secret Book. Profits, debtor and creditors' statements, merchandise, household goods, cash to balance the various "reasons' are recorded.

Table 1. Translation of the household goods account, Prato, AS, D. №357, c. 332r

1383		
Here below, we will write all the expenses made for our supply of household goods For many goods Nicolò and Lodovico di Bono purchased in Florence on our behalf, they must give as in the Memorial B at c. 3 is clearly stated		f. 55 s. 6 d. 1
For 1 bale of feather and For 2 pieces of liners for quilts	Of those goods we made 3 quilts for our house	f. 32 s. 18
We got them from Lorenzo Bianciardi and Co as it can be seen in the Memorial of the merchandises A at c. 178 of the old *ragione*		
For 50 rafters for the warehouses as in the Uscita B at c. 109		f. 3 s. 14
For 1 case and other goods as in the Uscita B at c. 109		f. 2 s. 18
For 1 pair of average scissors for the house as in the Uscita B at c. 109		s. 2 d. 10
For 1 chest and tables and perches and small desks and 2 benches and 1 case and other wood goods, it all can be seen in Uscita B at c. 109		f. 9 s. 6
For 10 ½ *canne* [rods] of scarlet *baracano*[a] for 1 mattress for us as in the Uscita B at c. 109		f. 8 s. 2
For 6 ¾ *canne* of Burgundy towel for 1 mattress for the maid as in the Uscita B at c. 109		f. 1 s. 13
For 1 cover and 1 white blanket Stoldo di Lorenzo purchased, as in the Uscita B at c. 109		f. 10 s. 10
For 2 pairs of large bedclothes for the bed of the boy and the maid, Stoldo di Lorenzo purchased as in the Uscita B at c. 109		f. 6 s. 6 10
Here below we will write down matters and quilts and feather we purchased from a *coltriciaio*[b] living in [*di Landi?*] as in the Memorial B at c. 1 where he has given where he must have f. 22 for his activity		f. 25 s. 5 d. 5
For lbs. 200 of wool for our mattress at f. 3 ¼ per 100	f. 6 s. 35	
For making 2 quilts, one for us and the other for the maid, all in all is	f. 1 s. 10	
For making 2 mattresses, one of our bed and one for the maid's bed	f. 1 s. 20	
For 1 quilt and 1 mattress for the shop boy	f. 3 s. 60	
For lbs. 100 of feather for the said quilts and pillows	f. 2	
For dyeing in blue of a towel for our bedroom	f. – s. 64	
For 1 mattress of br. 4 ½ long and 3 ½ wide for the shop boy's bed	f. 4 s. 40	
The total sum is f. 25 s. 5 d. 5 *d'oro* as in the Memorial B at c. 1 he must give, and he gave		
And for custom duties of 2 coffers of goods, that is household goods, Nicolò and Lodovico di Bono purchased in Florence on our behalf as in the Quaderno of the *Ragione* B at c. 7		f. – s. 14 d. 4
For many wooden objects and writing desks and 3 sleeping beds and benches and many other necessary goods for the house as in the Memorial B at c. 3 at the *Maestragio*		f. 11 s. 4 d. 3

(*continued*)

Table 1. (*continued*)

1383	
For many household goods Niccolò and Lodovico di Bono purchased in Florence on our behalf as in the Memorial B at c. 16	f. 33 s. 10 d. 5
For 1 bed and wooden objects we got from Parentario woodworker as in this book at c. 27 he had where he should have had	f. 1 s. 13 d. 3
For 31 rafters we purchased on September 10th to be put under the wool, as in the Uscita B at c. 114	f. 2 s. 2 d. 10
For 1 brass bowl 1, Matteo, purchased in Florence at f. 6 s. 14 and 1 sent to Pisa what remained by letter as in the Uscita B at c. 114	f. 2
For 5 Napolitan barrels we purchased on October 10th from Naples […] as in the Uscita B at c. 115	f. 7 s. 3
For lbs. 41 of tin in 4 small dishes and 12 bowls and 12 bowls [sic] for s. 10 per pound, we purchased from Antonio Borsaio as in Uscita B at c. 116	f. 5 s. 17 d. 2
220.6.8	
Put here further at c. 336	

[a] Rough goat or camel hair cloth.
[b] The medieval name of the master of blankets and pillows.

On October 26, 1383, the first folio (Prato, AS, D. №357, c. 332r) of the household goods account was fully filled in. The amount accumulated on it (f. 220 s. 6 d. 8) has been transferred from the credit of the closing account to the debit of the continuing account Prato, AS, D. № 357, c. 336r in the same Ledger (photocopy to Fig. 2, transfer to Table 2).

Fig. 2. The household goods account, Prato, AS, D. №357, c. 336r

On the credit account Prato, AS, D. No. 357, c. 336r, seven entries were placed. They are due to the removal of household goods from a sole proprietorship. Incoming money from the sale of household goods in two cases was registered in Entrata B at c. 76, B. In four instances, buyers "gave money to Francesco di Marco himself as stated in this book at c. 451". And, only one entry indicates the reflection of income from sales (f. 6 s. 16 d. 4) in the buyer's account: "They have given on this day f. 6 s. 16 d. 4 for many household goods sold to Stefano di Barto middleman as in this book at c. 403 into his account". The amount for all credit records is f. 47 s. 10 d. 4.

Table 2. Translation of the household goods account, Prato, AS, D. No. 357, c. 336r

1383		
Household goods *must give* as in can be seen back here at c. 332 f. 220 s. 6 d. 8 *d'oro*		220 s. 6 d. 8
For 1 large wooden case we purchased on November 3rd as in Uscita B at c. 116,		f. 1 s. 5 d. 8
For 1 2-case chest and 1 barrel *di meno* and 1 cask *da cietto*	Those goods were purchased on December 12 from Salvestro Barduci and co as in the Uscita B at c. 119	f. 3
For Expenses for an exposition of cloths in the house where we live and other expenses as it can be seen in the Uscita B at c. 133		f. 14 s. 4 d. 7 *a oro*
238.16.11		
They *have given* on August 16 f. 3 s. 5 d. 8 *a oro* for many household goods we sold to Zanobi di Lacopo as in Entrata B at c. 76, B		f. 3 s. 5 d. 8
They *have given* on August 18 f. 3 for 1 chest we sold to Antonio middleman as in Entrata B at c. 76, B		f. 3
They *have given* on this day f. 6 s. 16 d. 4 for many household goods sold to Stefano di Barto middleman as in this book at c. 403 into his account		f. 6 s. 16 d. 4
They *have given* on August 27th f. 6 s. 7 for 8 barrels we sold to Primo Tanner and on our behalf, he gave the money to Francesco di Marco himself as in this book at c. 451 Francesco *must give*		f. 6 s. 7
They *have given* on September 11th f. – s. 15 for 1 pitcher and 1 brass bowl we sold to Manno del Migliore and on our behalf he gave the money to Francesco di Marco himself as in this book at c. 451 Francesco *must give*		f. – s. 15
They *have given* on September 16th f. 11 s. 2 for many household goods given to Agnolo di Lupo of Siena and on our behalf, he gave the money to Francesco di Marco himself as in this book at c. 451 Francesco *must give*		f. 11 s. 2
They *have given* on September 27 f. 16 s. 4 d. 4 *a oro* for […] 4 ½ of bed cloths and 5 tablecloth [*tovaglie*] and 6 cupboards [*guarda nape*] and 2 hand-towel [*sciugatoi*] and 8 dishtowels [*canovaccio*] for knives, all was given to Francesco di Marco as in this book at c. 277 Francesco *must give*		f. 16 s. 4 d. 4 *a oro*
47.10.4		
191.6.7		

As it follows from the transfer of the account (Prato, AS, D. №357, c. 336r) (Table 2) and the diagram (Fig. 3), there is a difference between the debit of the account (top part - f. 238 s. 16 d. 8) and the credit of the account (bottom part - f. 47 s. 10 d. 4). 191 s. 6 d. 7. The photocopy of the account (Fig. 2) shows this amount is recorded separately from all entries in the lower-left corner of the page. The researchers made an unsuccessful attempt to find this amount in other accounts and in second proprietorship accounts.

Figure 3 presents a diagram of the formation of the Household goods account.

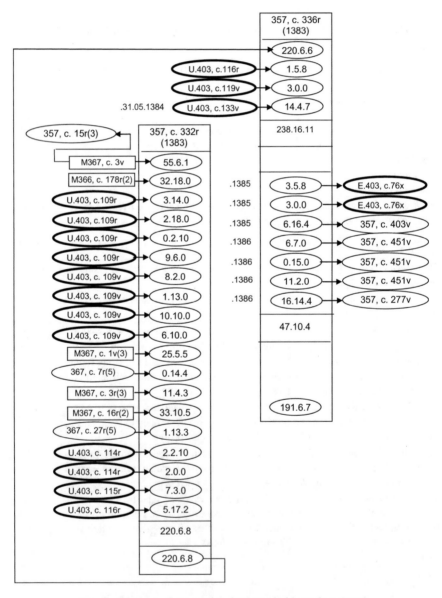

Fig. 3. Diagram of accounting for household goods account

Let's try to explain it. The tip of the proprietorships accounting system Francesco Datini in Pisa was the "Profits on Merchandise" account. Trial Balance hasn't been applied yet.

Officially, the first proprietorship was closed in 1386. It meant that this sole proprietorship stopped purchasing goods. At the same time, the sale of merchandise purchased before 1386 continued. At the same time, entries were made in previously opened and current and newly opened personal accounts of customers. In 1406 a "profit on goods" account was prepared, when all goods were sold, and all merchandise accounts were closed.

A carryover eventually closed the Merchandise accounts from balance to Merchandise accounts, which reflected gains and losses on the sale of goods. Personal accounts of suppliers and buyers remained open, with debit or credit balances denoting the amount of active or passive debt. The balance of accounts remained in the accounts, as there was no Trial Balance to carry them over.

The situation was similar with the household goods account. Its balance f. 191 s. 6 d. 7 should have transferred to Trial Balance, but it did not exist.

In 1386, a second proprietorship was established on the same production base (household goods of the first proprietorship), which operated in parallel with the first. For this reason, no new household goods account was opened in the second proprietorship accounting system. Officially, the proprietorship was closed in 1392, although the sale of inventory continued in 1393. In 1393, when the second proprietorship was really closed, for the first time in proprietorships and companies of Francesco Datini a "Profits on Merchandise" account was established.

In 1392, the First Company was founded. When creating an accounting system (parallel to the two already in operation), the company opened a household goods account, which was periodically subject to asset impairment procedures. Neither asset impairment nor depreciation has been charged to the household goods account of the first sole proprietorship.

Years of research at the Datini companies in Pisa (1392–1400) have confirmed the practice of asset impairment procedures in all companies. Initially (1393), asset impairment was charged to the household expenses account. At the closure of the company in 1394 "losses on the household goods were charged to the Profits on Merchandise account" [16].

De Roover followed the common practice of presenting "losses on the household goods" as "Depreciation". Describing the "Profits and losses" account of Francesco di Marco Datini (11 July 1397-31 January 1399)" in Barcelona [4, p. 144], he mentioned:

"Depreciation on office equipment, lb. 16 s. 17".

A more detailed study of the claim that we were interested in the fact of depreciation led us to the account Prato, AS, D. № 801, c. 389v:

"Per danno di masarizie di casa in XVIII months in questo c. 379 masarizie di casa lb. xvi s. xvii"

For loss of household goods for 18 months, as in this book on carta 379 [the account for] household goods £ 16 s. 17.

Unfortunately, de Roover has not paid due attention to the entry in credit side of the household goods account (Prato, AS, D. №801, c. 379r):

"Masarizie di chasa de' dare a dì XXXI di gienaio 1398 per danno **di masarizie in XVIII mesi abatuto a ragione di X per cento l'anno in questo** a c. 389, lb. xvi s. xvii".

It means: "Office equipment should give on January 31, 1398 "wear and tear of furniture" **calculated over 18 months based on 10 percent annually**, on page 389, lb. 16 s. 17".

Note that an algorithm for distributing the value of household goods over their useful lives is highlighted in bold.

Thus, the authoritative researcher has left the historical creation of Simone Bellandi without attention, who was an accountant at that time in Datini's company in Barcelona.

This was the only depreciation practice we discovered at the end of the 14th century. Bellandi did it only once, in 1399. Before that, he used the asset impairment. In the beginning of the next century, Simone Bellandi left the company due to illness. A new manager, who had arrived to replace him, refused to continue the practice started by Bellandi and returned to the asset impairment practice [17].

4   Conclusion

The paper looks at early examples of the use of a household goods account in sole proprietorships of Francesco Datini. The accounting mechanism was simple. The receipt of household goods was reflected in the debit of the household goods account. Entrata e Uscita accounts or personal accounts of suppliers or buyers of household goods were used as contra entries.

Special attention is paid to the presence of an unbalanced balance in the household goods account. Household goods account, as well as personal accounts, were not closed as there was no Trial Balance.

Apparently, the second sole proprietorship used household goods that belonged to the first sole proprietorship. The authors arrived at this conclusion because there was no such account in the accounting system of the second sole proprietorship.

We point out, that asset impairment or depreciation procedures have not been applied to household goods in the sole proprietorships, despite the presence of the "Profits on Merchandise" account.

In this paper, it is emphasized that for the preserved archival materials in the Middle Ages, only the procedure of asset impairment was applied, which gave a reason to combine its name with depreciation. However, the authors recalled that they have established that in 1399 in the company Datini in Barcelona, its accountant Simone Bellandi performed the standard procedure of distributing the value of household goods in its useful life, which corresponds to the algorithm of descriptions in the credit of the household goods account.

References

1. Martinelli, A.: The origination and evolution of double entry bookkeeping to 1440. ProQuest Dissertations & Theses Global, p. n/a (1974)
2. Zerbi, T.: Le Origini della partita dopia: Gestioni aziendali e situazioni di mercato nei secoli XIV e XV. Marzorati, Milan (1952)
3. Melis, F.: Documenti per la storia economica dei secoli XIII—XVI, Leo S. Olschki, Firenze (1972)
4. de Roover, R.: The development of accounting prior to Luca Pacioli according to the account-books of Medieval merchants. In: Littleton, A.C., Yamey, B.S. (eds.) Studies in the History of Accounting, London, pp. 114–174 (1956)
5. Sokolov, Y.V.: Accounting: From the Beginnings to the Present: A Manual for Schools. Audit Unity, Moscow (1996)
6. Chatfield, M., Vangermeersch, R.: History of Accounting: An International Encyclopedia. Garland Publishing Inc., New York (1996)
7. Rihll, T.E.: Depreciation in vitruvius. Class. Q. **63**(02), 893–897 (2013)
8. Lee, G.A.: The coming of age of double entry: the Giovanni Farolfi ledger of 1299–1300. Acc. Hist. J. **4**(2), 79–96 (1977)
9. Macve, R.: Some glosses on "Greek and Roman accounting". Hist. Polit. Thought **6**(1/2), 233–264 (1985)
10. Smith, F.: The influence of Amatino Manucci and Luca Pacioli. Br. Soc. Hist. Math. Bull. **23**(3), 143–156 (2008)
11. Sangster, A.: How double entry was taught before Luca Pacioli. In: Levant, Y., Trébucq, S. (eds.) Théorie comptable et sciences économiques du xve au xxie siécle: Mélanges en l'honneur du Professeur Jean-Guy Degos, pp. 109–120. L'Harmattan, Paris (2018)
12. Woodward, P.D.: Depreciation – the development of an accounting concept'. Acc. Rev. **31**(1), 71–76 (1956)
13. Mellis, J.: A Brief Instruction and Manner How to Keep Books of Accounts (1588). John Windett, Reprinted Scholar Press, London (1980)
14. Melis, F.: Storia della Ragioneria. Cesare Zuffi, Bologna (1950)
15. Melis F.: Aspetti della Vita Economica Medievale. Studi nell'Archivio Datini di Prato, Siena (1962)
16. Kuter, M., Gurskaya, M.: Accounts of household expenses in the medieval companies. In: Antipova, T., Rocha, A. (eds.) DSIC 2018. AISC, vol. 850, pp. 286–295 (2019). https://doi.org/10.1007/978-3-030-02351-5_33
17. Kuter, M., Gurskaya, M., Andreenkova, A., Bagdasaryan, R.: Asset impairment and depreciation before the 15th century. Acc. Hist. J. **45**(1), 29–44 (2019)

Digital Economics

Multi-agent Systems as the Basis for Systemic Economic Modernization

Gennady Ross$^{(\boxtimes)}$ and Valery Konyavsky$^{(\boxtimes)}$ ⓘ

Plekhanov Russian University of Economics, Moscow, Russia
ross-49@mail.ru, konyavskiy@gospochta.ru

Abstract. This paper considers the mechanisms that give multi-agent systems the ability to self-organize and develop, and the manifestations of these abilities are studied in relation to the systemic modernization of the economy. The analysis of financial bubbles management in the development of new methodology of forecasting economic processes is carried out. The necessity of new methodology is determined by the increasing role of financial bubbles in digital economy. The proposed methodology is based on the theory of equilibrium random processes, which includes a complex of theoretically substantiated evolutionary and simulation models, methods and methods of decision making. The method of evaluation of financial bubbles sizes is proposed, the concept of financial bubble share and the method of its calculation is introduced. An example of calculating financial bubbles using the Decision tool system is considered.

Keywords: Digital economy · Equilibrium random process theory · Evolutionary simulation model · Risk of overstatement · Risk of understatement · Chaos · Type 2 collapse · Multi-agent systems · Self-organization · Financial bubbles · Instrumental system decision · Share of financial bubble

1 Introduction

It is now clear that the global financial sector has "broken away" from reality, turned into a global bubble and started to put pressure on the economies of all countries. The transition to a digital economy was not improving but rather exacerbating the situation by increasing the rate at which bubbles were being blown and diversifying the way in which they were stored. This leads to an economy that is not manageable and chaotic, unfair. The question arises: what to do? The answer to this question can only be given by a measure in the mathematical sense: one has to learn how to measure financial bubbles (FB). Only knowing the measure, it is possible to create a monitoring system and management institutions adequate to the digital economy. In our opinion, the necessary scientific and methodological groundwork for solving this problem, namely, the theory of equilibrium random processes (EP), the methodology of mathematical modeling of EP, and the instrumental system Decision have already been created [1–4, 27, 28].

© Springer Nature Switzerland AG 2021
T. Antipova (Ed.): ICIS 2020, LNNS 136, pp. 177–187, 2021.
https://doi.org/10.1007/978-3-030-49264-9_16

When we look at economics as a complex of multi-agent systems (MAS), we find that economics cannot exist either as an organization (planned economy) or as a self-organization (market economy). These statements are a consequence of the theorems [1, 2] that the planned economy has shown the unviability of organization, while the market economy shows the collapse of the 2nd kind and chaotic self-organization. Among the various and numerous manifestations of the collapse of the 2nd kind one can observe such as the detachment of the nominal value of money from any reality. The cause of the collapse of the 2nd kind and chaotic economy are objective regularities of its emergence, which lead to the distortion of information, lies, incompetence and stupidity of elites. Blowing up an FP is the most effective investment. To be more precise, investments in financial turnover have the highest average specific profitability taking into account the risk. Therefore, capital flows from the real sector to the financial sector. One more fundamentally important problem related to FBs comes from the fact that money is a carrier of economic information. All derivative economic indicators (income, profit, GDP, etc.) are built on prices. False information is transferred from economy to marketing, PR technologies, politics and all other spheres.

In economic science, ignoring the fundamental laws of MAS leads to the fact that research, including those awarded the Nobel Prize, becomes helpless. In the article [3] it is illustrated by concrete examples of the Nobel laureates works such as F.A. Hayek, G. Murdahl, (1974), G.S. Becker (1992), R. Merton, M. Sholes (1994). All these studies are unable to explain the most significant, large-scale negative processes: capital flow from production to financial sphere, transformation of financial sphere from service to bubble, transformation of financial sphere into bubble, and business into rentier, etc. The regularities peculiar to the MAS allow explaining why the attempts to overcome the problems of economy through planning have not been successful. Only an economy based on managing self-organization and development can be viable. At the same time, the state must intervene in the economy in a timely manner to switch various market segments from market self-organization to planned management and back.

All the results obtained by MAS are derived from the evolutionary simulation model (EMM) of the economic agent (EA) [24]. Economic movements towards collapse of the 2nd kind appear in different kinds: transformation of capital into FBs; decrease in reliability of economic information; chaos; increase in social injustice. All these are manifestations of the natural aging process of the MAS [11]. Only the state for survival and prosperity can resist this tendency.

The works [5–9] are devoted to the comprehensive and thorough research of FBs, among which the work of Smirnov [6] is particularly worth mentioning, where such mechanisms of FB formation as financial leverage and others are disclosed. Kaurova [10] investigates the problems of the level of openness of the economy; Kobyakov A. and Khazin M. [11] thoroughly investigates the role of high technologies in the world economy; Soros [12] analyzes some sources of global threats resulting from the motives of the market participants' behavior. The increasing role of psychological factors and related threats is the subject of research by Becker et al. [13]. In the article by Abebe [14] considers the role of electronic commerce for business orientation; in the articles Beamish [15] and Beck. The articles Beamish [15] and Beck. [16] consider the role of contractual relations and financial aspects of international trade, respectively.

The research on creation of conditions for FB are devoted to the work [17, 18] and methods of possible fight against them in [19–31]. A good review of literary sources is given in the book that was written by Chirkova [22]. However, the ideas expressed by her are not sufficient for a full-fledged theory of FB management.

2 Self-organization and Development - Emergent Properties of Multi-agent Systems

For all the achievements in the field of FB [23, 24] the central issues remain unclear: "What is Financial Bubles (FB)?" and "What is an economic agent (EA)?" In our opinion, it is self-organization that is the obligatory, systemic, emergent property that serves as an indispensable attribute of FB. The definitions of economic agent, FB and self-organization are interconnected and mutually complementary. In formulating these definitions, it is necessary to select the necessary and sufficient set of conditions under which the mechanism of self-organization is capable of functioning, and in the absence of at least one of them - is inoperative [4]. The difference between self-organization and organization is especially evident in the economic community, where it is presented as a distinction between free market economy and planned economy [25, 26]. Self-organization and organization are combined at different hierarchical levels: an agent may and should even be an organization within himself, but interaction of agents is possible either only on the principles of organization or only on the principles of self-organization. If neither is present, chaos reigns. Organisation and self-organisation are alternative, they mutually exclude each other, because in self-organisation, the agent derives from his own interests, and in organisation there is a plan common to all agents, where the common interest takes precedence over the interests of individual agents. The existence of a plan is an indispensable feature of the organisation, and its absence is a necessary, albeit insufficient, condition for self-organisation. It is in this sense that organization and self-organization are alternative. At the same time, they can coexist physically. For example, marketing sales plans may be established in the firm, but it is allowed to select suppliers (also agents), thus leaving the freedom for self-organization.

It should be noted that only active EAs can self-organize. Activity is the ability to formulate individual goals, develop plans and achieve them independently. In this case, the agent is under the influence of random factors or in conditions of uncertainty. The presence of a random environment is fundamentally important. In its absence, the whole set of agents can be organized, and this organization will obviously be more effective than self-organization. The combination of randomness and planning means that an agent's functioning is an equilibrium random process (ESP) [1]. Thus, an integral property of an agent is its activity, i.e., its ability to plan and execute its plans, and to interact with other agents. In this case, the PL plan of an MAS agent is the solution to the agent's ESM, which is proposed in [1]. It is established that MAS functioning naturally and inevitably leads to information collapse and chaos. In this case, there comes the collapse of the 2nd kind, at which the information becomes unreliable, distorted or simply false. Idealized this situation is described in the fairy tale of G.S. Andersen about the naked king: the servants praised the beauty of the royal clothes while the king was naked! Lying information makes communication, self-organization, planning and management impossible. To characterize the value of information owned by agents, we introduce the reliability index P^0. In the process of self-organization, the overwhelming majority of agents' P^0 is steadily decreasing, approaching zero. Chaotization and information collapse of the 2nd kind are two sides of the same coin; they are different manifestations of the same process. In order to preserve self-organization, slow down or prevent these tendencies, external interference in the management of self-organization is necessary. It can be said that management of self-organization is management of interaction between agents, while management of development is management of structural and functional rearrangements, in particular, input or shutdown of communication channels.

Thus, looking at the economy as an ICA, we can find regularities that determine the content of the most important economic, political and social trends today and for the foreseeable future.

3 Evolutionary Simulation Model of an Economic Agent

3.1 A Subsection Sample

We take into account the main definitions for describing of an economic agent (EA) by considering the structural formulation of Evolutionary simulation model (ESM) of an economic agent (see below (1)–(7) equations). It is a model of economic agent, for example, a firm to which we will appeal for convenience and specificity of presentation.

Any firm is in a random environment, i.e. under the influence of various random factors, such as exchange index f_1, volatility f_2, etc. Let us consider that each factor f_i is a scalar random value and refers to a certain moment t of the future, not to the time interval. We can consider any number of f_i factors, with any laws of probability distribution of values and any covariance relations between them. Let's assume that $t = 0$ is the present moment, $t > 0$ is the moment of the future, and $t < 0$ is the moment of the past. In the past, all random factors have specific implementations, which we will mark with the upper index "e". So f_{it}^e is the stock index in the specific conditions that were in the past $t < 0$, and f_{it} is possible options for the stock index at the time of the future $t > 0$ with the appropriate probabilities. In general, thus, a random environment in which EA is immersed at the moment $t > 0$ is a random vector $\bar{f}_t = \{f_{1t}, \ldots, f_{lt}\}$.

For any economic agent, it is possible to select several most important integral characteristics and accept them as phase coordinates. For example, a company working in the market sector of a particular product, say, a grain, is characterized by coordinates at any moment of time: x_1 - price and x_2 - sales volume. Another obligatory coordinate is time. The phase space is called the Euclidean space with the named coordinates. The point of phase space at the moment t, that is a vector $\{x_{1,t}, x_{2,t}, \ldots, x_{n,t}\}$, let's designate Fa_t (from the word "fact"). For any moment $t > 0$ of the future, Fa_t is a random vector since depends on a random environment $\bar{f}_t$. Besides, Fa_t depends on initial indicators characterizing EA (for example, on production capacity of the firm p_1, on external conditions of its existence, say, the rate of the profit tax p_2, etc.). The vector of initial indicators denotes as $\bar{p} = (p_1, \ldots, p_J)$. Equation (1) shows that with the help of a simulation model $\rho\left(\overline{f_t}, \bar{p}\right)$ it is possible to obtain sales Fa_t through the implementation of factors $\overline{f_t}$ at given input indicators $\overline{p_t}$. The content of simulation model algorithms $\rho, \rho_1, \rho_2, \rho_{2+k}, k = 1, \ldots, K$ that included in (1)–(7) are presented in general form and their specific implementation depends on the specifics of the task [30].

Examples of specific formulations of ESM for different categories of economic agents are given in [6–9].

$$Fa_t = \left\{x_{1,t}, x_{2,t}, \ldots, x_{n,t}\right\} = \rho\left(\overline{f_t}, \bar{p}\right) \tag{1}$$

$$\Psi_1\left(PL_t, Fa_t\right) = \rho_1\left(PL_t, Fa_t, \overline{f_t}, \bar{p}\right), если \quad PL_t \succ Fa_t \tag{2}$$

$$\Psi_2\left(PL_t, Fa_t\right) = \rho_2\left(PL_t, Fa_t, \overline{f_t}, \bar{p}\right), если \quad PL_t \prec Fa_t \tag{3}$$

$$\min_{PL_t}\left\{\max_{L \in \{1;2\}} \{M\{\Psi_L(PL_t, Fa_t)\}\}\right\} \tag{4}$$

$$P_t^0 = p(PL_t \succ Fa_t) \tag{5}$$

$$3/3_t = \lim_{\Delta \to 0} \frac{M\{\Psi_1\left(PL_t + \Delta, Fa_t\right)\} - M\{\Psi_1\left(PL_t, Fa_t\right)\}}{M\{\Psi_2\left(PL_t, Fa_t\right)\} - M\{\Psi_2\left(PL_t + \Delta, Fa_t\right)\}} \tag{6}$$

$$r_{k,t} = \rho_{2+k}\left(PL_t, P_t^0, 3/3_t, \bar{p}, \bar{f}\right), k = 1, \ldots, K \tag{7}$$

As it was shown above, the activity of EA is expressed by such criteria as maximizing the firm's profit or income, or increasing capitalization, or at least not going bankrupt. Based on this goal, the firm is forced to react to the impacts of a random environment $\overline{f_t}$. Incentives that determine this or that reaction of EA to the external environment are managing impacts. Active behavior in a random environment leads to the fact that the functioning of the firm and any EA is "an equilibrium random process - a process whose trajectory in the phase space is determined by a combination of random factors and control actions, the direction and force of which are determined by the size and direction of deviation of the actual trajectory from the smoothed one" [1]. The meaning of the RSP is obvious on the example of the market of a specific product. In this case, phase coordinates: x_1- price and x_2 - volume of sales; the smoothed trajectory is the equilibrium price and volume of sales; the actual trajectory is the instantaneous values of price and volume of sales; managing (stimulating EAs) effects are profit (for the supplier) and costs (for the buyer).

In order to clarify the meaning of "managerial impact", it will be necessary to introduce the concept of "plan". On any set of points $Fa_t^e, t \leq 0$, i.e. on the actual trajectory up to the present moment $t = 0$, it is possible to construct a smoothing trajectory of EA in phase space and to find on it a point PL_t (from the word "plan") corresponding to any moment of the future $t > 0$. Here, PL_t does not matter to us whether there is a real possibility to build a smoothing trajectory. Moreover, is not necessarily the position to which the process will reach "by inertia" if it is not managed. On the contrary, PL_t is the position that EA would like to reach by the time $t > 0$, despite random influences by $\overline{f_t}$. The motive for reactions from EA is a mismatch of fact Fa_t and plan PL_t, of what is and what one would like to have. For example, the sales marketing plan $PL_t = \{x_{1t}, x_{2t}\}$ indicates the desired sales volume x_{1t} and price x_{2t} at the time $t > 0$. If there were no random environment $\overline{f_t}$, the fact Fa_t would be coincidence with the plan PL_t and efforts are not needed to influence what is happening. But at $t > 0$, Fa_t- a random, and the plan PL_t - a deterministic value and the probability of the event $PL_t = Fa_t$ is close to zero.

Incentives for this or other variant of EA behavior are not in itself a mismatch Fa_t and PL_t, but the negative consequences that will come from this mismatch. For example, if Fa_t turns out to be bigger than the plan PL_t, there will be lost profits, and if less - not efficient use of resources. To quantify these effects, it is necessary to be able to compare Fa_t and PL_t, to determine the direction of deviation Fa_t from PL_t and have a quantitative measure of this deviation.

Ratio (2) shows that the simulation model $\rho_1\left(PL_t, Fa_t, \overline{f_t}, \overline{p}\right)$ when $PL_t \succ Fa_t$ it is possible to calculate a quantitative measure of the entire set of negative consequences of deviation Fa_t from PL_t. Negative consequences are denoted as $\Psi_1(PL_t, Fa_t)$ and called overestimation costs which, at the same time, are also the controlling influence. So, if Fa_t not better than PL_t, there are negative consequences (managing influence) as $\Psi_1(PL_t, Fa_t)$.

Ratio (4) expresses the minimal strategy of EA behavior. Its essence is that EA seeks to minimize the greater of the risk of overestimation $M\{\Psi_1(PL_t, Fa_t)\}$ and the risk of undervaluation $M\{\Psi_2(PL_t, Fa_t)\}$, which are mathematical expectations of the costs of overestimation $\Psi_1(PL_t, Fa_t)$ and undervaluation costs $\Psi_2(PL_t, Fa_t)$, respectively. For example, the firm tries to draw up such a sales plan and set such prices in order to minimize the greater of the risks.

Formulas (1) to (4) unambiguously define the optimal plan PL_t, while the equality (5) shows that the mutually unambiguously PL_t related to indicator P_t^0, called reliability, expresses the probability that the optimal plan will be as good as the fact. Equality (6) defines an indicator $3/3_t$ (from "overvaluation/lowering") that expresses the risk of overvaluation to the risk of underestimation around the optimum. Estimated indicators $r_{k,t}, k = 1, \ldots, K$ can be calculated on the basis of $PL_t, P_t^0, 3/3_t, \overline{p}, \overline{f}$ what the equality (7) indicates.

4 Old and New Economic Paradigm

The above notions describing the model (1)–(7) simultaneously give a formulation of the paradigm of the digital economy. The mathematical expression of the new paradigm is condition (4). Equivalently, the same condition can be written down as follows:

$$\text{Risk of overstatement} = \text{Risk of understatement}$$

The equality of risks of overvaluation and undervaluation determines the behavior of economic agents of all categories (low, medium and high yield). Even if EA has no resource limitations, if it has variable preferences and uncertain intentions, it still, in all circumstances, is guided by its subjective risks as it understands them. EA always proceeds from how not to do something superfluous (risk of overstating) and, at the same time, not to miss something necessary (risk of understating). This is true regardless of the horizon on which EA focuses: a momentary situation, a medium-term perspective or a distant strategic goal.

At the same time, it is a classic economic paradigm, which is expressed by equality:

$$\text{Demand} = \text{Supply}$$

is a special case of condition reflected in formula (4).

There is strict mathematical proof that the equilibrium of supply and demand is a special case of equilibrium of risks of overstatement and understatement, i.e. the classical paradigm is a special case of a new paradigm - paradigm of digital economy.

5 Estimation of the Financial Bubble Share (SFB)

The proposed SFB (share of financial bubble) measure the share of the secured portion in the nominal value of a currency or security. At the same time, SFB expresses the probability of the reliability of an economic indicator, such as profit, capitalization or price. SFB can be likened to the indicator "Strength of alcoholic beverage", which is a ratio of "… the volume of dissolved anhydrous alcohol to the volume of all the beverage multiplied by 100%"[1]. The statement that the SFB of the dollar is 97% means that the secured portion and the real price of green paper is 3 cents. The other 97% is bubble, financial nothing, emptiness. However, dollar liquidity, expressed as an indicator (reliability), is still quite high ($P^0 \approx 1$): today dollars are easily exchanged for other currencies and commodities. At the same time, SFB expresses the probability of the reliability of a price or other economic indicator. For example, if the capitalization of some firm is equal to $N, then SFB is a probability that the figure N is true. SFB is a level of distrust to the indicator, and (1 - SFB) is a level of trust. Except for the price of goods, SFB can be attributed to the firm's capitalization and, therefore, to EA, which is especially important when assessing the business value.

---

[1] https://dic.academic.ru/dic.nsf/ruwiki/425252.

At the same time, SFB is an expression of social injustice of various origins: the difference between SFB and 0 indicates the presence of theft, or attachments, or monopolization, or frenzy, or shortage, or imbalance, or financial leverage, or over-heating of the economy, etc. The only exception is the increase in SFB due to tech-nological know-how, which temporarily distorts the balance. In order to take this into account, it is necessary to set the standard (reference, permissible) $SFB^h > 0$ for a country, or for a region or for a certain category of EA within a certain period of time. It is clear that SFB is not a replacement of liquidity (reliability) indicator or its specifi-cation, but a new additional characteristic.

To calculate the SFB of some EA (in its capitalization, in the quotes of its secu-rities, in the price of goods issued by it, etc.) is necessary:

1. To take as a reference average EA from the category under consideration, except for 10% of the most profitable and 10% of the least profitable EA.
2. Calculate for the benchmarks and research EA at least one of the characteristics $PL, P^0, 3/3$ either directly on the basis of available data or on the basis of indirect data $\bar{f}$ using the Decision tool system.
3. Use a mutually unambiguous correspondence of indicators $PL, P^0, 3/3, D$ and uni-versal comparability of indicators $3/3, P^0, D$ to calculate $3/3$ and $D_3 = \dfrac{P^0}{3/3}$ for the benchmark.
4. Calculate $D_\mu = \dfrac{P^0}{3/3}$ in the same way for the studied EA.
5. Calculate SFB for the studied EA by the formula: $SFB = \dfrac{Du - D_3}{Du} * 100\%$.

If SFB > 0 - there is a usual financial bubble associated with the price deviation to the right, similar to the one that occurs in the fussy demand.

If SFB < 0 - there is a "bubble on the contrary", for example, overheating the economy, or inefficient, not paying off the excitement, for example in construction. For practical realization of the suggested method of SFB calculation the Decision system [27, 28] can be used. The easiest way to collect input data is to collect data for evaluation P^0. In many cases P^0 is possible to estimate on the basis of statistical reporting or on the basis of special, limited, confidential sociological surveys. This is not difficult to do even when EAs tend to hide information.

As a reference EA it is possible to choose an agent that works in conditions of close to perfect competition and has reached the technological level - the industry average, or among the considered category of EA. In practice, they try to create conditions for perfect competition on exchanges. At the same time, however, it turns out that not every commodity is an exchange. The requirements of homogeneity and separability actually make us talk about a rather narrow sector of the market of a particular com-modity, for example, grain, and tied to place and time. Therefore, as a reference EA should be taken an average or model EA, which is presented as a specific ESM with average input data.

Proximity of the reference EA to the studied EA means only the similarity of the basics of organizational and financial activities, while the direction of activity or ownership can be any. The similarity of the benchmark and the EA under study is calculated using a formula: $SFB = \dfrac{Du - D_3}{Du} * 100\%$.

The fight against FBs implies the creation of two mutually complementary systems: a system for monitoring financial bubbles and a system for ensuring quality economic growth. The objectives of management are to equalize SFB and D in different EAs, but not to achieve strict equality SFB = 0 or $D_i = D_{i'}, \forall i \neq i'$. There is only necessary to ensure, on the one hand, the aspiration of SFB $\rightarrow$ SFB^h where SFB^h > 0 is a normative value and, on the other hand, to ensure approximate (not strict) equality of returns: $D_i \approx D_j, i \in L$, where L is a set of EAs similar to this benchmark. These requirements ensure fairness. At the same time, they provide the direction of bubble management to eliminate unwanted anomalies in size and stability of bubbles.

Governance of FPs differs from ordinary public economic governance in that the moment, purpose, size and direction of public intervention is determined by the dynamics of the ratio of values, and SFB for specific EAs or categories of EAs. For example, if some EAs have increased from 0.29 to 0.35 during the year and the SFB has not changed, the result will be achieved due to an increase in productivity, or improved administration, or increased production, etc. If SFB also increased, for example, from 0.21 to 0.27, then most likely the increase is a consequence of bubble inflation. This can be clarified in various ways, for example, by comparing capitalization growth with bubble growth in rubles. If it turns out that the increase is due to SFB, then the state should intervene, and checks should be done (whether there are any additions, money laundering, etc.).

Financial bubble (FB) management consists in the reaction of the state to deviations of the result of functioning of some category of EA or even of a single sufficiently large EA from a given norm. Management of FBs does not replace other methods of economic management, but is a necessary complement, creating the necessary conditions for ensuring the effectiveness of other types of management. Model (1)–(7) is implemented in a processor device for which a patent [31] has been received.

6   Conclusions

As a result of the above, the main findings and results can be summarized briefly:

1. The analysis of the existing scientific sources has shown the absence of concrete researches in the field of creation of scientific and methodical support of FP management in conditions of uncertainty.
2. It is reasonable to consider the economy as MAS, formed by the aggregate of EA. In this case, each EA, being under the influence of random factors, is able to enter into the exchange of goods, as well as endowed with the activity, which is expressed in the planning of its behavior and ability to self-organize and develop.
3. The proposed scientific and methodical support of FI management allows:

- systematically solve a whole complex of interrelated tasks for all categories of EA, providing the increase of reliability and reliability of economic information, reduction of disproportions, increase of efficiency of economy;
- to ensure the possibility of effective state intervention in the economy without significant limitations of market self-regulation mechanisms;
- provide a reliable guarantee against making irrational decisions.

The proposed methodological approach makes it possible to comprehensively take into account the whole set of reasons distorting information, including the psychology of EA behavior, conscious corruption distortions, ensuring the reduction of uncertainty due to the system approach.

References

1. Liechtenstein, V.E., Ross, G.V.: Equilibrium random processes: theory, practice, info business, 424 p. Finance and Statistics, Moscow (2015)
2. Liechtenstein, V.E., Ross, G.V.: Mathematical proof of necessity of changes in economics. Inf. Commun. 1 (2013). http://www.decision-online.ru/
3. Liechtenstein, V.E., Ross, G.V.: Nobel Prize and ideology of economic justice. Econ. Hum. Sci. 5(292) (2016)
4. Liechtenstein, V.E., Konyavsky, V.A., Ross, G.V., Los, V.P.: Multi-agent systems: self-organization and development. Financial Statistician, Moscow (2018)
5. Batkovskiy, A.M., Konovalova, A.V., Semenova, E.G., Trofimets, V.Y., Fomina, A.V.: Risks of development and implementation of innovative projects. Mediter. J. Soc. Sci. 6(4), 243–253 (2015)
6. Smirnov, A.D., Macrofinance, I.: Methodologies for modeling bubbles and crises. HSE Econ. J. 3, 275–310 (2010)
7. Dorofeev, M.L., Samarsky, G.V.: Modeling of financial bubbles processes on the Russian stock market. Financ. Credit 15, 275–310 (2016). 7, 45–62
8. Grebenyuk, E.A., Malinkina, A.V.: Application of econometric data analysis methods for identification and dating of "bubbles" on the financial markets. Control Sci. 5 (2014)
9. Methods of financial bubbles testing. http://allbest.ru
10. Kaurova, N.K.: Financial and economic security in conditions of national economy openness (theoretical and methodological aspects). IE RAS, Moscow (2013)
11. Kobyakov, A., Khazin, M.: The Sunset of the Dollar Empire and the end of "Pax Americana". Veche Publishing House, Moscow (2003). 368 pages
12. Soros, D.: Alchemy of Finance. INFRA-M, Moscow (1997)
13. Becker, G.S., Kominers, S.D., Murphy, K.M., Spenkuch, J.L.: A Theory of intergenerational mobility. MPRA Paper No. 66334, August 2015
14. Abebe, M.: Electronic commerce adoption, entrepreneurial orientation and small- and medium-sized enterprise (SME) performance. J. Small Bus. Enterp. Dev. 21(1), 100–116 (2014)
15. Melnik, M., Antipova, T.: Organizational aspects of digital economics management. In: Antipova, T. (ed.) Integrated Science in Digital Age. ICIS 2019. LNNS, vol. 78, pp. 148–162. Springer, Cham (2020). https://doi.org/10.1007/978-3-030-22493-6_14
16. Beck, T.: Financial development and international trade: is there a link? J. Int. Econ. 57(1), 107–131 (2002)
17. Amihud, Y., et al.: Stock liquidity and cost of equity capital in global markets. J. Appl. Corp. Financ. 27(4), 68–74 (2015)
18. Arslanalp, S., et al.: Chinas growing influence on Asian financial markets. J. Issue 173 (2016)
19. Gali, J., Gambetti, I.: The effects of monetary policy on stock market bubbles: some evidence. Natl. Bureau Econ. Res. (19981) (2014)
20. Gerding, E.F.: Law, Bubbles, and Financial Regulation, vol. 18. Routledge, London (2013)

21. Chirkova, E.V.: Prerequisites of a financial bubble appearance, no. 1, pp. 79–88 (2012). (in Russian)
22. Chirkova E.V. Financial Bubble Anatomy. https://e-reading.mobi/bookreader.php/1007597/ Chirkova_Anatomiya_finansovogo_puzyrya.html
23. Skobelev, P.O., Mayorov, I.V.: Multi-agent technologies and self-organization of the networks connected by schedules for management of resources in real time. IU Marine Syst. 1(17) (2015)
24. Bashminov, A., Mingaleva, Z.: The use of digital technologies for the modernization of the management system of organizations. In: ICIS 2019. LNNS, vol. 78, pp. 213–220 (2020). https://doi.org/10.1007/978-3-030-22493-6_19
25. Mingaleva, Z., Oborina, A., Esaulova, I.: Use of information technologies for managing executive compensations in network companies. In: Antipova, T. (ed.) Integrated Science in Digital Age. ICIS 2019. LNNS, vol 78. Springer, Cham (2020). https://doi.org/10.1007/978-3-030-22493-6_21
26. Avdiysky, V.I., Moneyless, V.M., Liechtenstein, V.E., Ross, G.A.: Economic justice and security of economic agents, 272 p. Finance and Statistics, Moscow (2016)
27. Lakatos I.: Evidence and disproof. How the theorems are proved. Nauka, Moscow (1967). http://www.bourabai.kz/dm/logic/txt10.htm
28. Liechtenstein, V.E., Ross, G.N.: New approaches to economics. Finance and Statistics, Moscow (2013)
29. Liechtenstein, V.E., Ross, G.V., Los, V.P.: Economic safety: management of financial bubbles, 100 p. Finance and Statistics, Moscow (2019)
30. Liechtenstein V.E., Ross G.V. (2008) Information technologies in business. Workshop: application of the Decision tool system in micro- and macroeconomy: a training manual. Moscow: Finance and Statistics, 2008. 512 pgs
31. Lalaev, S.G.: Depreciation award as a factor of the economic growth (in Russian). J. Financ. 7 (2008)
32. Ross, G., Konyavsky, V.: Models and methods of identification of threats related to the uncontrollability of capital flows. In: Integrated Science in Digital Age, ICIS 2019. LNNS, vol. 78, pp. 242–250 (2020). https://doi.org/10.1007/978-3-030-22493-6_22
33. Liechtenstein, V.E., Ross, G.V.: Way of optimal control of equilibrium random process. Patent for invention No. 2557483, 25 June 2015

Financial Costs Optimization for Maintaining Critical Information Infrastructures

Gennady Ross[1(✉)] and Valery Konyavsky[2(✉)] 🆔

[1] Financial University under Government of Russian Federation,
Moscow, Russia
ross-49@mail.ru
[2] Plekhanov Russian University of Economics, Moscow, Russia
konyavskiy@gospochta.ru

Abstract. This paper considers the complex of economic and mathematical models of financial expenses' optimization for preventive maintenance of critical information infrastructure of the enterprise. The offered models are focused on information protection of the automated control systems of technological processes. The analysis of possible risks and their measurement at rendering services in service of critical information infrastructure of the enterprise is resulted. Simulation models for optimization of service costs and models for optimization of distributed in time service costs are developed, which are based on the evolutionary and simulation methodology. Methodological proposals on the choice of options of insourcing, sourcing or outsourcing are formulated.

Keywords: Critical information infrastructure · Preventive maintenance models · Evolutionary simulation model · Instrumental system "decision" · Outsourcing · Insourcing · Sourcing · Technological processes · Data security · Hazard analysis

1 Introduction

One of the most important tasks of successful functioning of the enterprise's critical information infrastructure (CII) is organization of periodic preventive maintenance of its software and hardware complex [1–5]. These complexes provide information security in the computer network of industrial objects.

The subnetwork of the enterprise computers sometimes is designed isolated and does not assume external connections and processing of the data of not checked up sources. However, in practice those subnetworks have communication (constant or temporary) with a general network of the enterprise which knots have an Internet access. In addition, there is a human factor due to the operator's maintenance of each computer. Also, software errors are possible in the control system functioning that may occur due to software imperfection, incorrect memory allocation, etc. All the above factors may cause a disruption of normal computer network functioning, which leads to stoppages in shops, accidents, uncontrolled development of the production process, disasters.

© Springer Nature Switzerland AG 2021
T. Antipova (Ed.): ICIS 2020, LNNS 136, pp. 188–200, 2021.
https://doi.org/10.1007/978-3-030-49264-9_17

For practical realization of the above problem in the environment of information support calculation algorithms based on adequate mathematical models of real processes taking place at operation of CII are necessary.

Development of methodical approaches to estimation, forecasting and optimization of means for CII service include: economic and mathematical models of cost optimization; sources and methods of reception of the initial information; instrumental means for program realization; definition of directions of expediency of additional researches.

The main scientific and methodical problems connected with the development of methodical approaches consisted in what should be taken into account:

- objective limitations on obtaining cost data;
- "littering" of data from departments and inability to obtain data that would allow the use of statistical methods;
- risks in the operation of CII, as well as the dependence of other operating costs;
- specific features of CII of different classes (telecommunication systems, distributed CII, as well as CII that provide information protection);
- various conditions for financing the operation of CII (own funds, borrowed funds, outsourcing, etc.).

The derivative goals of developing methodological approaches are to identify reserves for optimizing CII and taking appropriate managerial decisions.

2 Hazard Analysis in Critical Infrastructure Maintenance

All the main aspects of establishing the cost of servicing enterprises' CII become most obvious, assuming that its owner uses outsourcing. In this case, he acts as a Customer who hires a certain Contractor to provide services to the CII. In this case, the maintenance costs are equal to the Contractor's price, and the real cost of CII maintenance is determined as a reasonable price compromise acceptable both for the Client and the Contractor. With this approach, the real complexity of reaching a compromise becomes evident. This is due to the fact that taking the position of one or another party, one has to operate with different components of the notion of "cost" and different factors determining its value. With the compromise approach, the "cost" of the Contractor is acceptable to the Client, if the dependence of consumer qualities on the Contractor's costs is justified.

In order to find a compromise price, it is possible to proceed from a comparison of the risks of the Client and the Contractor, who have 2 alternative risks: the risk of overstating and the risk of undervaluation [6, 7]. For a Client who owns an CII, the following risks make sense:

- The risk of overstating the client is the risk of suffering losses due to the fact that too much money is spent on maintaining the FII, some of which is irrevocably lost;
- Under-customer risk is the risk of losses due to insufficient funds being spent on maintaining CII, resulting in malfunctions, loss of information and other effects, the monetary equivalent of which measures this risk.

For an executor servicing the CII, risks make different sense:

- Performer risk is the risk of incurring (or being ready to incur) unclaimed services and, therefore, unsustainable costs;
- underperformance risk is the risk of incurring losses due to the impossibility (or unpreparedness) to perform services for which there is a solvent demand.

In this case, the price of the Customer is determined by the condition:

risk of overstating the customer = risk of undervaluation of the customer.

The price of the contractor is determined by the condition:

risk of overvaluation of the contractor = risk of underestimating the contractor.

The compromise price is determined by the condition:

risk of overcharge by the client = risk of undercharge by the contractor.

The latter condition leads to the setting of a compromise and, at the same time, an optimal service price for the CII in the form of a game between the client and the contractor. This game, in its turn, is reduced to the evolutionary simulation model (ESM) [8, 18, 19].

In order to develop a methodical approach allowing to find a compromise price of CII servicing, it is necessary to develop algorithmic models of risk calculation based on the factors that determine these risks. For this purpose, it is necessary to identify the factors, establish their interrelationships and determine the methods of quantitative evaluation.

Risks, as follows from the above wording, are a monetary expression of irrational costs and under expenditure. By its structure, the costs of maintenance of CII are similar to the costs of scientific and technical support of complex information telecommunications systems (which are part of CII), including the costs of: regulatory work; modernization; repair of technical facilities; support; development.

Each of these items includes equipment, usually consisting of serial computer equipment, telecommunication and video equipment, and programming. The basic expenses of time, as experience shows, are required basically for works with hardware-software at all stages of life cycle of CII. Therefore, except for expenses for technical means, the main emphasis is made on determination of hardware-intensive nature [9–20].

3   Simulation Models for CII Maintenance Cost Optimization

The analysis of existing methods for estimating various types of costs on CII at different facilities allows us to consider that there is at least a fundamental possibility of creating procedures for collecting information on the following factors:

$f_{1,t}$ - material costs include the cost of basic and auxiliary materials required for the work;

$f_{2,t}$ - fuel and energy resources include the cost of fuel and energy resources as a share related to the maintenance of CII;

$f_{3,t}$ - labor costs include labor costs of employees involved in the maintenance of CII;

$f_{4,t}$ - capital expenditures include expenses, if necessary, on preparation (construction, redevelopment) of premises, purchase of vehicles, purchase of licenses, etc.;

$f_{5,t}$ - overheads include incidental expenses related to the main production, which are related to the maintenance of CII (in the amount of 70–150% of labor costs, i.e. from the factor $f_{3,t}$), direct costs in accordance with the regulations of the Ministry of Finance of the Russian Federation on business trips, and so-called non-productive expenses.

$f_{6,t}$ - profit margin includes industry average profit margin expressed in %;

$f_{7,t}$ - taxes in cost of sales include total tax deductions from all items included in cost of sales (wages and salaries, etc.) expressed in %.

$f_{8,t}$ - income tax includes income tax rate expressed in %.

$f_{9,t}$ - mandatory contributions to funds include social contributions, the amount of the unified social tax and contributions for insurance against accidents and occupational diseases

In this case, t - indicates the period of time to which we attribute all factors (this may be, for example, or the entire life of the project, say, 5 years), and each factor we consider as a random value, characterized by some law of probability distribution of values. Since we are in a situation of uncertainty, not risk, when estimating the cost of CII services, we usually do not have the possibility to collect the necessary array of statistical information about the factor. The source of information is some or other methods of expertise, the results of which should be: selection of the type of the law of probability distribution of the factor and assessment of the parameters of this law. For example: material costs of maintenance of research institutes in the period t (factor $f_{1,t}$) have a uniform distribution law in the interval [A,B] (or a normal distribution law with mathematical expectation M and average square deviation K; or a piecewise linear law of probability distribution with corresponding parameters).

It is possible to calculate using the factors entered:

$A = f_{1,t} + f_{2,t} + f_{3,t} + f_{4,t} + f_{5,t}$ - cost of maintenance of research institutes;

$B = A * (100 + f_{7,t} + f_{9,t})/100$ - cost price in the sum of taxes;

$C = (A + B) * (f_{6,t}/100)$ - planned profit;

$Fa_t = (A + B) + C * f_{8,t}$ - the cost of servicing the CII.

In general, thus, we have the following imitation model of forming the price of CII services based on cost estimates:

$$
\begin{aligned}
A &= f_{1,t} + f_{2,t} + f_{3,t} + f_{4,t} + f_{5,t} \\
B &= A * \left(100 + f_{7,t} + f_{9,t}\right)/100 \\
C &= (A + B) * \left(f_{6,t}/100\right) \\
Fa_t &= (A + B) + C * f_{8,t}
\end{aligned}
\tag{1}
$$

Let's consider alternative imitation models of the cost of servicing the CII. There are situations when it is possible to allocate and estimate the factors reflecting expenses depending on technology of work of CII and its parameters: $f_{10,t}$ - number of jobs; $f_{11,t}$ - average salary; $f_{12,t}$ - database volume; $f_{13,t}$ - standard database maintenance costs; $f_{14,t}$ - number of transactions; $f_{15,t}$ - unit transaction costs; $f_{16,t}$ - fault tolerance; $f_{17,t}$ - standard expenses for elimination of failures; $f_{18,t}$ - throughput; $f_{19,t}$ - standard expenses for channel service; $f_{20,t}$ - functional completeness; $f_{21,t}$ - standard expenses for service of function; $f_{22,t}$ - complexity; $f_{23,t}$ - standard expenses for service.

Factors $f_{10,t}$; $f_{12,t}$; $f_{14,t}$; $f_{16,t}$; $f_{18,t}$; $f_{20,t}$; $f_{22,t}$ are assumed to express the corresponding CII characteristic in natural units, while factors $f_{11,t}$; $f_{13,t}$; $f_{15,t}$; $f_{17,t}$; $f_{19,t}$; $f_{21,t}$; $f_{23,t}$ specify the monetary equivalents of these units. Estimating the values of these factors is the most difficult. However, since there are quite a number of different operating CIIs with different characteristics, it can be assumed that some statistical data on the values of factors $f_{i,t}$, where $i \in \{10, 12, 14, 16, 18, 20, 22\}$ and the cost of servicing CIIs are available. Such information may be available from organizations specializing in the design, installation and maintenance of CIIs.

If $f_{i,t}^e$ - value of the factor realized in any concrete operating CII, and X^e - expenses for its service $f_{i+1,t}^e = \frac{X^e}{f_{i,t}^e}$ will be realization of values of the factor $f_{i+1,t}^e$. Having an array: $f_{i+1,t}^e, e = 1, 2, 3, \ldots$ it is possible to approximate the law of distribution of probability values of the factor $f_{i+1,t}$. In this case, we get the following simulation model of CII service price formation:

$$
Fa_t = \sum_{i \in G} f_{i,t} * f_{i+1,t}
\tag{2}
$$

where $G = \{10, 12, 14, 16, 18, 20, 22\}$.

One more variant of construction of a simulation model of formation of expenses for service of CII consists in allocation of functional subsystems of CII and estimation of standard expenses for their service depending on parameters. In [2] variants of classification of information systems on different signs which in aggregate completely enough characterize any concrete CII are offered. Let's result these classifications and we will consider each element of functional structure as the specific factor, and elements of other classes - as values of parameters. Let's consider following functional elements of CII: $f_{23,t}$ - automation of technical preparation of manufacture; $f_{24,t}$ - marketing and strategy of development of the enterprises; $f_{25,t}$ - technical and economic planning; $f_{26,t}$ - the finance (accounting, the financial analysis); $f_{27,t}$ - material and technical maintenance; $f_{28,t}$ - operative and calendar management of manufacture; $f_{29,t}$ - management of sale of finished goods; $f_{30,t}$ - personnel management.

Parameters that take into account other features of CII are shown in Table 1. The formula for calculating the CII price is as follows:

$$Fa_t = \prod_{j=1}^{5} p_j * \sum_{i=23}^{30} f_{i,t} \qquad (3)$$

The methods of calculating the service price of CII (1), (2) and (3) are alternative, approximate and rather crude. They give an unacceptably large error of Δ in the estimation of F_{at} price. Error Δ can be reduced by 1–2 orders of magnitude with the help of ESM; besides, its application allows to use the Decision tool system as a necessary software [4–7, 18–20].

Table 1. Structurally appropriate coefficients

Organizational structure		
$p_1 =$	k_1	Automated workstation (AWP) for managerial staff
	k_2	Complex of interrelated ARMs
CII limitation		
$p_2 =$	k_3	CII of enterprise (organization)
	k_4	CII industry research institutes
	k_5	State CII
	k_6	International CII
Degree of integration		
$p_3 =$	k_7	Local CII (isolated information space)
	k_8	Partially integrated CII (common information space)
	k_9	Fully integrated corporate CII
Information and technology structure of CII		
$p_4 =$	k_{10}	CII of centralized architecture of construction (one data center)
	k_{11}	Distributed architecture FRI (computer networks, multiple processing and storage centers)
CII specialization		
$p_5 =$	k_{12}	Management institute (or organizational and economic management, information management system - IMS)
	k_{13}	Information retrieval systems (InformationRetrievalSystem - IRS)
	k_{14}	Automated learning systems (EducationInformationSystem - EIS)

The structural formulation of the model is defined by relations (4)–(8) [6]. In this case:

t	- moment or period of time; $f_{1,t,...,fN,t}$ - factors (random values);
$p_1,...,p_5$	- initial indicators (conditionally constant values);
$\rho_1,...,\rho_4$	- simulation models; M - mathematical expectation sign;

F_{at} - random realization of CII value;

$F_{1(PLt,\ Fat)}$ - overstatement costs arising at $PLt > Fat$;

$F_{2(PLt,\ Fat)}$ - understatement costs arising at $PLt < Fat$;

rt()   - vector of calculated indicators, i.e. indicators that depend on PLt:

$$Fa_t = \rho_1\left(f_{1.t}, \ldots, f_{n,t}, p_{1.t}, \ldots, p_{m,t}\right) \tag{4}$$

$$F_1(PL_t, Fa_t) = \rho_2\left(PL_t, Fa_t, f_{1.t}, \ldots, f_{n,t}, p_{1.t}, \ldots, p_{m,t}\right), npuPL_t > Fa_t \tag{5}$$

$$F_2(PL_t, Fa_t) = \rho_3\left(PL_t, Fa_t, f_{1.t}, \ldots, f_{n,t}, p_{1.t}, \ldots, p_{m,t}\right), npuPL_t < Fa_t \tag{6}$$

$$\min_{PL_t}\left\{ \max_{q\in\{1,2\}} \left\{ M\{F_q(PL_t, Fa_t)\} \right\} \right\} \tag{7}$$

$$r_t() = \rho_4\left(PL_t, p_{1.t}, \ldots, p_{m,t}\right) \tag{8}$$

As a simulation model ρ1 in (4) can be used algorithm (1), or algorithm (2), or algorithm (3). Condition (7) is equivalent to the requirement:

risk of overstating the client = risk of underestimating the performer.

Condition (8) can be used to calculate additional CII service parameters such as profit. To complete the wording of the model, we are left to develop models ρ2 and ρ3 in (5) and (6), i.e. to develop algorithms for calculating the risk of overestimating the customer and the risk of underestimating the performer respectively.

As at overestimation of expenses on service of CII the excessively spent means irrevocably disappear, ρ_2 has the form:

$$F_1(PL_t, F_{at}) = \rho_2\left(PL_t, F_{at}, f_{1,t}, \ldots, f_{n,t}, p_{1,t}, \ldots, p_{m,t}\right) = PL_t - F_{at}$$

For calculation of expenses of underestimation of the executor, that is for construction of model ρ3 from (6) it is necessary to consider the factors defining the sizes of losses from failures in work of EFI because of insufficiency of expenses for service, in other words from that PLt < Fat. Let's enter some more initial indicators and factors which allow to connect the size of shortage of means (PLt-Fat) with the sizes and probabilities of negative consequences.

Let:

− p_5 and p_6 - linear regression coefficients, connecting the probability of failure V_1 in the work of CII with the value of funds for CII maintenance $(PL_t\text{-}F_{at})$, that is:

$$V_1 = p_5 + p_6 * (PL_t - F_{at}) \tag{9}$$

- p_7 and p_8 - linear regression coefficients, which connect the duration of failure V2 in the work of FRI with the value of defects of funds for maintenance of FRI (PLt-Fat), that is:

$$V_2 = p_7 + p_8 * (PL_t - F_{at}) \qquad (10)$$

- p_9 and p_{10} - linear regression coefficients linking the value of loss V3 with the idle time or loss of functionality of KIIV2, i.e.:

$$V_3 = p_9 + p_{10} * V_2 \qquad (11)$$

Using (9)–(11) you can create a loss calculation formula:

$$F_2(PL_t, F_{at}) = \rho_3\left(PL_t, F_{at}, f_{1,t}, \ldots, f_{n,t}, \ p_{1,t}, \ldots, p_{m,t}\right) = V_1 * V_2 * V_3$$

Hence:

$$F_2(PLt, Fat) = (p_5 + p_6 * (PL_t - F_{at})) * (p_7 + p_8 * (PL_t - F_{at})) * (p_9 + p_{10} * (p_7 + p_8 * (PL_t - F_{at})))$$
$$(12)$$

Ratios (1) to (12) have several alternative formulations of the model for calculating the cost of maintaining an FDI based on risk weighting. There are no methodological or other problems to implement these models in the Equilibrium module environment of the Decision tool system.

In cases when there are difficulties in setting the values of $p_7, \ldots, p_{10}$ - you can use the estimation of the cost price of S maintenance of CII and their price C.

Besides, it is necessary to note those directions on which the given methodical approach can be specified and improved.

The simulation model of forecasting the actual costs of CII maintenance $\rho 1$ in the period t can be specified on the basis of:

(a) Processing statistical data on similar CII (histograms and trend forecast);
(b) Clarification of the sectoral, functional characteristics of CII, its structure and scale;
(c) Technical standards.

At a formulation of the simulation model of costs of overestimation $\rho 2$, arising at superfluous investments we have assumed, that costs in the size $(PL_t\text{-}F_{at})$ simply disappear. It is the simplest case. More difficult consequences are also possible, for example, underfunding of any other articles and occurrence of corresponding negative consequences.

The simulated model of the underinvestment costs arising at lack вложений$\rho 3$ can be specified at the expense of more detailed consideration of consequences: failures in work of the equipment, idle times, losses at technologically connected systems etc. The specification of ways of gathering the initial data with application of expert estimations can have the important value.

In the presence of algorithms of calculation of integral indicators ρ_4 in period t (such as the income, profit, reliability) depending on PL_t and indicators additional criteria of optimality can be formulated.

At program implementation of models in Equilibrium environment, it is possible to provide a dialog mode of choice of this or that model variant.

Further development of methodical approach supposes the development of forms and procedures of data collection, methods of preliminary data processing, in particular, methods of forecasting the values of factors and indicators for the period t depending on the CII lifetime. As a result of use of dialogue procedures of system Decision values of plan PL_t, its reliability, revealing of the basic risks, construction of dependences PL_t from factors and initial indicators (values, schedules) can be received.

4   Optimization of Time-Distributed CII Maintenance Costs

There may be situations when the cost of maintenance of CII in the period t + 1 from the cost in the previous period t. This means that PL_{t+1} is some function from PL_t that is not explicitly defined. For example, at time t, we may spend different amounts, k = 1, ...,K, which is different in that we add options for updating to the service.

Then a consequence tree appears, which is illustrated in Fig. 1. At that, and - variants of cost sizes in the period t = 2. A lot of numbers of these variants $G_2 = \{1, 2, 3\}$.

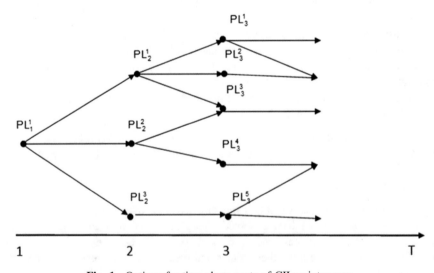

Fig. 1. Options for time-share costs of CII maintenance.

The task is to find a path in the figure where the total cost is minimal, that is:

$$\min_{\forall k} \sum_{t=1}^{T} PL_t^k * \alpha_{k,t} \tag{13}$$

$$\alpha_{k,t} = \begin{cases} 1 \\ 0 \end{cases}, \forall k, \forall t \tag{14}$$

$$\sum_{k \in G_t} \alpha_{k,t} \leq 1, \forall t \tag{15}$$

$$\sum_{k \in G_t} \alpha_{k,t} \geq 1, \forall t \tag{16}$$

Thus, the conditions for choosing the optimal path in the oriented graph in Fig. 1 are expressed by the relations (13)–(16). At the same time, condition (13) expresses the requirement to minimize the total costs for the whole period of operation of the CII, condition (14) introduces a Boolean variable corresponding to each variant of costs k at each moment of time t, condition (15) means that at each moment of time only 1 variant of costs can be realized, and condition (16) means that one of the variants of costs at each moment of time must necessarily be made.

The formulation of the task (13)–(16) belongs to the class of Boolean programming tasks and can be implemented programmatically in the Combinatorics module of the Decision tool system [7]. The complex method of establishing the optimal size of the expenses distributed in time for the maintenance of CII implies the following main stages:

1. Schemes of possible CII maintenance are designed in the form of a graph (Fig. 1).
2. For each point of the chart by means of any of the model variants (1)–(12) in the dialogue with Equilibrium are calculated.
3. In the dialogue with Combinatorics the path on the chart in Fig. 1 is found, which forms the optimal cost plan for the whole period of the CII servicing. And this plan also contains variants of updates at each moment of the planned period.

5 Methodological Approaches to the Selection of Options for Insourcing, Sourcing or Outsourcing in CII Maintenance

The selected option for servicing CII usually covers a fairly long period of time [1,T], comparable to the life of an CII. To solve the problem of whether to sourcing, insourcing, or outsourcing [13, 14], the unit cost of the service is more preferable. Let's call this index as the index of the service cost (ICS) and denote its value indices, and the associated total costs for the service total costs. Other conditions being equal: levels of trust; reliability; functionality, the producer of the service is better, which has less ICS.

We will consider the ICS as a criterion for selecting an option. According to this criterion, the task of selection is solved as follows:

- sourcing - the service provider that has the minimum ICS is selected;
- sourcing - is necessary for those services for which the owner has a smaller ICS;
- outsourcing - is necessary for those services, on which the business has a smaller ICS than the owner.

One of the variants of the Evolutionary Simulation Model (1)–(12) is the main one for calculation of the ICS.

$$V_I = TS/(tT = 1\,PL_t)$$

Taking into account the expected difficulties in assessing $p_7, ..., p_{10}$, which are necessary to calculate the cost of underestimating the performer, we will consider a different approach.

With the help of the model (1), (2) or (3) it is possible to estimate the limit values of the maintenance cost of CII Fat. Let's enter the following designations: minimum value of Fat; maximum value of Fat; S - the cost of CII maintenance; C - the cost of CII maintenance.

The price differs from the cost price in this case by the value of the industry average profit margin, that is $C = S * (1 + D/100)$. In which the mathematical expectation $f_{6,t}$.

In this case:

risk of customer overvaluation $= M\{S * (PL_t - F_{at}), PL_t > F_{at}\}$;

the risk of underestimating the performer $= M\{(C - S) * (F_{at} - PL_t), PL_t < F_{at}\}$.

Condition: risk of overestimating the client = risk of underestimating the performer means that:

$$M\{S * (PL_t - F_{at}), PL_t > F_{at}\} = M\{(C - S) * (F_{at} - PL_t), PL_t < F_{at}\} \qquad (17)$$

You can get it out of here:

$$PL_t = Fa_t^{min} + \left(Fa_t^{max} - Fa_t^{min}\right) * \frac{1}{1 + \sqrt{\frac{S}{C-S}}} \qquad (18)$$

Reliability, in other words, the probability that the PL_t will be exceeded, can be calculated by the formula:

$$P^0 = 1 - \frac{PL_t - Fa_t^{min}}{Fa_t^{max} - Fa_t^{min}} \qquad (19)$$

6 Conclusions

When developing specific methodological recommendations, there may be a need for additional research [14–17], the content of which we will formulate on the example of information and telecommunication technologies:

1. Usually, a telecommunication service provides a whole set of services. In this connection there is a necessity to compare not separate services, but complexes of services and to develop methods of data collection on complexes of services.
2. Services produced by different manufacturers may have differences in functionality and customer coverage. These differences need to be taken into account in the list of factors and, at the same time, how to ensure the universality of methods of data collection on the factors ($f_{1,t}$, ..., $f_{N,t}$), so that the methodology is also universal and does not require improvement at each specific application.
3. When assessing the services' efficiency, as in a number of other cases, it is inevitable to use expert assessments. On the one hand, it may decrease the credibility of the methodology and therefore expert evaluation procedures should be well developed. On the other hand, in order to make the technology of sourcing, insourcing and outsourcing extremely easy and convenient to use, it is necessary to conduct almost all or even all expert assessments not in the process of a dialogue in a particular application of the methodology, but in advance. In other words, it is desirable to present the results of pre-examined examinations in the methodology itself in the form of ready-made tables (or charts, or nomograms, or service programs). In other words, it is desirable to present the results of pre-examined examinations in the methodology itself in the form of ready-made tables (or charts, or nomograms, or service programs).

The offered technique of preventive maintenance of CII is based on the economic and mathematical model which allows:

1. To calculate expenses for service of CII taking into account various ways of gathering the initial information (from the statistical data, on the basis of expert estimations or technical standards).
2. To carry out calculations for any given variant of modernization.
3. Develop modernization programs that are most effective in terms of maintenance costs (provided that the modernization options are given).

Software implementation of the model is carried out in the Equilibrium module environment of the Decision tool system and uses the solution dialog procedures.

References

1. Federal Law "On security of critical information infrastructure of the Russian Federation" of 26.07.2017 N 187-FZ
2. Categorization of critical information infrastructure objects (CII). Practical examples. https://rtmtech.ru/articles/kategorirovanie-obektov-kii-primery/

3. Security of objects of critical information infrastructure of the organization General recommendations (version 2.0) Moscow (2019). http://aciso.ru/files/docs/metodichka_2.0.pdf

4. Mikhailovsky, N.: Comparison of methods of estimation of cost of projects on development of information systems. https://www.cfin.ru/management/practice/supremum2002/15.html

5. Lipaev, V.V.: Feasibility Study of Complex Software Projects. Sinteg, Moscow (2004)

6. Liechtenstein, V.E., Ross, G.V.: Information technologies in business. Application of tool system Decision in micro- and macroeconomy. Finances and Statistics, Moscow (2008)

7. Lichtenstein, V.E., Ross, G.V.: Information technologies in business. Application of tool system Decision in the decision of applied economic problems. Finances and Statistics, Moscow (2009)

8. Liechtenstein, V.E., Ross, G.V.: Equilibrium random processes: theory, practice, infobusiness. Finances and Statistics, Moscow (2015)

9. Federal unit prices for equipment installation. Estimated norms of the Russian Federation, Moscow (2003)

10. Federal Unitary Prices for Start-up and Adjustment Works. Estimated norms of the Russian Federation, Moscow (2003)

11. Order of the Ministry of Labor of the Russian Federation dated July 23, 1998 N 28 "Interindustry standard time standards for service maintenance of personal electronic computers and organizational equipment and software maintenance"

12. Extended time standards for development of computer software (approved by the Resolution of the USSR State Labour Committee, Secretariat of the All-Union Central Council of Trade Unions dated 24.09.1986). As of 12 October 2006. http://www.consultant.ru/document/cons_doc_LAW_100176/

13. Ross, G.V., Nikiforova, S.V.: Perfection of the archive organizational structure oriented to the outsourcing services (in Russian). Adm. Acc. J. (4) (2008)

14. Melnik, M., Antipova, T.: Organizational aspects of digital economics management. In: Antipova, T. (eds.) Integrated Science in Digital Age. ICIS 2019. Lecture Notes in Networks and Systems, vol. 78, pp. 148–162. Springer, Cham (2020). https://doi.org/10.1007/978-3-030-22493-6_14

15. Chernyavskiy, G.I.: Adaptive control of the technical state and operation safety of the complex technical systems under the resource restrictions. Energy Saving (2) (2006)

16. Mingaleva, Z., Deputatova, L., Starkov, Y.: Management of organizational knowledge as a basis for the competitiveness of enterprises in the digital economy. In: Integrated Science in Digital Age. ICIS 2019. Lecture Notes in Networks and Systems, vol. 78. Springer, Cham (2020). https://doi.org/10.1007/978-3-030-22493-6_18

17. Akatov, N., Mingaleva, Z., Klackova, I., Galieva, G., Shaidurova, N.: Expert technology for risk management in the implementation of QRM in a high-tech industrial enterprise. Manag. Syst. Prod. Eng. 27(4), 250–254 (2019). https://doi.org/10.1515/mspe-2019-0039

18. Alekseev, A.A.: Identification and diagnostics of systems. Academy, Moscow (2009)

19. Ross, G., Konyavsky, V.: Models and methods of identification of threats related to the uncontrollability of capital flows. In: Integrated Science in Digital Age. ICIS 2019. LNNS, vol. 78, pp. 242–250. Springer (2020)

20. Ross, G., Liechtenstein, V.: Management of financial bubbles as control technology of digital economy. In: Antipova, T., Rocha, A. (eds.) Information Technology Science. AISC, vol. 724, pp. 96–103. Springer (2018)

Creation of a System for Climate-Related Risks Disclosures in Companies' Reporting

Olga Efimova[1](✉) ⓘ, Olga Rozhnova[1](✉) ⓘ,
and Elena Zvyagintseva[2](✉) ⓘ

[1] Department of Accounting, Analysis and Audit,
Financial University under the Government of the Russian Federation,
Leningradsky Pr., 49, Moscow 125993, Russia
{Oefimova, ORozhnova}@fa.ru
[2] Language Training Department, Financial University under the Government
of the Russian Federation, Leningradsky Pr., 49, Moscow 125993, Russia
EZvyagintseva@fa.ru

Abstract. The article explores the way of disclosing issues related to climate change in companies' reports as well as these issues' impact on the business functioning and business response to them. The conducted study has analyzed the financial and corporate reports (integrated, environmental, on sustainable development) drawn up by the leading Russian oil and gas companies, that consider climate change influence or environmental impacts as the most significant ones. Based on linguistic, statistical and qualitative analyses, it was possible to obtain the following conclusions. Firstly, there is no comprehensive approach to revealing information on the climatic aspects of business activities in various types of company's reports with the corresponding cross-references. Secondly, there is almost no confirmation of disclosures in non-financial reports supported with financial reporting data. Thirdly, there is hardly traced mutual influence, both direct and reverse, between a climate and a business entity in all types of reporting. Thus, the unifying factor for both non-financial and financial reporting is, first and foremost, the risk zones under climate change for the company on the one hand, and the companies for the environment on the other hand. Nowadays, one of the tools to turn risk zones into indicators of success, including progress of a company, lack of harm to the climate, can and should be systematized climate disclosures in reports. The article makes recommendations on improving the logical alignment and quality of disclosures on the climate-related risks impact on the company's strategy and its most important financial indicators. The implementation of the proposals suggested will allow increasing the validity of investment decisions and make the financial market more transparent.

Keywords: Financial and non-financial reporting · Climate-related risks disclosures · Financial consequences of climate-related risks

© Springer Nature Switzerland AG 2021
T. Antipova (Ed.): ICIS 2020, LNNS 136, pp. 201–211, 2021.
https://doi.org/10.1007/978-3-030-49264-9_18

1 Introduction

In the annual report of the World Economic Forum on Global Risks at the end of 2019, systemic threats associated with climate change and environmental degradation are considered as the most significant and destructive factors that will threaten the world over the next decade [1].

A great number of participants in capital markets are increasingly becoming aware of the risks and their significance as well as new opportunities that are inextricably linked to climate change. As a result, there is an increase in information connected with decision-making. At the same time, numerous surveys and interviews with representatives of investment communities [2, 3] indicate the absence of a unified approach to disclosing information on climate-related risks. Therefore, it leads to the incompatibility of reporting data regarding to the impact of climate change on the results of companies' operations and prospective cash flows. These factors are becoming the main obstacles for accounting the climate change risks when making decisions by investors, lenders and insurers.

Studies show that investors need to be provided with two types of information: how companies and their businesses affect the environment and how climate change influences the company's strategy, its business model and ability to create value [2, 4]. Today, the investment and analytical community is the leader in the developing the process associated with the consideration of climatic factors and their disclosure in the financial statements [5, 31, 32].

The main objective nowadays is to build a system for the formation and disclosure of data related to climate change. The need for such disclosures is in demand by both the expectations of investors and also by the desire of companies to ensure the long-term business functioning.

In 2015, the Financial Stability Board at the request of the G20 leaders set up Task Force on Climate-Related Financial Disclosures (hereinafter TCFD), that is a methodological platform, aiming, on the one hand, to assist companies in revealing climate-related risks and their possible financial consequences, and on the other hand, to help financial markets integrate these risks into key instruments [6, 7]. Such an approach allows investors to make financial investments more effectively.

The recommendations of TCFD suggest a consideration of two separate aspects of the problem: 1) the risks posed by climate change (for example, risk of raising the oceans' level for companies located in the coastal zone, thus being vulnerable to natural disasters); 2) the risks of transition to a low-carbon economy of a new type (that is of great importance, for instance, to representatives of the oil industry) [7, 8].

One of the core differences between the TCFD recommendations is included in the approach itself: this is the disclosure of climate impacts *on* the company, rather than the environmental impact *of* the company, that reporting compilers present in most cases currently. In accordance with the recommendations, disclosure of the climate impact on the company's activities involves reporting data in four key areas: management (how the responsibility for managing climate-related risks is shared in a corporate structure); strategy (identifying risks and opportunities over time periods and explaining how they affect strategic and financial planning); climate risk management in the frame of the company's risk management; key and target indicators (i.e. quantitative indicators of the climate influence degree on company's activities, monitoring and analysis of their conduct).

This approach may be called as universal and applicable in any organization.

Nevertheless, the recognition of the climate issues relevance for business strategies by various stakeholders does not guarantee the strengthening of particular actions among different companies. Many of business units are currently refusing to put into practice the recommendations on climate disclosures, as they beware of an increased pile of work on reporting and the outflow of significant resources despite the fact that the TCFD recommendations are logically integrated into the system of non-financial disclosures [9, 31–33].

As a result, in December 2017 the Climate Action 100+ initiative[1] was launched in order to improve the disclosure of data on the financial consequences connected with climate-related risks. Besides, the project will also focus on how the TFCD recommendations are implemented by companies. The list of those closely monitored includes 100 companies that systematically emit greenhouse gases and have a significant impact on the environment, as well as more than 60 companies with great potential to reduce their emissions and switch to clean energy. Among Russian companies that have come to the focus of investors' attention are Gazprom, Rosneft, LUKOIL and Norilsk Nickel.

In June 2017, TCFD issued official climate disclosure guidelines [8], which later were used as the basis for changes to existing non-financial reporting systems, such as GRI, IIRC and SASB, ESG indexes and ratings.

The Institute of Chartered Accountants in England and Wales (ICAEW) in collaboration with The Carbon Trust, provided guidance [10] to assist companies in implementing the TCFD-recommended disclosure approaches, highlighting the following issues as the most significant: What kind of climate information is the company reporting on? How is the information exchange between financial, strategic, and managerial functions within the organization carried out in order to obtain the necessary data? To what extent are factors of long-term climate change considered at the board level? Are the general expectations about qualitatively made climate disclosures known among investors and other market participants? Is there opportunity to use other internal sources of information, for example, the results of materiality assessments? Can the current sustainable development strategy be superimposed on the long-term risks and opportunities in order to assess progress?

Another complex problem is the lack of climate-related risks standardization, the methods for risk determination and assessment, the integration of modeling results, connected with possible climate change scenarios and the consequences into the financial analysis relating to investment and financial decisions.

Both companies and interested users of financial statements face various issues and potential obstacles when transferring climate scenarios into a comprehensive financial analysis system. Different types of climate scenarios and a wide variety of results call forth uncertainty in choosing appropriate climate scenarios. Most existing climate

---

[1] http://www.climateaction100.org.

scenarios have been developed for economic and scientific purposes, not for financial ones. Companies will also need to decide how to integrate this analysis into their current strategy and scenario planning, or risk assessment. Finally, companies will need to associate the impact of the scenario with prospective business performance.

2 Literature Review and Current State of the Issue Assessment

J. Richard et al. (2017), EY survey (2017) outlined the progress made by governments, stock exchanges, non-governmental organizations and other parties in the field of disclosure systems [5, 22]. However, despite the achievements the disclosure of information related to climate change does not provide the required completeness.

There is a demand in an assessment of the climate change impact on the reporting companies' activities from a financial standpoint. Such an approach will serve as a milestone for developing recommendations on the voluntary disclosure of relevant information in the framework of mandatory financial statements. In general, the disclosure of information related to climate change should be guided by the processes of effective corporate governance. It also should let on the significant factors and their impact on the company's activities, as well as fully reflect the strategy used by a company to manage the relevant risks.

Currently, most companies are still at an early stage of introducing recommendations into practice. TCFD forecasts it will take about five years of voluntary use before companies learn to be aware of their investors' expectations. However, in many cases it is not necessary to get a new start, since the environmental disclosures, that the report compilers have made so far, may be seen as the basis. The EU has adopted an approach to corporate social responsibility from the internal and external standpoints of responsibility among economic entities [16–18].

M. Kuter et al. (2017) argued that historical analysis of accounting development and its particular branches demonstrate the significance and particular impact on it with the economic conditions and stakeholders' requirements [11]. Y.V. Sokolov (2011) identified five paradigms in accounting theory including sociological. The issues of environmental responsibility are related to the sociological theory in the broad sense [12].

The issues of environmental responsibility are mainly perceived as a part of the sociological theory of accounting, which was developed by such scientists as Littleton and Zimmerman (1962), Tinker (1984), Bedford (1973) - [13–15].

An analysis of scientific works on this topic reveals the ambiguity of interpretations, the presence of different approaches to the rules for reporting corporate social and environmental responsibility, which is largely due to the existence of various standards for corporate reporting (GRI (Global Reporting Initiative) standards and the conceptual framework of IR (Integrating Reporting), Sustainability Accounting Board Standards).

Moser D.V., Martin P.R. (2012) argued that the presence of different standards impedes mutual understanding between different categories of utilizers: economic entities themselves, these entities and users of their reporting. On the other hand, reporting in this area expands its utilizers' range [19]. In addition, mutual understanding also requires confidence in the reliability and high quality of reporting. Such an issue is very significant and is covered in the scientific literature, for example, in the works by Edgley et al. (2014), Trotman A. (2015) and Trotman K. [20, 21].

There are some scientific papers devoted to the study of disclosure content regarding to specific issues of environmental responsibility [22–24]. However, such studies are only the basis for a more serious analysis of the further improving the disclosure requirements for climate impacts, risks and opportunities. For example, there are a number of papers on an analysis of investors' needs in presenting information on the risks connected with the impact of climate factors on business both in the short and long-term prospective [26, 27]. A profound scientific analysis on the conceptual and methodological issues of environmental accounting based on more than 80 scientific sources was carried out by Richar P. and Altukhova Yu. (2017). The authors proposed CARE model, which is a transposition of a traditional capitalism accounting model into natural and human capital [28]. The authors suggest the idea "to establish a new accounting environmental law that ratifies a new vision of both capital and profit concepts for environmental and social management". Jones M. (2010) propounds a multi-level theoretical model that underlies environmental accounting and reporting [29]. This theoretical model assumes a reorientation to sustainable development as the goal of any business. It considers the need for a new comprehensive accounting that would reflect the environmental impact of the enterprise.

In general, existing laws and regulations already require disclosing the information related to climate change risks in the cases where such information is considered material (IAS 1 Presentation of financial Statements, IFRS 7 Financial Instruments, IAS 16 Property, plant and equipment, IAS 36 Impairment of Assets, IFRS 3 Business Combinations and some other financial reporting standards).

According to the Central Bank of the Russian Federation[2], the work on analyzing and accounting a climate-related risks impact on the financial market and financial stability is on the early stage in Russia. Such work can be carried out through: development of "green" financing instruments; stress testing of financial institutions that are most exposed to climate-related risks; a number of risk assessments in the real sector of Russian economy which are associated with the transition to low-carbon production methods and the spread of new ecological standards in order to include these scenarios in stress testing.

3   The Main Body

3.1   Statement of Basic Materials

The study has analyzed financial and corporate reports of 13 largest Russian oil and gas companies (PJSC Gazprom, PJSC Gazprom Neft, PJSC Lukoil, PJSC Tatneft, PJSC Novatek, PJSC Rosneft Oil Company, Sakhalin Energy Ltd, PJSC Bashneft, PJSC

---

[2] http://www.insur-info.ru/press.

Zarubezhneft, PJSC Surgutneftegas, Severneftegazprom, PJSC Transneft, Gazprom EnergyHolding Ltd), for which the impact of climate change and environmental impact is the most significant. An interim conclusion is following: despite the fact that the performed linguistic analysis has not revealed the mention of such terms as "climate" or "climate-related risks" in their reporting, the financial statements of these companies reflect and disclose the estimated obligations associated with environmental activities.

The analysis of non-financial reports was based on a study of 9 environmental reports, 30 sustainability reports and 8 integrated reports compiled by oil and gas companies that were presented on the RSPP website in 2015–2018[3]. All analyzed reports reflect the environmental impact of these companies and include the indicators that characterize emissions, wastes generation and their usage, energy efficiency aspects.

Some Russian companies (Gazprom[4], Rosneft[5], Surgutneftegas[6]) in their ecological reports reveal the information on current innovative projects related to climate change. The environmental reports of these companies reflect topics of environmental management system, goals and programs, financing, negative environmental impact fee, environmental impact indicators, efforts to prevent environmental impact, environmental assessment of projects, innovation in environmental protection.

In general, according to the National Register of Non-Financial Reports[7] of Russian Union of Industrialists and Entrepreneurs, 52% of all non-financial reports registered on the website in 2018 identified the Goal of "taking urgent measures to combat climate change" as a priority.

Despite disclosing financial data related to the costs of maintaining and protecting the environment, an assessment of the financial consequences due to climate change and their impact on critical financial statements is not presented in any report. At the same time, the influence of climate-related risks on the most important financial indicators is becoming increasingly significant (Table 1).

As Table 1 demonstrates, the impact of climate-related risks on key financial indicators becomes very significant and their ignoring in the nearest future may lead to the disability of financial statement to fulfill its main function, i.e. to provide reliable presentation of data on financial position, financial results and cash flows required for decision making. The relationship of climate-related risks and their financial consequences for the most important indicators of financial statements is presented in Fig. 1.

Both Fig. 1 and Table 1 provide the opportunity to forecast negative scenarios of a climate change impact on the financial reporting indicators.

---

[3] http://rspp.ru/simplepage/157.

[4] http://media.rspp.ru/document/1/5/f/5f027f91643685a8c60801bb6dce5b12.pdf.

[5] https://www.rosneft.ru/upload/site1/document_file/Rosneft_CSR18_RU_Book.pdf.

[6] http://media.rspp.ru/document/1/0/5/05aac601a3054eb5a170ef5dd30d8d71.pdf.

[7] http://rspp.ru/simplepage/475.

Table 1. Climate-related risks and impacts.

Risks	Impact	Financial Impact
Physical and social		
Changing weather patterns Floods, fires Epidemics Recourses availability and Quality	Asset damage Health, safety, death rate Transportation interruptions	Increased costs and capital expenditures Asset impairment Changes in the useful life of assets Changes in the fair valuation of assets Asset write-offs
Regulatory		
Current changing regulations	Impact on market demand Restriction of licenses, availability and use Market restrictions	Compliance Increased costs Capital expenditures Asset valuations
Business model and technology		
Changes in demands for products/services Transition to renewable energy Growing demand for green products and a negative attitude towards products and technologies associated with increased emissions and waste	Lower demand Higher costs for transition Changing or adapting business models to new requirements	Lower revenues Increased costs Changes in the useful life of assets and therefore the amount of depreciation/amortization recognized Changes in the fair valuation of assets Potential provisions and contingent liabilities arising from fines and penalties
Reputational		
Stakeholder (clients, suppliers, investors and creditors) attitudes	Reduced availability of capital Litigation/penalties Reduced demand for goods/services	Increased costs Reduced revenues Asset valuations

For mining companies and some other sectors mostly affected by climate change, it is possible to predict attributes of asset impairment, which will require annual (or more frequent) asset impairment verifications. An impairment test when a loss is identified in a cash generating unit (CGU) will lead to a decrease or even loss of goodwill, which will inevitably affect an investment attractiveness. In addition, CGUs may be depreciated due to the lack of required resources for the activities and demand for the products.

3.2   Recommendations

Even though IFRS does not contain direct indications of the need for climate disclosures, information on the effects of climate change on assets value and future cash flows for companies in certain industries is invaluable. Therefore, such statistics should be reflected in the financial statements. Since investors, lenders, insurance companies pay more attention to the financial consequences of climate change, while making decisions on the capital distribution, such information becomes more and more significant for them.

These data are also important for the development of models while assessing the impairment of long-term assets, the justification of depreciation rates and their useful lives, accruals creation, the forecast of financial results and cash flows. Companies need to provide all interested parties with the information on the principles and assumptions used in their financial statements, as well as disclose those factors and risks that have an essential impact on the value creation process and a business model. The disclosure of risks associated with climate change and opportunities is also required.

Based on the development of TCDF, the results of investment research and analysis of both corporate and financial reports, we consider it appropriate to give the following recommendations to companies that form the initial experience of climate disclosures.

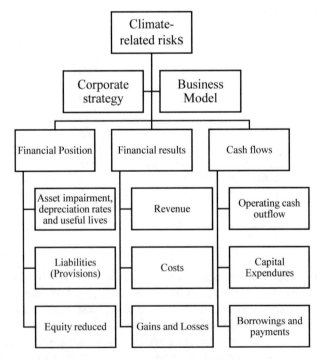

Fig. 1. Links of disclosures connected with the impact of climate change on financial reporting indicators.

The disclosed information should respond to the following questions:

- whether climate issues are relevant to the company; what mechanisms are used by a company management to evaluate and consider climate-related issues; who is responsible for climate-related issues; what information and indicators are monitored by management in this regard;
- how the business model can be affected by climate issues; whether it will remain sustainable and able to make a profit; what changes in strategy and business model the company may need; what opportunities and risks are most relevant for the company's business model and strategy;
- how risks and opportunities associated with climate change are reflected in the financial statements (for example, assumptions used in testing for impairment, depreciation rates, decisions on decommissioning, additional investments necessary for their restoration, and other similar obligations);
- which systems and control processes are used for identifying, assessing and managing climate-related risks; what are the results of a scenario analysis on the impact of climate change on key financial indicators;
- which methodology was used to work out the system of indicators and how these indicators are comparable with the ones of other companies the same industry.

One of the key conditions for success in this is to ensure the coherence and comparability of the disclosed financial and non-financial information, which becomes possible using modern digital technologies [30].

4 Conclusion

Improving the quality of climate risk disclosures in financial statements will increase the feasibility of investment decisions and make the financial market more transparent, thereby reducing the likelihood of a significant and unexpected adjustment of assets valuation, which could lead to destabilization of financial markets. In addition, the companies themselves will be able to assess both their own risks and the risks of their suppliers and competitors more reasonably. Such information could help in some issues. Firstly, in making more justified financial and investment decisions regarding specific companies. Secondly, in getting by stakeholders an awareness of the carbon dioxide-related assets' extent in the financial sector and realizing the degree to which the financial system is exposed to climate change risks.

Financial reporting is an integral element of a capital market and an economic system. In contemporary challenges, financial reporting should be considered as a system and as a basis in successful accounting for external and internal factors that pose a danger to a business model and ability to generate cash flows in view of capital providers and their requirements.

References

1. World Economic Forum. The Global Risks Report 2020. http://www3.weforum.org/docs/WEF_Global_Risk_Report_2020.pdf2
2. Eccles, R.G., Kastrapeli, M.: The Investing Enlightenment: How Principle and Pragmatism Can Create Sustainable Value through ESG, Environmental, Social and Governance (ESG) (2017)
3. Kahn, R.N.: The future of investment management. The CFA Institute Research Foundation (2018). https://www.cfainstitute.org/en/about/press-releases/2018/future-of-investment-management
4. London Stock Exchange Group: Guidance for issuers on the integration of ESG into investor reporting and communication (2018). https://www.lseg.com/sites/default/files/content/images/Green_Finance/ESG/2018/February/LSEG_ESG_report_January_2018.pdf
5. EY. Reporting climate change risk. A step-by-step guide to implementing the Financial Stability Board Task Force Recommendations for disclosing climate change risk (2017). ey-cfd-paper-final.pdf
6. Climate Disclosure Standards Board: Task Force on Climate-related Financial Disclosures (2018). https://www.cdsb.net/what-we-do/task-force-climate-related-financial-disclosures/commit-implement-recommendations-task
7. Climate Disclosure Standards Board. First Steps. Corporate climate and environmental disclosure under the EU Non-Financial Reporting Directive (2018). https://www.cdsb.net/sites/default/files/cdsb_nfrd_first_steps_2018.pdf
8. The Task Force on Climate-related Financial Disclosures: 2018 Status Report [PDF]. https://www.fsb-tcfd.org/wp-content/uploads/2018/09/FINAL-2018-TCFD-Status-Report-092618.pdf
9. Holder, M.: Mind the gap: are corporates translating climate risk disclosure into business action? (2019). https://www.greenbiz.com/article/mind-gap-are-corporates-translating-climate-risk-disclosure-business-action
10. ICAEW in association with the Carbon Trust. Reporting on climate risks and opportunities. A practical guide to the recommendations of the task force on climate-related financial disclosures. https://www.icaew.com/-/media/corporate/files/technical/financial-reporting/tcfd.ashx
11. Kuter, M., Gurskaya, M., Andreenkova, A., Bagdasaryan, R.: The early practices of financial statements formation in medieval Italy. Acc. Hist. J. **44**(2), 17–25 (2017)
12. Sokolov, Y.V., Sokolov, V.Ya.: History of Accounting: A Textbook. Master, Moscow (2011)
13. Littleton, A.Ch., Zimmerman, V.K.: Accounting Theory, Continuity and Change, p. 292. Prentice-Hall (1962)
14. Bedford, N.M.: Corporate accountability. Management Accounting (November), pp. 41–44 (1973)
15. Tinker, A.: Social Accounting for Corporations: Private Enterprise versus the Accounting Interest. M. Wiener, New York and Manchester University Press, Manchester (1984)
16. European Commission. Communication from the Commission to European Parliament, the Council, The European Economic and Social Committee and the Committee of the Regions. A renewed EU strategy 2011–14 for Corporate Social Responsibility, Brussels, 25 November 2011

17. The European Parliament and The Council of The European Union. Directive 2014/95/EU of the European Parliament and of the Council of 22 October 2014 amending Directive 2013/34/EU as regards disclosure of nonfinancial and diversity information by certain large undertakings and groups (2014). http://eur-lex.europa.eu/legal-content/EN/TXT/?qid=1421410689280&uri=CELEX:32014L0095

18. European Commission. Communication from the Commission—Guidelines on nonfinancial reporting (methodology for reporting non-financial information) (2017/C 215/01) (2017). https://eur-lex.europa.eu/legal-content/EN/TXT/?uri=CELEX:52017XC0705(01)

19. Moser, D.V., Martin, P.R.: A broader perspective on corporate social responsibility research in accounting. Acc. Rev. **87**(3), 797–806 (2012)

20. Edgley, C., Jones, M.J., Atkins, J.: The adoption of the materiality concept in social and environmental reporting assurance: a field study approach. Br. Acc. Rev. **47**(1), 1–18 (2014)

21. Trotman, A.J., Trotman, K.T.: Internal audit's role in GHG emissions and energy reporting: evidence from audit committees, senior accountants and internal auditors. Audit.: J. Pract. Theory **34**(1), 199–230 (2015)

22. Fernandez-Feijoo, B., Romero, S., Ruiz, S.: Commitment to corporate social responsibility measured through global reporting initiative reporting: factors affecting the behavior of companies. J. Clean. Prod. **81**, 244–254 (2014)

23. Gamerschlag, R., Moller, K., Verbeeten, F.: Determinants of voluntary CSR disclosure: empirical evidence from Germany. Rev. Manag. Sci. **5**(2–3), 233–262 (2011)

24. Hassan, A., Ibrahim, E.: Corporate environmental information disclosure: factors influencing companies' success in attaining environmental awards. Corp. Soc. Responsib. Environ. Manag. **19**(1), 32–46 (2012)

25. Jose, A., Lee, S.M.: Environmental reporting of global corporations: a content analysis based on website disclosures. J. Bus. Ethics **72**(4), 307–321 (2007)

26. Qiu, Y., Shaukat, A., Tharyan, R.: Environmental and social disclosures: link with corporate financial performance. Br. Acc. Rev. (2014). http://www.sciencedirect.com/science/article/pii/S0890838914000705

27. Marsat, S., Williams, B.: Does the market value social pillar? 2014. SSRN 2419387 (2014). http://www.efmaenn.org/0EFMAMEETINGS/EFMA%20ANNUAL%20MEETINGS/2014-Rome/papers/EFMA2014_0296_fullpaper.pdf

28. Richar, J., et al.: Proposals on reforming the fundamental principles of the enterprise, joint-stock company and social interest through environmental accounting (part I). Int. Acc. **20**(22), 1318–1335 (2017)

29. Jones, M.J.: Accounting for the environment: towards a theoretical perspective for environmental accounting and reporting. Acc. Forum **34**(2), 123–138 (2010). https://ideas.repec.org/a/eee/accfor/v34y2010i2p123-138.html

30. Efimova, O., Rozhnova, O.: The corporate reporting development in the digital economy. In: Antipova, T., Rocha, A. (eds.) Digital Science. DSIC 2018. Advances in Intelligent Systems and Computing, vol. 850, pp. 150–156. Springer, Cham (2019). https://www.springer.com/gp/book/9783030023508

31. 2degrees investing initiative. Transition Risk Toolbox: Scenarios, Data and Models (2016a). http://2degrees-investing.org/IMG/pdf/2ii_et_toolbox_v0.pdf

32. 2degrees investing initiative. Climate Disclosure: How to make it fly (2016b). http://2degrees-investing.org/IMG/pdf/make_dislosure_fly_v0.pdf?iframe=true&width=986&height=616

33. 2degrees investing initiative. Trails for Climate Disclosure (2016c). http://2degrees-investing.org/IMG/pdf/2ii_trails_v1.pdf?iframe=true&width=986&height=616

Assessment of the Factors' Impact on Innovation Activity in Digital Age

Irina Toropova[1] , Anna Mingaleva[1]([✉]) , and Pavel Knyazev[2]

[1] GSEM, Ural Federal University named after the first President of Russia
B. N. Yeltsin, Yekaterinburg 620002, Russia
mingaleva.ann@yandex.ru
[2] FSBEI of HE "Ural State Economic University",
Yekaterinburg 620219, Russia

Abstract. Digitization as a global phenomenon, covering an increasing number of aspects of human life, has a direct and indirect impact on many processes. The innovation activity of enterprises and organizations is not an exception. Moreover, the impact of digitization on various aspects of innovation is developing in different ways. The article presents the results of econometric modeling of socio-economic and scientific-technological factors impact on the output of innovative product. Six factors that currently have the greatest impact on the innovative development of the regions were identified and analyzed by creating a logarithmic model with fixed effects. The influence of the digitalization process on these factors is shown.

Keywords: Digitization · Innovation activity · Macroeconomic model

1 Introduction

One of the most complex methodological and methodic problems in the study of innovative development issues is the selection and evaluation of indicators that reliably assess the impact of various socio-economic, political, scientific, technical, demographic, national and other factors on innovative development [1–4]. At the same time, the level of innovation activity and effectiveness is measured both with the help of quantitative indicators and qualitative indicators. Among the quantitative indicators, the most frequently used are such indicators as the "volume of innovative products", "the number of patents and inventions", "the number of innovative enterprises", "the number of researchers" and others [5, 6]. The list of quality indicators is more restricted and among them there are such indicators as innovation competitiveness [7–9] and innovative susceptibility [10]. Moreover, the choice of factors and indicators of their assessment naturally differs for different economic entities: for countries it will be one [11, 12], for regions and clusters another [13, 14], for enterprises and organizations the third [15–17]. Innovative factors and indicators in management are separately investigated [18, 19].

Many scientific works of recent years are devoted to the substantiation of the choice of these indicators, as well as the assessment of the relationship between various factors at different levels: the relationship between capital accumulation and innovation in

© Springer Nature Switzerland AG 2021
T. Antipova (Ed.): ICIS 2020, LNNS 136, pp. 212–220, 2021.
https://doi.org/10.1007/978-3-030-49264-9_19

long-run growth [20, 21], the relationship between the indicators of innovative and investment-resource potentials in different regions [22]. Numerous studies are also devoted to the analysis of the relationship levels between innovation and knowledge and to the selection of indicators to assess their impact [11, 23–26]. Especially many researches are devoted to the selection and evaluation of environmental innovation indicators at all levels: national, regional, corporate [27–31].

A special area of research is the development of a research scheme for innovative processes for different entities, the analysis of methods for their assessment [32–34].

At the same time there are many methods for measuring innovative development, including the regression method, the taxonometric method (Euclidean distance method), the iterative method of cluster analysis of k-means, the rank correlation analysis method, nonparametric analysis methods and others [29, 35–37].

Method of integrated rating analysis of innovative assessments, the functional model for assessing innovative activity and competitiveness of regions using a system of statistical indicators are also widely used while conducting research and evaluating the results of innovative activities. The choice of an assessment method is largely determined by the purpose and objectives of the study, as well as reliable statistics for the calculations.

2 Research Method and Database

Econometric analysis has been used as a main research method. An official statistical data for 83 regions of the Russian Federation for the period from 2010 to 2016 were used to build an econometric model. The city of Sevastopol and the Republic of Crimea were excluded from the analysis, since the development of official statistical information for these regions has been carried out only since 2014. Regression model was built in «Stata» program.

The research method includes the following stages of regression building.

1. *Choice of indicators for assessing innovative development*

Determining key indicators for assessing the innovative development of Russian regions is the most complex methodological aspect of the study. To determine the key indicators, it is necessary to exclude their mutual influence on each other, therefore, before calculating the level of innovative potential of the region, a correlation analysis was carried out to check the presence of interconnections between indicators in order to avoid the problem of multicollinearity.

The indicator "volume innovative products output by enterprises" was taken as a measure of innovative development. This indicator was chosen due to the fact that it allows to measure innovations not through patent activity and R&D costs but using index of market success from the implementation of innovative activities. Moreover, both state and private organizations, as well as research institutes that produce innovative products, are considered as subjects of innovative activity. Explanatory variables that were used in the regression are listed in Table 1.

Table 1. Variables used in the model and their designations.

Designation	Variable	Measuring
ia	Innovative activity of organizations	%
aa	Average age of the population	Years
r_d	Costs of one enterprise for research and development	Thousand rubles
ti	Costs of one enterprise for technological innovations	Thousand rubles
ewhe	Share of employees with higher education	%
crisis	Crisis	1 – yes, 0 - no

Source: compiled by the authors

2. Econometric modelling

A logarithmic regression was constructed, where the indicator "volume of innovative products produced by enterprises" was used as an explained variable. Panel data was used in the research to construct the regression. Data are presented in the form of tables "object - sign". The peculiarity of the panel data is that it adds another dimension - time. The use of panel data due to the increasing number of observations allows to provide greater efficiency in estimating the parameters of the regression model and to increase the objectivity of the study.

A model with fixed effects was used to build a regression that considers the individual characteristics of each region. In the following study, the model is built to determine the weighting coefficients of indicators of innovative potential, therefore, usual linear model with a constant was considered:

$$Y = a + a_1 x_1 + a_2 x_2 + \ldots + x_n \tag{1}$$

In the compiled model, the explanatory variables were also logged in addition to the explained variable. The exception was variables whose values are measured in per cent and variables whose values are less than one, as well as the crisis variable, that is a dummy variable, i.e. takes values 0 or 1.

In general the database are no strong emissions in the sample, since its size is quite large. Many variables in the process of selecting the correct specification of the model were excluded from the analysis as they were insignificant. Displayed equations are centered and set on a separate line.

Next, a correlation matrix was constructed to determine the tightness of the relationship between the variables. After correlation analysis, various regression options were constructed: pooled regression, regression with fixed and random effects. Comparison of pooled regression, models with fixed and random effects using the tests of Wald, Broysch-Pagan and Hausman showed that the best model is fixed effects. The results of constructing a regression with fixed effects are presented in Table 2.

Table 2. Econometric model for "volume of innovative products produced by enterprises".

Average age of the population	22,27***
	(7,92)
Costs of one enterprise for research and development	0,004*
	(0,002)
Innovative activity of organizations	0,031*
	(0,019)
Share of employees with higher education	−0,080***
	(0,030)
Crisis	0,293**
	(0,140)
Costs of one enterprise for technological innovations	0,188***
	(0,067)

Source: compiled by the authors
***Variable is significant at 1% level
**Variable is significant at 5% level
*Variable is significant at 10% level

Model validations with fixed effects on multicollinearity and heteroscedasticity were performed. Next, model was tested for presence of multicollinearity and heteroscedasticity. There is no multicollinearity in the model (VIF = 1, 25). R^2 also has not changed and amounted to 35, 48%.

According to White's test, there is presence of heteroscedasticity in the model. Standard errors were corrected to fix heteroscedasticity. When adjusting, all variables remained significant, only the significance level changed. At the 1% level, the indicator "research and development costs" turned out to be significant. At a 5% level the following variables are significant: the logarithm of the "average age of the population" and the logarithm of "technological innovation costs". At a 10% level "innovative activity of organizations", "share of employees with higher education" and "crisis" are significant.

3 Research and Results

The results of the study revealed those socio-economic and scientific-technological factors that currently have the greatest impact on the innovative development of the regions. Some of the identified factors are traditional and their strong influence on innovative development is noted in many other studies. These factors are "innovative activity of organizations", "costs of research and development", "costs of technological innovation" and "crisis".

The model showed that with an increase in all these factors, the output of innovative products increases by a certain amount. This relationship is logical and traditional, only the strength of influence changes - i.e. rate of increase in the volume of production of innovative products. The currently existing power of influence used for the analysis of factors is given in Table 3.

Table 3. The contribution of factors.

	Increase in the output of innovative products, %
Innovative activity of organizations	3.13%
Average age of the population	22.27%
Costs of one enterprise for research and development	0.38%
Costs of one enterprise for technological innovations	0.19%
Share of employees with higher education	−7.97%
Crisis	29.34%

Source: compiled by the authors

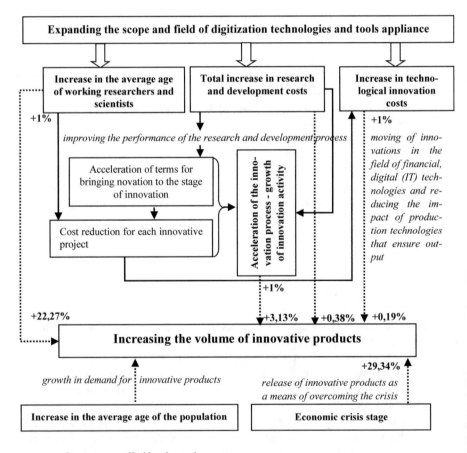

Source: compiled by the authors

Fig. 1. Scheme of the influence of the main scientific, technical and socio-economic factors on the volume of innovative products taking into account digitization processes

Such a force of influence of factors is explained by the fundamental impact of the digitalization process on all elements of the life of society and business and their change. Figure 1 shows the direct and indirect impact of the digitalization process on the factors of innovative development of the regions as a whole.

A feature of the innovation process and the creation of innovative products in many industries is that the success and speed of bringing research to practical implementation in the form of innovative products largely depends on the experience of the leaders of scientific groups (project groups) involved in the development of a specific project. Practice has shown that many projects led by old experienced managers have fewer mistakes and failures and are more quickly brought to the finished product stage than younger managers (although in the latter case, innovation may be more creative).

Emergence of the possibility of remote access to discussions on the progress of research, control of experiments using video equipment and online modes, etc. significantly extended the age and term of intellectual and creative longevity of many researchers with tremendous experience, which allows them to immediately see errors in experiments or options for solving problems that young researchers cannot find due to lack of experience.

In addition, if we talk about the effect of an increase in the average age of the population as a whole, in modern conditions this is also reflected in an increase in demand from older people in more convenient and multifunctional household appliances, in devices that make life easier for older people, in new, more effective medicines etc. This, in turn, stimulates scientific research in these areas and increases the production of innovative products specifically for this category of the population.

4   Conclusion

Scientific, technical and socio-economic factors affecting the production of innovative products in the Russian regions were identified, analyzed and evaluated during the study considering the digitalization process. The main directions and consequences of the impact of digitalization on factors affecting the expansion of enterprises' innovative activity in Russia were also analyzed.

Six significant variables that had the greatest impact on the growth of innovative products were identified using an econometric analysis based on a logarithmic model with fixed effects. These are five variables with a positive influence - "innovative activity of organizations", the logarithm of "average age of the population", "research and development costs", the logarithm of "costs of technological innovation", "crisis" and one variable with a negative impact - "share of employees with higher education". It should be noted that the traditional influence of the "higher education" factor on such indicators of society's development as "innovative development" and "competitiveness" is positive [38, 39].

A preliminary analysis showed that there is no simple and obvious explanation for this phenomenon, therefore, the study of the influence of this factor will require additional research. In particular, a more in-depth study of the sectoral structure of the regional economy and the corresponding professional qualification structure of the labor potential of the regions, indicators of the presence of research workers in the region and their areas of activity (technical sciences, humanitarian, medical, etc.), as well as the presence of among them researchers with advanced degrees. A preliminary analysis at the data collection stage showed that in some Russian regions data on research workers are generally absent. Checking the reasons for the lack of such data—either there are no research workers at all, or official data are not presented for them—is also the subject of future research.

References

1. Arundel, A., Kemp, R., Parto, S.: Indicators for environmental innovation: what and how to measure. In: The International Handbook on Environmental Technology Management, pp. 324–339 (2007)
2. Clemente, C., Civiero, P., Cellurale, M.: Solutions and services for smart sustainable districts: innovative key performance indicators to support transition. Int. J. Sustain. Energy Plan. Manag. **24**, 95–106 (2019)
3. Blind, K.: The influence of regulations on innovation: a quantitative assessment for OECD countries. Res. Policy **41**(2), 391–400 (2012)
4. Acs, Z.J., Audretsch, D.B., Lehmann, E.E., Licht, G.: National systems of innovation. J. Technol. Transfer **42**(5), 997–1008 (2017)
5. Acs, Z.J., Anselin, L., Varga, A.: Patents and innovation counts as measures of regional production of new knowledge. Res. Policy **31**(7), 1069–1085 (2002)
6. Gorshkov, A.P.: Aggregate demand for innovative products in international trade. Bull. PNIPU. Socio-Econ. Sci. **1**, 210–218 (2019)
7. Mingaleva, Zh., Mingaleva, A.: Assessing innovation susceptibility of regions and municipal districts. Procedia-Soc. Behav. Sci. **81**, 595–599 (2013)
8. Klenner, P., Husi, S., Dowlin, M.: Ex-ante evaluation of disruptive susceptibility in established value networks-when are markets ready for disruptive innovations? Res. Police **42**(4), 914–927 (2013)
9. Fonseca, L., Lim, V.: Countries three wise men: sustainability, innovation, and competitiveness. J. Ind. Eng. Manag. **8**(4), 1288–1302 (2015)
10. Belyakova, G.Y., Belyakov, G.P., Sumina, E.V., Badyukov, A.A.: Project-based approach to formation of innovative region receptivity. Reg. Sci. Inq. **9**(2), 119–130 (2017)
11. Mingaleva, Zh., Mirskikh, I.: On innovation and knowledge economy in Russia. World Acad. Sci. Eng. Technol. **66**, 1032–1041 (2010)
12. Balzat, M., Hanusch, H.: Recent trends in the research on national innovation systems. J. Evol. Econ. **14**(2), 197–210 (2004)
13. Buesa, M., Heijs, J., Pellitero, M.M., Baumert, T.: Regional systems of innovation and the knowledge production function: the Spanish case. Technovation **26**(4), 463–472 (2006)
14. Noskov, A.A.: Methodological directions for assessing the innovative development of regions and the scientific and innovative activities of universities. Bull. PNIPU. Socio-Econ. Sci. **4**, 363–372 (2018)
15. Deng, X., Cheng, X., Gu, J., Xu, Z.: An innovative indicator system and group decision framework for assessing sustainable development enterprises. Group Decis. Negot. (2019)
16. Gorshkov, A.P.: Cartoon effects induced by the private sector in the production of innovative products. Bull. PNIPU. Socio-Econ. Sci. **2**, 156–165 (2018)
17. Mingaleva, Z., Mirskikh, I.: Small innovative enterprise: the problems of protection of commercial confidential information and know-how. Middle East J. Sci. Res. **13** (SPLISSUE), 97–101 (2013)

18. Rybin, M.V., Stepanov, A.A., Morozova, N.V.: The system of key performance indicators of innovative activity as management innovation in oil and gas companies. Lecture Notes in Networks and Systems, vol. 115, pp. 605–612 (2020)
19. Melnik, M., Antipova, T.: Organizational aspects of digital economics management. In: Antipova, T. (eds.) Integrated Science in Digital Age. ICIS 2019. Lecture Notes in Networks and Systems, vol. 78, pp. 148–162 (2020)
20. Arranz, N., Arroyabe, C.F., de Arroyabe, J.C.F.: The effect of regional factors in the development of eco-innovations in the firm. Bus. Strategy Environ. **28**(7), 1406–1415 (2019)
21. Korchagina, I., Korchagin, R., Howitt, P., Aghion, P.: Capital accumulation and innovation as complementary factors in long-run growth. J. Econ. Growth **3**(2), 111–130 (1998)
22. Emelyanova, O.S., Tolstykh, V.I., Toropova, I.V.: Using the methods of economic and statistical analysis in the study of the relationship between indicators of innovative and investment-resource potentials in the federal districts of Russia. In: The Collection: Russian Regions in the Focus of Changes. Collection of Reports from Special Events of the XII International Conference, pp. 507–511 (2018)
23. Asheim, B.T., Coenen, L.: Knowledge bases and regional innovation systems: comparing Nordic clusters. Res. Policy **34**(8), 1173–1190 (2005)
24. Barrutia, J.M., Echebarria, C., Apaolaza-Ibáñez, V., Hartmann, P.: Informal and formal sources of knowledge as drivers of regional innovation: digging a little further into complexity. Environ. Plan. A **46**(2), 414–432 (2014)
25. Golovchanskaya, E.E.: Intellectual resources as a factor of innovative economic growth of the national economy. Bull. PNIPU. Socio-Econ. Sci. **3**, 183–193 (2019)
26. Hui-Boon, T., Hooy, Ch.-W., Islam, S.M.N., Manzoni, A.: Relative efficiency measures for the knowledge economies in the Asia Pacific region. J. Model. Manag. **3**, 111–124 (2008)
27. Antonioli, D., Borghesi, S., Mazzanti, M.: Are regional systems greening the economy? Local spillovers, green innovations and firms' economic performances. Econ. Innov. New Technol. **25**(7), 692–713 (2016)
28. del Río, P., Peñasco, C., Romero-Jordán, D.: Distinctive features of environmental innovators: an econometric analysis. Bus. Strategy Environ. **24**(6), 361–385 (2015)
29. Hoff, Ph.: Greentech Innovation and Diffusion: A Financial Economics and Firm-Level Perspective, p. 258. Springer, Heidelberg (2012)
30. Sheshukova, T.G., Mukhina, E.R.: Problems of accounting for eco-innovation in the system of environmental accounting and auditing. Bull. PNIPU. Socio-Econ. Sci. **2**, 147–155 (2018)
31. Vukovic, N., Pobedinsky, V., Mityagin, S., Drozhzhin, A., Mingaleva, Z.: A study on green economy indicators and modeling: Russian context. Sustainability (Switzerland) **11**(17), 4629 (2019)
32. Barnhoorn, F., McCarthy, M., Devillé, W., Alexanderson, K., Voss, M., Conceição, C.: PHIRE (Public Health Innovation and Research in Europe): methods, structures and evaluation. Eur. J. Public Health **23**, 6–11 (2013)
33. Bergek, A., Jacobsson, S., Carlsson, B., Lindmark, S., Rickne, A.: Analyzing the functional dynamics of technological innovation systems: a scheme of analysis. Res. Policy **37**(3), 407–429 (2008)
34. Proksch, D., Busch-Casler, J., Haberstroh, M.M., Pinkwart, A.: National health innovation systems: clustering the OECD countries by innovative output in healthcare using a multi indicator approach. Res. Policy **48**(1), 169–179 (2019)
35. Stevens, J.P.: Applied Multivariate Statistics for the Social Sciences. Lawrence Erlbaum Associates, Mahwah (1986)

36. Toropova, I.V., Merkulova, L.A., Perminov, E.A.: Nonparametric methods in the analysis of innovation and economic activity of the constituent entities of the Russian Federation. Econ.: Yesterday Today Tomorrow **8**(12A), 67–77 (2018)
37. Toropova, I.V., Yemelyanova, O.S., Tolstykh, V.I.: Use of the rank correlation analysis in the investigation of the interaction of innovative potential and investment attractiveness of the federal districts of the Russian Federation. Manag. Issues **3**(46), 78–82 (2017)
38. Stanišić, T., Leković, M., Stošić, L.: Relationship between the quality of higher education and Balkan countries' competitiveness. Int. J. Cogn. Res. Sci. Eng. Educ. (IJCRSEE) **7**(3), 49–59 (2019)
39. Krstić, B., Petrović, J., Stanišić, T.: Influence of education system quality on the use of ICT in transition countries in the age of information society. TEME **39**(3), 747–763 (2015)

Strengthening the State's Economic Security in the Tax Sphere: Problems and Prospects of the Russian Federation

Irina Korostelkina(✉) ⓘ, Marina Vasilyeva ⓘ, Lyudmila Popova ⓘ, and Mikhail Korostelkin ⓘ

Orel State University, Naugorskoe Highway, 40, Orel 302020, Russian Federation
cakyra_04@mail.ru

Abstract. At the state level, society is provided with various benefits in the form of law enforcement, social guarantees, education, healthcare, and environmental safety, which directly depends on the level of budgets and the tax potential of the regions. The effectiveness of the state and regions in terms of fulfilling their functions, which are financially provided by the tax system, should be taken into consideration in order to evaluate the stability and sustainability of the socio-economic development and the ability to ensure economic security of the state and regions. Therefore, a study dedicated to finding ways to strengthen economic security in the tax sphere is regarded as relevant. The purpose of this article is to consider the factors and threats to economic security in the tax sphere with the application of world best management practices and to identify government measures of taxation efficiency. General and particular research methods, methods and tools of graphical interpretation, comparative analysis, and related changes are used as methodological tools for this study. The article presents a theoretical analysis of the categorical apparatus and elements of security, threats and factors of their occurrence in the tax sphere, explores the indicators of the underground economy, forms a mechanism for ensuring economic security and identifies ways to strengthen tax security from the perspective of using tools and methods of the state tax policy. The implementation of measures to prevent and neutralize threats to economic security in the tax sphere ensures state stability and an increase in competitiveness and living standards of the population.

Keywords: Economic security · Tax policy · Tax mechanism · Risks and threats to tax security

JEL classification: F52 · H21 · H30 · G28

1 Introduction

The most important condition for the development of any society in modern realities is the formation of a particular condition withing the state in which its security would be fully ensured. The security of the state is the protection of its vital interests from

© Springer Nature Switzerland AG 2021
T. Antipova (Ed.): ICIS 2020, LNNS 136, pp. 221–232, 2021.
https://doi.org/10.1007/978-3-030-49264-9_20

external and internal negative influences, in which normal conditions for its functioning and the possibility of its stable development are created.

The term «security» arose in philosophical science in 1190 and was defined as calm state of mind of a person who considers himself protected from any danger until the XVII century. During XVII–XVIII centuries, the judgment that the main goal of the state is general welfare and security dominates in almost all countries. Since that time, the term «security» has acquired a new interpretation as a situation of calm, the absence of real danger (physical, moral), as well as material, economic, political conditions; relevant institutions and organizations contributing to the creation of this state [1]. Modern approaches to the interpretation of the concept of «security» appeared relatively recently - in the twentieth century, which is largely due to the ongoing wars, revolutions, and the increased danger of technological and environmental disasters. As a result, in the concept of «security», there was a shift in emphasis from a single individual to society, to the state, and to world space. Moreover, this put the problem of security in the forefront of the life of any state, and its provision became a priority of a state policy.

The term «economic security» arose at the turn of the 19th and 20th centuries at the intersection of economics and political science. This term was introduced into the domestic economy from foreign studies defining a different combination of categories of economic sovereignty, economic independence, as well as the stability of economic interests. Ensuring economic security should help achieve a balance of interests of all subjects of economic relations, reduce uncertainty and prevent risks in the framework of socio-economic development [2].

Abroad, scientists in the early 90 s of the twentieth century began to show interest in the study of security issues, including problems on the international level. Edward Luttwack (consultant to the National Security Council and the US Department of State) in his work «From Geopolitics to Geo-economics» noted the increasing role of economics and information in resolving geo-economic conflicts [3]. Research by Western scholars on security issues mainly relates to geo-economics and global financial crises. In particular, Fulcheri Bruni Roccia, in his work «The Geo-economic Factor in Financial Relations with Foreign Countries», considers financial instruments and methods as a weapon in an economic war [4].

Russian experts are actively engaged in the process of studying the problems of economic security in general and its individual areas (economic, environmental, military, etc.), starting in the 1990s of the twentieth century. This was facilitated by various crisis situations associated with the formation of the new state and its political and socio-economic areas of development. Russian economists of that time considered economic security through the concepts of «sustainability» (L.I. Abalkin, V.S. Pankov), «interests» (V.K. Senchagov, E.V. Pridius), and «independence» (A.E. Gorodetsky, A.A. Illarionov).

In particular, L.I. Abalkin defines economic security as a special state of the economic system in accordance with which tasks of a social nature are dynamically developing and effectively being resolved, as well as the state in which the state is given the opportunity to develop and implement an independent economic policy [5]. V.S. Pankov also considers economic security as a state of the national economy, characterized by its stability and immunity to internal and external factors that disrupt the normal functioning of the process of social reproduction, undermine the achieved standard of living of the population and thereby cause increased social tension in society, as well as a threat to the existence of the state [6].

A.A. Illarionov [7] understands economic security as a combination of economic, political and legal conditions that ensure the long-term sustainable production of the maximum amount of economic resources per capita in the most efficient way. A.E. Gorodetsky has a similar opinion, a decade later, he designated economic security as reliable and secured by all necessary means and institutions of the state (including power structures and special services), the protection of national-state interests in the economy from internal and external threats, economic and direct material damage [8].

By the beginning of the twenty-first century, opinions appeared defining economic security as the need to protect interests. So, E.V. Pridius considers economic security as a qualitative characteristic of the economic system, which determines its ability to maintain normal living conditions of a particular enterprise, industry, population, sustainable provision of resources for the development of the national economy, as well as the consistent implementation of regional and state interests [9]. And according to V. K. Senchagov economic security is an economic condition that provides, under conditions of stability (uncertainty) of internal and external processes, guaranteed protection of national interests, social policy, and sufficient defense potential [10]. Currently, an integrated approach is used when economic security is defined as a system of qualities and criteria characterizing a certain state of security. Now in Russia, national security is ensured by the Federal Law on December 28, 2010 No. 390-ФЗ «On Security» [11] and the National Security Strategy of the Russian Federation, approved by Decree of the President of the Russian Federation on December 31, 2015 No. 683 [12].

An important segment of the state's economic security is tax security. In this case, the object of security is the monetary resources that form the budget tax flows, and the subjects are the participants in tax relations. Taxes are a link between the budget, financial, and tax systems, the main source of financial support for the state's activities, an element of the budget revenues, which form economic relations, interactions and monetary relations and, accordingly, they are one of the primary elements of economic security. The state performs its functions in an appropriate manner on condition of having a proper fullness of budgets, which directly depends on the size and structure of tax revenues. That is why tax security at the present stage of economic development presents a fundamental importance in ensuring national security.

Bogdanov A.S. [13], Selyukov M.V. [14], Kolmakov V.S. [15], Vorontsov B.V. [16] give significant attention to the study of tax security as part of financial and economic security in general, and the identification of threats and tax risks in their works. Bogdanov A.S. considers tax security through the categories of tax policy and tax administration, exploring the objective prerequisites for ensuring tax security through tax control and regulation [13]. A similar opinion is presented in the work of V.S. Kolmakov [15]. Selyukov M.V. understands tax security as the state of the tax system, ensuring the harmonious development of the tax sphere, the possibility of using tax instruments to guarantee the protection of state interests, maintaining socio-economic stability in society, as well as the formation of a system of financial resources that successfully counteracts predictable and unpredictable internal and external threats

[14]. Moreover, the author characterizes issues of strengthening tax security through the prism of the process of making managerial decisions (tax management), control and verification activities and the system of state tax planning and forecasting. Vorontsov B.V. [16] explores the risks arising in the process of tax activity of the state and economic entities.

The modern economy creates new challenges and poses threats to economic security in the tax sphere. Scientists and economists are trying to solve the problem of ensuring economic security in the tax sphere by creating the optimal set of tools that takes into account the balance of interests of all subjects of tax relations. Therefore, the study of objective prerequisites and current trends in ensuring the economic security of the state in the tax sphere, as well as assessing the activities of tax institutions from the perspective of identifying threats to the economic security of the state and measures to strengthen it, is important and necessary at the present stage of economic development.

The aim of the research is to study the world and domestic experience of strengthening economic security in the tax sphere, to make the analysis of factors and causes of threats and the level of the underground economy, to identify the problems of effective implementation of the state tax policy at the present stage of economic development, and to suggest a mechanism for strengthening economic state security through the interaction of tools and elements of tax policy.

2 Materials and Methods

The methodological basis of the study consists of a set of general scientific approaches and methods: obtaining results and formulating conclusions, methods of comparison, analysis and classification were used while writing an article. The provisions of the study are also argued using a systematic approach and particular scientific methods (formal, comparative and functional methods, concretization, etc.). In the process, tools were used for graphical interpretation, comparative analysis, and for the method of concomitant changes. The study of the concepts of «economic security», «tax security» and theoretical analysis of factors and threats was based on the method of analysis, comparison and deduction. The mechanism for ensuring the economic security of the state is based on the method of focus group discussions. The identification of threats to economic security in the tax sphere and the proposal of ways to eliminate them is justified by using empirical methods and expert methods.

3 Results

Economic security in the tax sphere ensures a level of development of the tax system in which the necessary conditions are created for the socio-economic stability of the state. The efficiency of the main parts of the economy and the satisfaction of the basic needs of the state and population depend on the soundness and adequacy of the tax system.

Taxes and tax policy are factors of «such an economic development in which acceptable conditions are created for the life and development of the individual, the socio-economic and military-political stability of society and the preservation of the integrity of the state, successful opposition to the influence of internal and external threats» [15]. Taxes as factors of economic security should ensure the stability of opposition to threats to the country's security from the outside, as well as internal threats to the social, financial, budgetary and other systems of Russia.

Tax policy in the system of ensuring tax security plays a dual role. On the one hand, it acts as a resource and a tool for ensuring security in the hands of the state, and on the other hand, it is a risk factor and, in this aspect, is a source of threats from the point of view of the course of socio-economic processes in society [17]. Figure 1 presents the main factors contributing to the emergence of threats directly related to the implementation of tax policy.

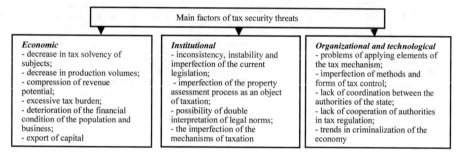

Fig. 1. Main factors contributing to threats to tax policy implementation

Threats to economic security are such phenomena and processes that negatively affect the economic state of the country, restrict the economic interests of the individual, society, and the state, and create a threat to national values and national way of life. Threats that arise in the tax sphere have a negative impact on the system of ensuring economic security in general, since there is a possibility of losses in the tax process. We agree with M.V. Selyukov [14] and V.B. Vorontsov [16] that all security threats can be defined as threats within the tax system:

- legal (associated with the imperfection of the tax legislation),
- institutional (related to the problems of interaction of elements of the tax mechanism-planning, regulation, control, management)
- and essential (related to non-payment of tax payments, the shadow economy) nature.

To strengthen economic security in the tax sphere, it is necessary to form a mechanism for ensuring it, the purpose of which is to create a set of conditions for ensuring the financial, social and political stability of society [18]. The mechanism for ensuring economic security must perform not only protective and regulatory functions, but also preventive, innovative and social ones, in other words, provide the most optimal state of the state's economy for its effective progressive development [19]. We believe that the mechanism itself should institutionally include organizational and managerial and legal, financial and budgetary, investment and innovation, and insurance support (Fig. 2). Each element of the mechanism for ensuring economic security should reflect the directions of mutual influence and dependence of the developed tools and methods, both in statics, dynamics, and in the perspective of identifying strategic challenges and risks in the functioning of the socio-economic system.

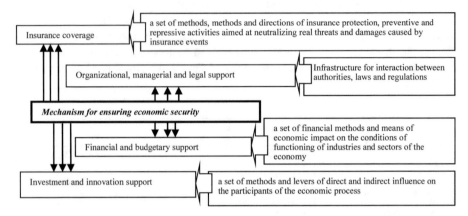

Fig. 2. Elements of the mechanism for ensuring the economic security of the state

Measures to ensure the economic security of the state in the tax sphere belong to the financial and budgetary direction. The assessment of the level of security requires a system of indicators (criteria), which, given the existing tax threats and factors include indicators of efficiency of tax system (tax collection level, indicators of tax burden and tax capacity, tax arrears, the level of shadow economy and dynamics of tax offences, indicators of efficiency of tax control and regulation and others) [20].

The study of problems related to the shadow economy occupies an important place in economic science, since shadow business and the criminalization of the economy cause significant damage not only to economic security, but also to the stability of the state as a whole. According to the World Bank, the scale of the shadow sector in the world is increasing year after year. The shadow economy in Russia is estimated by various studies to be between 23 and 80% of GDP.

According to a study by the Association of Chartered Certified Accountants (ACCA), in 2016 Russia was among the five largest shadow economies, taking fourth place in the ranking, which includes 28 countries. Its volume is 33.6 trillion. Ukraine (46% of GDP), Nigeria (48% of GDP) and Azerbaijan (67% of GDP) have the largest share of this indicator [21].

The International Monetary Fund (IMF) analyzed the shadow economy in 158 countries. In 2015, in developed countries (the United States, the Netherlands, Japan, Switzerland, and Singapore), the level of the shadow economy ranges from 7% to 15% of GDP [22]. In Russia, this figure amounted to 33.7% of GDP in 2015. In 2017, it reached 38.3% of world GDP (Fig. 3), so this phenomenon is of increasing interest to experts. In 2017 the scale of the shadow economy in Europe ranged from 6.5–22.4% of GDP, in the US-5.9%, and in Latin America-40%, in Japan-8.4%, in Greece-22.4%, in Germany-12.2% (average value), in the UK-9.4%. According to data published by the

World Bank for 151 countries, in Russia, the share of the shadow economy was about 50% of GDP [23]. Having analyzed the World Bank reports, experts in this industry have concluded that the opaque sectors in Russia are service, trade, food production, agriculture, as well as illegal activities, including those with a high corruption component.

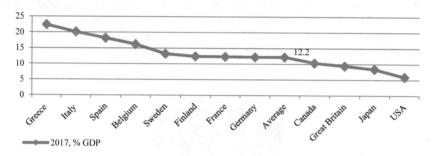

Fig. 3. The level of the shadow economy of countries in 2017, % [23]

The most important criteria that characterize the state of financial stability of a country and its individual subjects are the level of tax collection and tax debt, and their causal relationship. The indicator of tax collection in the Russian Federation is shown in Fig. 4. During the reviewed period, the tax collection rate generally decreases, despite an increase in the collection of basic tax payments (for example, corporate income tax and corporate property tax).

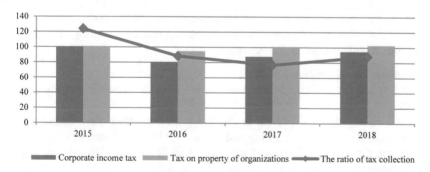

Fig. 4. Dynamics of the tax collection index in the Russian Federation for 2015–2018, % (calculated on the basis of official statistics of tax authorities https://www.nalog.ru)

These figures confirm the existing threats to economic security due to significant non-payment of tax payments. The amount of tax debt, its structure and dynamics characterize the level of tax offenses, which are characterized by significant opportunities for latency. Statistics show an increase in the number of tax offenses: they depend on ways of understating sales (41.3%), hiding revenue (18%) and overstating expenses (17%).

The distribution of tax offenses by industry in 2017 is shown in Fig. 5. The largest number of tax offenses in large and especially large amounts is observed in sectors such as: industry (23.3%), trade (17.4%), construction (17.1%), and agriculture (11.6%). The share of violations of tax legislation in the transport sector amounted to 4.7%, and in the credit and financial sector - 2.1% of all detected violations. According to the statistics from the Ministry of Internal Affairs of the Russian Federation, a fairly large number of economic crimes are committed in the country every year, including crimes in the tax sphere. In 2017, 105,087 economic crimes were registered, but there is a negative trend in 2016. The number of tax crimes also decreased by 6.8%. The share of economic crimes in the total number of registered crimes in 2017 was 5.1% [24].

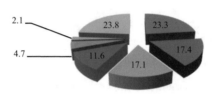

■ industry ■ trade ■ construction ■ agricultural industry ■ transport ■ financial and credit sphere ■ other sectors

Fig. 5. Dynamics of tax offenses by sectors of the national economy (compiled according to [24])

It is possible to ensure the economic security of the state by adopting an appropriate strategy focused on the implementation of information and analytical measures (monitoring and forecasting of socio-economic development), regulatory measures (improving the regulatory framework, improving the quality of tax administration, increasing the socio-economic and budgetary efficiency of tax policy instruments).

When applying the mechanism for ensuring economic security, the following algorithm should be used in the tax sphere:

– identification of national priorities;
– defining a system of criteria for assessing the level of economic security in the tax sphere;
– search for external and internal threats to tax security, their identification;
– search for directions for warning and preventing identified threats;
– formation of organizational and managerial bases for ensuring security and their adjustment in accordance with possible changes.

To determine the common threats from the tax system, it is necessary to pay attention to the competitiveness and profitability of the region and the industry, which are manifested in the assessment of production assets and import dependence, as well as the level of budget (financial) security. These indicators, having a growing synergistic effect, ultimately have a significant impact on the economic security of the state. Common threats also have a direct impact on the effectiveness of tax policy implementation in the region, since budget revenues and the region's ability to finance various social, investment and infrastructure projects depend on their impact. In addition to common threats, there are specific tax threats that have a direct impact on the level of economic security in the tax sphere. In particular, it is the imperfection of the tax administration system, the presence of the shadow economy and offenses, inter-budgetary irregularity and problems of tax federalism, the shortcomings of the system of tax benefits and tax sanctions [25]. We offer the following solutions to the identified problems:

Problem 1 - improving the quality of tax administration. To optimize tax administration in order to increase the economic security of the state:

- it is advisable to consolidate the indicators of evaluating the effective work of the tax inspector from the position of motivational-stimulating, control and service-oriented components at the legislative level. There are the two well-known models of tax administration that determine the nature of interaction between participants in tax relations (aggressive and partner), and it is necessary to focus on the latter, since its implementation will evolutionarily contribute to the formation of the tax culture of all participants in tax relations and, as a result, strengthen the economic security of the state. In this model budget revenues are formed under the influence of the assessment of the quality and performance of tax authorities;
- develop information services and electronic systems, which is relevant in the light of the widespread digitalization of the economy. The use of tax technologies provides important benefits for tax administration, including reducing costs, improving the quality of taxpayer service, and preventing corruption;
- introduce a legal norm in the tax code of the Russian Federation that determines the possibility of clearing tax payables with an expired maturity (this is now indefinite). Reduction of accounts payable allows the company to acquire the status of an investment-attractive business entity. In this case, the possibility of additional participation of taxpayers in the formation of the revenue part of the budget increases.

Problem 2 - the presence of the shadow economy. The activities of tax authorities should be adjusted to take into account the economic situation. To do this, you must:

- eliminate the practice of fighting for formal indicators of collected taxes, optimize the number of field inspections, and maximize the effectiveness of in-house examination of the financial condition of taxpayers;
- improve the procedures for administration, maintenance and submission of really popular tax reports, eliminating duplicate indicators.

The greatest damage from the functioning of illegal business is borne by the state authorities, the population, entrepreneurs and enterprises that conduct their business legally. Therefore, as a rule, they are the initiators of the fight against shadow business.

The solution to the problem of the shadow economy in modern conditions is also provided by:

- improving the principles of taxation of profits of controlled foreign companies;
- equalizing tax conditions for domestic and foreign e-commerce operators by introducing customs duties on such operations;

- clarification of the procedure for calculating excise taxes and VAT on certain goods and services;
- establishing a tax regime for self-employed citizens (tax on professional income);

Problem 3-inter-budgetary inequality in the redistribution of tax revenues. We believe that the current economic conditions do not fully ensure the effectiveness of fiscal federalism. This is related to the financing of national projects as the main sources of expenditure. We can refer the following features to the areas for improving the setting up of inter-budgetary relations and tax federalism:

- allocation of regional priorities for targeted inter-budget transfers;
- taking measures to expand the revenue base of the regions (increase the tax potential), expand the powers of regional and local authorities in establishing tax benefits and using other tax policy tools.

Solving these problems in the interests of improving the country's economic security will ensure the achievement of social (preservation of jobs, stabilization of the social situation, improvement of living standards), economic (ensuring the level of financial stability and solvency, increasing competitiveness and stabilization of financial indicators) and budgetary effects (financial support of state functions).

4   Conclusion

Thus, in the course of research of the conceptual apparatus and theoretical aspects of strengthening national interests and economic stability, it was determined that economic security should be considered as a certain ability of the state to ensure and maintain the existing status of socio-economic development, regardless of the existence of external and internal threats.

Economic security in the tax sphere is a guarantee of the progressive development of the state, since without proper financing, which is formed from tax revenues by more than 80%, it is impossible to perform state functions. Thus, taxation acts as a link between the redistribution of monetary resources between taxpayers and the state.

The determining role of tax security in the system of economic security of the state is determined by the significance of tax functions in general. One of the indicators of a high level of economic security is the stability of the tax system. Tax security is functionally aimed at ensuring the implementation of the interests of the national economy, as well as the social function of tax policy. We have analyzed common and specific threats to economic security in the tax sphere, identified objective prerequisites and directions for strengthening security. The study of international experience in strengthening economic security in the tax sphere has allowed us to determine the mechanisms and tools for effective implementation of the state's tax policy at the current stage of economic development.

Proposals for improving tax instruments to ensure economic security will improve tax system, and their implementation will contribute to the stabilization and further development of the country's economy and market structures, which together ensures economic security. The goal set at the beginning of the study was successfully achieved.

References

1. Latov, Yu.V.: The Russian shadow economy in the context of national economic security. Terra Economicus **1**, 16–27 (2007)
2. Vechkanov, G.S.: Economic Security: A Textbook for Universities. Petropolis, St. Petersburg (2015). 534 p.
3. Luttwak, E.N.: From Geopolitics to Geoeconomics. Natl. Interest **6**, 17–23 (1990)
4. Fulchery Bruni Roccia: Geo-economic factor in financial relations with foreign countries. Carlo Jean, Paolo Savona. Geoeconomics, Moscow (1997)
5. Abalkin, L.I.: Economic security of Russia: threats and their reflection. Issues Econ. **12**, 4–5 (1994)
6. Pankov, V.S.: Economic security. Interlink **3**, 114–120 (1992)
7. Illarionov, A.A.: Criterion of economic security. Issues Econo. **10**, 35–58 (1998)
8. Gorodetsky, A.E.: Economic security in a crisis. Bull. Econ. Secur. **5**, 49–57 (2010)
9. Prudius, E.V.: On the concept and system of economic security. Bus. Law **1**, 66–70 (2008)
10. Senchagov, V.K.: Strategic goals and mechanism for ensuring economic security. Probl. Manag. Theory Pract. **3**, 18–23 (2009)
11. Federal Law «On Security» dated 12.28.2010 № 390-FL [Electronic resource]. http://www.consultant.ru/cons/cgi/online.cgi?req=doc&base=LAW&n=187049&fld=134&dst=100008.0&rnd=0.6891267646889752#0
12. Decree of the President of the Russian Federation of December 31, 2015 № 683 «On the National Security Strategy of the Russian Federation» [Electronic resource]. http://www.consultant.ru/cons/cgi/online.cgi?req=doc&base=LAW&n=191669&fld=134&dst=1000000001.0&rnd=0.07444807760803052#0
13. Bogdanov, A.S.: Analytical support of the formation of the tax policy of the region. Manag. Acc. **11**, 44–51 (2014)
14. Selyukov, M.V.: The concept of the influence of the elements of the tax mechanism on the tax security of the state. Econ. Humanit. Sci. **9**, 65–71 (2014)
15. Kolmakov, V.S.: Tax policy of the Russian Federation and its role in ensuring economic security. Bull. Saratov Socio-Econ. Univ. **5**, 25–28 (2011)
16. Vorontsov, B.V.: Tax security of the state and economic systems. Probl. Mod. Econ. **3**, 586–588 (2008)
17. Selyukov, M.V.: The main directions of the state tax policy and its role in ensuring the tax security of the region. Manag. Acc. **11**, 65–72 (2014)
18. Bart, A.A.: The mechanism for ensuring the economic security of Russia. Russ. J. Entrepreneurship **11**, 4–9 (2010)
19. Samoilova, L.K.: Structural elements of the state economic security system. Bull. Orenburg State Agrarian Univ. **1**, 211–214 (2014)
20. Korostelkina, I.A., Simonova, T.S.: Tax policy and tax mechanism in the system of strengthening the economic security of the state. Manag. Acc. **8**, 65–79 (2019)
21. Russia entered the top five countries with the largest shadow economy [Electronic resource]. https://www.rbc.ru/economics/30/06/2017/595649079a79470e968e7bff
22. The IMF has published an estimate of the size of the shadow economy around the world [Electronic resource]. https://www.newsru.com/finance/07feb2018/imfassess.html
23. Rybasova, M.V., Savina, A.B.: Analysis of the scale of development of the shadow economy in Russia. Collection of articles on the materials of the III international scientific-practical conference «Scientific Forum: Economics and Management», pp. 66–73. Publishing House: «ICNO» - Moscow (2017)

24. The state of crime in Russia/Collection of statistical observations (2016–2018). GIATS Ministry of Internal Affairs of Russia, Moscow (2016–2018). 52 p.
25. Korostelkina, I.A.: Investigation of the causes of tax evasion and determination of directions for their elimination. Econ. Humanit. Sci. **8**, 106–113 (2016)

Digital Education

Using Smart Technologies
at the Classes of Foreign Languages
at a Non-linguistic University

Natalia Ignatieva$^{(\boxtimes)}$ ⓘ, Irina Zhdankina ⓘ, Darya Bykova ⓘ,
and Yulia Sysoeva ⓘ

Nizhny Novgorod State Engineering and Economic University,
Knyaginino 606340, Russia
ngieikonkova@yandex.ru

Abstract. The article considers different approaches towards the using of smart technologies in education in general, and at the classes of foreign languages in detail. It also mentions the attitude and support given to smart technologies in education by Russian competent bodies. The elements of smart education as well as the methods appropriate to apply when learning are mentioned. The article also shows the practical experience of using information technologies in the learning process at Nizhny Novgorod State Engineering and Economic University. The opportunities of smart education for students learning foreign languages as well as the ways to use different resources are highlighted. The positive and negative aspects of using smart technologies in education are given.

Keywords: Smart education · Smart technologies · Information technologies · Electronic environment · E-learning · Digitalization

1 Introduction

In modern society, the word "smart" has long been fixed, which is identified in many areas of human existence. It is about such terms as smart town, smart village, smart farming and smart house. Education is no exception in this area. Smart education is a new paradigm in both global and Russian education. Currently, Russian education is undergoing significant changes: the transition from traditional forms of classes to advanced ones, from outdated teaching methods to innovative ones, from passive to active learning content. Smart technologies have become an integral part of the educational process. Foreign language classes are no exception [1].

2 Discussion

Smart education aims to encourage learners who develop knowledge and skills needed in the 21st century to meet the needs and solve the problems of modern society. Smart education includes such concepts as "digital environment" and "digital content", "digital literacy" and "smart technologies", "ecosystem" and "information technologies". Many

© Springer Nature Switzerland AG 2021
T. Antipova (Ed.): ICIS 2020, LNNS 136, pp. 235–240, 2021.
https://doi.org/10.1007/978-3-030-49264-9_21

publications note the positive role of using smart technologies in the educational process [2–5]. Each author gives the own interpretation of this combination, but, in general, the meaning can be expressed as follows (Fig. 1).

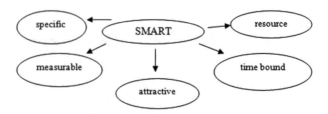

Fig. 1. Definition of the term "SMART" (Source: National project Education till 2024. https://edu.gov.ru/national-project)

The transformation of education has led to the fact that the changes taking place in this area are related not only to the use of innovative pedagogical technologies and teaching methods, but also to changes in teaching and learning principles and approaches, as well as in the concept of providing educational services in general.

The essence of smart education is to make the learning process more effective by transferring it into the electronic environment. There are three main elements of smart education: smart (intelligent) environment, smart (intelligent) pedagogy, and smart (intelligent) learners (Fig. 2). Smart pedagogy can have a significant impact on the smart environment where digital content resides. Smart pedagogy and smart environment support the development of smart learners in the context of developing new smart competencies [2].

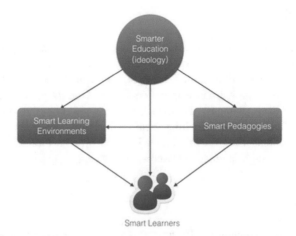

Fig. 2. Three main elements of smart education (Source: Marutschke, D.M. Smart Education in an Interconnected World https://doi.org/10.1007/978-981-13-8260-4_4)

By implementing smart learning approaches, educational organizations create the digital environment that includes research and teaching staff, students, and knowledge from around the world [6].

Thus, smart education is a flexible learning experience in an interactive educational environment using freely available digital content.

V. Tikhomirov gives the following definition of smart education – "it is the association of educational organizations and teaching staff to carry out joint educational activities on the Internet basing on common standards, agreements and technologies" [7]. At the same time, such principles as free access to knowledge, individual approach to learning for everyone, and mutual learning are implemented. Consequently, the role of educational organizations is changing from a knowledge provider to a creator of conditions for students to acquire new knowledge on their own [5].

The transition to smart education requires changes in the construction of the educational system in an educational organization:

- creating a digital intellectual educational environment;
- creating free mobile access to the digital intellectual educational environment;
- forming legal and scientific-methodological base, new educational materials and programs;
- improving the digital literacy of the educational process participants, developing and improving skills in using the advantages of digital technologies and services, etc.;
- creating of a virtual information portal that will allow the educational process participants to freely access methodological and analytical information;
- organizing and conducting events to exchange experience in the field of the educational environment digitalization as well as the introduction of smart education with partner universities (including foreign ones);
- creating a department performing a controlling function [4].

Many modern teachers in their practice apply innovative teaching methods in cooperation with new technical tools in practical classes.

Since foreign languages classes are held outside the country of the language being studied: its culture, customs, forms and ways of communication, modern technologies such as interactive textbooks and publications, smart boards, and the Internet resources help create a favorable environment for learning.

The implementation of smart education is impossible without the accumulated experience in e-learning and using smart technologies. Smart technologies play an important role in building an intelligent educational environment (smart environment), since learning in such an environment can take place at any time and place [8]. The smart technologies include:

- studies carried out in the form of remote (distant education);
- working in the virtual educational environment of an educational organization (working with digital materials, taking online courses, etc.);
- interactive textbooks and e-books (e-library systems such as Urait, IPRbooks, etc.);

- online materials (training sites such as www.native-english.ru, www.lengish.com, www.study.ru etc.), databases with video content (clips and films), apps for devices ("Anatomy 3D Pro" app, "Edmodo" app, "GeoGebra" app), virtual reality, virtual companions (chat bots), cloud technologies [2];
- creative and educational games;
- educational platforms created in social media (Facebook, Twitter), etc. [3, 4, 8].

3 Results

At Nizhny Novgorod state engineering and economic university, teachers use smart technologies at foreign language classes, such as interactive textbooks and e-books, presentation materials (material developed by the teacher and students), as well as the Internet materials, creative and educational games. Also, smart technologies used in education include working in the electronic educational environment of the University (working with digital material, taking online courses, etc.) [9]. It should be noted that language groups created in social media also provide assistance in learning a foreign language.

The use of smart technologies at foreign language classes contributes to the intensification of the educational process, as well as to the combination of different forms of work with students (group, individual) and types of educational activities [3]. Using this tool, students have the opportunity to work with authentic language materials, "become the object of real expectations and requirements related to language skills" [7].

It should be noted that students acquire the skills to search for useful information and materials about various spheres and aspects of life. The use of a variety of modern and accessible materials at foreign language classes contributes to the expansion of linguistic and cultural horizons, concentration when performing tasks. For example, a student with an insufficient vocabulary of foreign lexical units can independently continue learning the language at home or after classes on an online platform, performing lexical and grammatical tasks, listening to authentic monologs and dialogs, and pronouncing "difficult" words and expressions. As a result, the student has the opportunity to independently work out the studied material, analyze their mistakes and thus overcome the communication barrier [10].

The use of smart technologies methodically correlates with the requirements of the Federal state education standards (FSES) of the Russian Federation for secondary professional and higher education programs. Nowadays, not only the content a foreign language classes, but also the applied pedagogical technologies and methods must meet the implementation of the tasks set in the standards. Competence approach involves the development of universal (UC), general cultural (GCC) and general professional competencies (GPC) in the framework of the discipline "foreign language". For example, UC-4 and UC-5 in the Russian Federation FSES of new generation (on the example of "Agricultural Engineering" major, bachelor degree) assume the capacity of social interaction using a foreign language, as well as the perception of cultural and ethnic diversity. As for secondary professional education, the FSES of the Russian

Federation (on the example of "Mechanization Agricultural" major) includes GCC-5 and GCC-6, which assume the ability to communicate effectively in a team, as well as to use information and communication technologies, including in professional activities.

Using smart technologies not only allows to change the learning process qualitatively, but also to form the necessary competencies. For example, the principles of objectivity and clarity implemented through smart education contribute to the perception of diversity (working with authentic educational materials, using audiovisual means, direct foreign language communication via the Internet). Realized innovative approaches (for example, the project method) contribute to the development of students' soft skills, the ability to work in a team, and carrying out effective interaction. The principle of synergy, which is applied in smart technologies on the example of the integrated use of various learning tools (electronic information and educational environment, social media, training sites and blogs), can also contribute in these abilities. Students can also obtain skills to work with information and communication technologies using these tools both in the classroom and in the course of independent work.

The applying of smart technologies also contributes to the emergence of more effective tools for assessing the formation of competencies, such as electronic testing, summarizing and annotating authentic texts, research and creative projects, case studies, etc. The assessment of competence formation is carried out at the final stage of the discipline. To determine the degree of formation of various competencies, it is necessary to establish a relationship between the indicators of discipline development and the forms and criteria for evaluating competencies. In order to obtain more objective and complete results, it is preferable to use several evaluation tools, mainly based on interactive and innovative teaching methods [10].

Thus, the advantages of using smart technologies include increasing the effectiveness of teaching a foreign language to different groups of students (audio, visual, and audiovisual learning methods), quickly managing the course of the lesson, involving students in the process of searching for information, increasing the motivation and cognitive activity of students and interest in the subject, and removing the psychological barrier by students when using a foreign language as a means of communication [1].

Despite the positive aspects of using smart technologies at foreign language classes, both for teachers and students, it should be noted that such classes initially require a lot of time for the teacher to prepare, as well as special equipment, the Internet and all the accompanying "tools". Also, when working in a group, it is not always possible to correctly assess the contribution of each member of the group, since students have different backgrounds. Here the question of self-organization of students can be arisen when talking about individual task performing or about searching for additional information that is necessary for long-distance work in the classroom.

4 Conclusion

In general, it should be noted that smart technologies at foreign language classes are an auxiliary tool for all the participants of the educational process, and nothing will replace the "live, real" communication between the teacher and the student.

The use of smart technologies fully meets the requirements of the new educational standards for the formation, development and assessment of competencies, providing various tools and elements of discipline content that can be used in a comprehensive manner.

Smart education has a huge potential for the development of modern education. It is able to provide training for scientific and pedagogical staff that possess the necessary knowledge and skills, which is important in the development of modern advanced society, digital economy and innovative technologies.

References

1. National project Education till 2024. https://edu.gov.ru/national-project
2. Kacetl, J., klímová, B.: Use of smartphone applications in english language learning – a challenge for foreign language education. Educ. Sci. **9**(3), 179 (2019). https://doi.org/10. 3390/educsci9030179
3. Nguyen, T.T.H., Nguyen, T.M.: Information technology and teaching culture: application in classroom. In: Uskov, V., Howlett, R., Jain, L. (eds.) Smart Education and e-Learning 2019. Smart Innovation, Systems and Technologies, vol. 144. Springer, Singapore (2019). https://doi.org/10.1007/978-981-13-8260-4_32
4. Marutschke, D.M., Kryssanov, V., Chaminda, H.T., Brockmann, P.: Smart education in an interconnected world: virtual, collaborative, project-based courses to teach global software engineering. In: Uskov, V., Howlett, R., Jain, L. (eds.) Smart Education and e-Learning 2019. Smart Innovation, Systems and Technologies, vol. 144. Springer, Singapore (2019). https://doi.org/10.1007/978-981-13-8260-4_4
5. Li, K.: Visualization of learning activities in classroom blended with e-learning system. In: Uskov, V., Howlett, R., Jain, L. (eds.) Smart Education and e-Learning 2019. Smart Innovation, Systems and Technologies, vol. 144. Springer, Singapore (2019). https://doi.org/10.1007/978-981-13-8260-4_13
6. Strategies of development of Nizhny Novgorod region till 2035. https://strategy.government-nnov.ru/ru-RU/tilda/obrazovanie
7. Tikhomirov, V.P.: The world on the way to smart education. New opportunities for the development. Open Educ. **3**, 22–28 (2011)
8. Stepanek, J., Simkova, M.: Design and implementation of simple interactive e-learning system. Proc. Soc. Behav. Sci. **83**, 413–416 (2013). https://doi.org/10.1016/j.sbspro.2013. 06.081
9. Strategies of development of Nizhny Novgorod State Engineering and Economic University till 2035. www.ngieu.ru
10. Zhdankina, I.Yu., Ignatieva, N.N.: Assessment of competences formation on classes of foreign language at non-linguistic universities. KANT **2**(27), 41–47 (2018)

The Focus on Students' Attention! Does TikTok's EduTok Initiative Propose an Alternative Perspective to the Design of Institutional Learning Environments?

Markus Rach[1(✉)] and Marc Lounis[2]

[1] University of Applied Sciences and Arts Northwestern Switzerland,
Riggenbachstrasse 16, 4600 Olten, Switzerland
markus.rach@fhnw.ch
[2] Institute of Communication Sciences and Cognitions, University of Neuchâtel,
Pierre à Mazel, 2000 Neuchâtel, Switzerland

Abstract. This paper aims to provide a literature-based review of the current state of knowledge with regards to digital media and technology applications in learning environments. A systematic literature review of 27 sources was conducted to capture the current state of research and implementation of digital media and digital technology in learning environments. Findings exposed that scholars and learning practitioners alike apply a constructivist approach to shaping learning environments, yet almost all fall short to consider students' natural attention bias while assuming an exclusive focus on the learning environment. This paper follows to contrast the rapid growth of EduTok, TikTok's learning initiative to the digital media application of established learning institutions. The later provides a new perspective to re-evaluate the current state of digital media application in learning environments to propose an alternate framework for the application of digital media in learning environments by considering the so far neglected factor of attention arbitrage.

Keywords: Digital media · Digital technology · Learning environments · Social media · TikTok · EduTok · Technology adoption · Educational technology · Education

1 Introduction

Learning and learning environments have seen a strong impact of digital technologies over the last years, mostly referred to as educational technologies [1–4]. These technologies have grown in both quantity and impact, supported by the global digitalization trend and the influence of digital on our daily life [5]. By the end of 2019, the fast-growing educational technology (EdTech) market was estimated to reach USD 43bn [6]. With that, interest in EdTech applications and impacting factors has risen sharply, which is evidenced by both the number of scholarly articles but also commercially driven publications. Scholars have assessed the growth of technology applications as well as the influencing factors [7, 8], often with a focus on constructivist models

© Springer Nature Switzerland AG 2021
T. Antipova (Ed.): ICIS 2020, LNNS 136, pp. 241–251, 2021.
https://doi.org/10.1007/978-3-030-49264-9_22

serving to isolate acceptance factors [9, 10]. Ample research addresses the application disparity between educators and students, frequently focusing on factors impacting the EdTech adoption by instructors [11]. Through the transition of both authors from a marketing practitioner's role into an educator's role and the notion that much research in EdTech displays the addressed educators bias, this paper aims to answer the following research questions through a factor based literature review, followed by a marketing induced perspective on education, inspired by TikTok's EduTok initiative.

1. Do educational models and current research surrounding EdTech consider student's attention or is an exclusive attention bias assumed?
2. Does TikTok's EduTok initiative provide an alternate way to stimulate a new stream of research by considering student's flow of attention in EdTech application decision making?

2   Terms and Concepts

This chapter briefly introduces various key terminologies and concepts encountered in the literature review in order to provide the relevant background for this paper.

2.1   Learning Environments

Learning environments are classified as places where people can draw upon resources to make sense out of things and construct meaningful solutions to problems [12]. With that, a learning environment is made up of the learner and the setting or space where the learner acts, using tools and devices, collecting and interpreting information or interacting with others [12]. One theory to deconstruct learning, the constructivist theory of learning, seems especially relevant to mention as it aims to explain how people learn [13], taking into consideration how knowledge is shaped through understanding and experiences [14]. The seven pedagogical goals of constructivist learning environments serve as a reference point for the theory [12, 15]:

1. Provide experience with the knowledge construction process
2. Provide experience in an appreciation for multiple perspectives
3. Embed learning in realistic and relevant contexts
4. Encourage ownership and voice in the learning process
5. Embed learning in social experience
6. Encourage the use of multiple modes of representation
7. Encourage self-awareness of the knowledge construction process

With the emergence of digital technologies and their application to the design of learning environments, the complexity of designing learning environments has intensified and poised educators with new challenges in design, adoption and application [16]. Ample of research has been focused on this area, conversely also revealing the ill-defined nature of modern learning environments [17].

2.2   Educational Technology

Since the emergence of information technology, learning has been influenced and transformed through the availability of computer enabled tools to reshape the constructivist learning environment [16]. Parallel to the fast growth of computing technology, educational technologies emerged, shaping a commercially driven market [6]. Although various definitions of the term educational technology have been found, the authors apply the following definition as the most encompassing: "Educational technology is the study and ethical practice of facilitating learning and improving performance by creating, using, and managing appropriate technological processes and resources" [18]. This definition seeks to widen the constructivist understanding of learning environments and gathered wide attention through the focus on how technologies support the learning process. Some scholars have further engaged in discovering negative effects of technology in learning environments, by discovering an increase in distractibility on the one hand [19] or a design bias between educators and students on the other [20, 21].

2.3   TikTok

TikTok is a Social Media platform owned by the Chinese firm Bytedance. TikTok has risen to global fame since its takeover of musical.ly in 2017 and the following growth trajectory to a user-base of over 500 million active users with over 1 billion app downloads in under 2 years [22]. TikTok has become the predominant music and dance focused content app by teenagers, mainly facilitated by the ease of content creation [23]. Lately, scholars have started to engage in researching the effects leading to TikTok's meteoritic growth [23, 24] but also its effect on users and cultural attributes [25, 26]. Nevertheless, too little is known about the platform from a scholarly point of view with fewer than 1000 relevant articles on Google Scholar and no official analytics reporting by TikTok.

2.4   EduTok

Through a series of politically induced events in India, one of TikTok's prime markets by user numbers, TikTok launched its educational initiative EduTok in late 2019 [27]. EduTok's mission is to bring positive change to life and society by democratizing education, connecting learners and communities and fostering the growth of digital literacy [28]. It has to be noted how EduTok's mission aligns with most goals of the constructivist learning theory as outlined earlier.

Although primarily focused on the Indian market, EduTok marked content has received over 73bn views by 2020, as reported by TikTok [29]. Some very prominent sub-sections of EduTok include technology under the hashtag #edutoktech with 2.6 bn views, language under the hashtag #edutoklanguage with 1.9bn views, career advice under the hashtag #edutokcareer with 2.1bn views as well as more generalist sub-sections specific to political focus areas in India, such as #edutokmotivation or #edutoklifetips with 9.3bn and 8.4bn views.

By any means of measure, the reach and attention generated by TikTok's EduTok initiative in general and also by each of the mentioned sub-categories are staggering and worthy of more scholarly attention.

3 Method

This paper adopted a systematic literature review process designed to avoid authors' bias and pre-selection errors. To account for this, a structured literature review process was designed, as presented in Fig. 1.

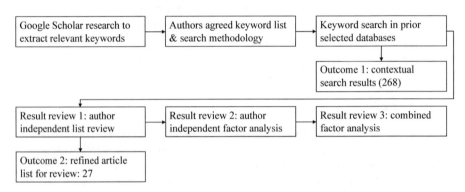

Fig. 1. Schematic visualization of the literature review process

3.1 Literature Review

The systematic literature review was conducted, based on the outline provided in Fig. 1, and with the following keyword search: "educational technology" or "digital technology" or "instructional technology" and "education" or "learning" for peer-reviewed academic journal entries ranging from 2011 to 2020.

The literature review was conducted in EBSCO Discovery Services, ScienceDirect, Elsevier Scopus and complemented by a search on Google Scholar.

A total of 268 peer-reviewed and contextually filtered academic journal articles were identified as a first outcome. These were then author independently reviewed to extract a list of the most relevant articles, as agreed through a mutual result review to condense the overall list of findings down to 27 articles. Following, each author reviewed the final article selection by the means of a qualitative factor analysis, which was yet again reviewed to agree on the most prevalent factors identified to impact the adoption of digital media and technologies in learning environments.

Due to the low number of findings and even lower number of observations per identified variable, the authors agreed to limit the review process to the qualitative review process.

4   Results and Discussion

Throughout the literature review of relevant articles, two clustering streams of research, with regards to EduTech and learning, emerged:

1. Focus on educators' adoption preferences
2. Focus on digital literacy as an adoption factor

Focus on Educator's Adoption

The dominant stream of research focuses on educators' adoption or implementation behavior of new technology [2, 9, 11, 30, 31]. The bulk of papers applies the Technology Acceptance Model (TAM) proposed by Fred D. Davis as a theoretical framework to examine how educators adopt new technologies, concluding that the individual's beliefs of usefulness, values and customs are amongst the higher determinants of technology adoption [31].

In many cases, educator's affinity to technology [30] is further being stated as a predeterminant for consideration, for example through educators' self-assessment of their skill and beliefs [32]. To further elaborate on educators' adoption of technology, it is important to expand on the notion of participation as an acceptance influencing factor [32]. Whether an educational technology is implemented through a top-down, bottom-up or a mixed method has been proven to be a success factor [32]. The later increases technology's utilization rate on both educators' and students' side.

Some studies include as additional factors students' motivation to learn; yet often addressed from an educator's point of view or a responsibility construct. Furthermore, although the TAM has evolved as a de-facto standard framework in this field of research, some scholars [11, 33, 34] have criticized its shortcomings, namely its lack of identifying the type of professional knowledge needed by teachers, conceptualized in the Technological Pedagogical Content Knowledge (TPACK) framework.

Others [33] consider further sociocultural factors on institutional levels and the predominant gap between information technology professionals in educational institutions versus educators. These studies have shown, that educators often lack application knowledge of current technologies. This leads educators to apply novel technologies based on their preconditioned knowledge of previous generation technologies.

In summary, the common determinant of most papers is, that the search, selection, adoption and implementation of technologies is being asymmetrically determined from the educator's point of view, giving little consideration to the perspective of students.

Focus on Digital Literacy as an Adoption Factor

Digital literacy is being identified from a student's as well as from an educator's perspective [35, 36]. Today, students have developed a strong understanding of digital technologies, in particular social media. This presents an opportunity for expanding the current state of research knowledge and should be taken into consideration when designing new learning environments.

Several authors have already suggested that pedagogy should be reconsidered in the digital age [37]. This, however, raises expectations towards educators [36] who tend to be "digital immigrants" and therefore show a legacy learning application bias [38].

Mitigating this potential gap requires more rigid educator training and updates of technological developments.

The TPACK framework [35], adopted in various papers, addresses this very training gap. The previously critiqued TAM model [11] might therefore benefit from a redefinition, accounting for the rapid growth of the digital chasm between educators and students.

Expanding the notion of literacy with the concept of range and depth of ICT embedding in all levels of the education ecosystem, Underwood's maturity matrix finds consideration [39]. Harrison et al. have used the linkage e-maturity model to highlight uneven adoption patterns to show that the linking of curriculum, learners and school's management system in a coherently connected way is an important factor for technology adoption [40]. This also provides a wider and more inclusive approach than the TAM model.

4.1 Discussion

It became apparent from the literature review that two factors were largely missing. One is the student's perception or perspective of technology as well as the effect of technology on the learning progress as measured on the students' level. It is argued by the authors, that students' learning should be considered as a performance measure in the design of educational environments and therefore flow into the adoption decision making of technologies.

The other is the inclusion of students' attention dispersion to other technologies which the authors argue should be considered under the pedagogical goals of constructivist learning environments.

Although it is accepted by the authors that an intrinsic level of motivation to learn should be expected in any learning environment, the current notion of assuming exclusive attention towards implemented learning technologies is considered to provide a false and misleading picture, as has been identified by some researchers already in predating work [19].

It is thus that the authors propose to apply a marketing perspective to the pedagogical goals of constructivist learning environments by considering where and how students' attention flows.

In other sectors, following attention level is a given and industry generalist method for target group interaction. Marketing serves as the prime example. Marketers engage in what is loosely defined as attention arbitrage [41]. A concept that assumes marketers must understand where the attention of target groups is focused on, to effectively reach these audiences with their marketing message.

Only little scholarly work has been found that supports this concept in education, yet with the increase of digital technologies being applied in the physical and virtual classroom, some scholars have already reasoned the need for technology and learning materials to garner students' attention to be effective [42].

As a proof of concept to this argument serves the social media platform TikTok, particularly its EduTok initiative. EduTok serves as a point of reference to showcase that attention is shifting from exclusive learning setups to democratized learning environments.

Contrasting TikTok's EduTok initiative to the seven pedagogical goals of constructivist learning environments [12, 15] provides further reason to assume the consideration of natural attention arbitrage by the student body. Social Media by default feeds into goal number two, four and five of the constructivist learning theorem by allowing conversations, multiple opinions, instant feedback and thus differing point of views or perspectives to one argument.

To examine this further, the authors have inspected a total of 50 EduTok postings related to mathematics, a subject not known for high engagement rates by students. Postings were selected based on the following hashtag search "edutok" and "mathematics" or "math", resulting in over 150 million combined views. The authors selected 20 random clips with the condition to be uploaded in 2019 to allow for a minimum time span of viewing. Out of this selection, the most prominent clip had over 800.000 likes compared the least prominent with 38.000. The average number of likes was counted around 380.000 with an average sharing volume of 48.000 and 2700 comments. To highlight the adoption effect, clips dating less than 4 weeks of upload date based on the time of writing this paper had a maximum view count of over 4.4 million views, 71.000 likes, 14.500 shares and over 100 comments. The same analysis could be done with a series of other educational subjects.

Although some scholars have already assessed the effect of social media to bridge formal to informal learning [43], or to enhance the effectiveness of learning [44], no prior research was found that assessed the identified gap from the aspect of attention grabbing or attention arbitrage. Research considering attention focused on measuring attention spans [45, 46], impact of learning methods on attention levels [47] or attention monitoring by instructors [48].

It is therefore that we propose to include the consideration of students' non-learning-based attention to the selection of technologies for the design of learning environments. The higher the non-learning-based attention level, the more likely a selected technology is within the attention levels of students and thus not negatively affected by attention arbitrage effects of other media or technology. Following, various venues for further research are being proposed to verify this hypothesis and create a more holistic framework for technology selection and adoption processes in learning environments.

5   Conclusions and Implications

Based on the literature review undertaken and the arguments presented, it is reasoned that the design of effective learning environments must factor in the non-exclusivity of education with regards to students' attention, both in classroom and virtual education settings.

Attention should be considered a determining factor for the selection of EduTech but also the design of learning environments to induce effective learning. Applying basic marketing tactics to education is therefore a possible alternative to increase learning effectiveness by not competing with attention grabbing media or technology in either classroom setting but by leveraging technology's natural attention levels for the benefit of the learning effort. This requires a more thorough analysis of the student body, which in the literature research undertaken was not found to be a considering factor. Through a deeper analysis of the student body, educators can then design learning efforts which are not only fulfilling pedagogical objectives but also meet the demographic needs of the learning target groups. Assumed implications for educators and scholars are manifold.

First educators need to consider the student population to a greater effect in the selection of technologies. As a derivative, this requires educators to consider a best fit of various technologies for heterogeneous student bodies, e.g. in classic universities with undergraduate, graduate and executive education programs.

Second, educators and institutions have to increase their speed of technology scouting and technology adoption.

Third, educators or educational institutes might need to engage with the prospect of applying certain technologies, such as TikTok's EduTok platform, without the ability to have an exclusive access or content control for the benefit of the earlier hypothesized learning benefit.

Finally, institutions must consider educators' literacy and in particular continuous and up to date knowledge about emerging digital technologies such as social media platforms.

5.1   Limitations

Despite all efforts undertaken by the authors of this paper, several limitations need to find consideration when applying the findings and arguments of this paper.

First, since research findings were selected on the basis of a multistep process, involving both scholarly keyword searches and human sorting; some relevant literature might not have been considered.

Second, this study assumed a holistic perspective and did not limit its application to a particular sector of education, e.g. higher education.

Third, this study did not consider cultural differences which might bias findings when contrasting EduTech across various cultural spheres.

Fourth, although all possible efforts have been undertaken to provide a robust literature review of identified documents, interpretation errors cannot be excluded due to the mix of terminology used.

5.2   Practical Implications for Educators

Whilst educators are being trained on pedagogical principles, a struggle for attention in the classroom in distance and virtual education setting remains. Some educators attempt to ban the use of non-class related media, yet since the wide adoption of the mobile devices, this option fades in applicability in most cases. It is therefore that this paper hopes to stimulate a reconsideration of attention flows to reconsider a utilization of these attention flows for the benefit of learning outcomes. Educators are advised to experiment with new technologies to increase their own technological literacy to support the finding of possible areas of application.

5.3 Future Research

Based on the papers analyzed, the discussion of findings presented and the contrast to the growth of EduTok, three avenues for further research are proposed.

First, the growing divergence of how technologies are understood and applied by educators and students is expected to broaden with the increasing rate of speed of platform emergence and adoption. More research needs to be done to cover this field to understand this divergence to derive applicable countermeasures.

Second, students' attention fields should be considered as a variable in learning models or the design of learning environments. Understanding where students' attention is focused and how EduTech could help to exploit this focus could proof very beneficial in creating high attention and performance driven learning environments.

Third, the above requires a reframing of the exclusivity paradigm of the learning environment. Since EduTech application drivers have been largely researched on the basis of instructors' willingness to apply; it might be necessary to also research application facilitators or hurdles on the institutional level of education.

References

1. Minoli, D.: Distance Learning Technology and Applications. Artech House Inc., Norwood (1996)
2. Yuen, A.H., Ma, W.W.: Exploring teacher acceptance of e-learning technology. Asia-Pac. J. Teach. Educ. **36**(3), 229–243 (2008)
3. Henry, P.: E-learning technology, content and services. Educ.+Train. **43**(4/5), 249–255 (2001)
4. Facer, K.: Learning Futures: Education, Technology and Social Change, 1st edn. Routledge, London (2011)
5. Mirrlees, T., Alvi, S.: EdTech Inc.: Selling, Automating and Globalizing Higher Education in the Digital Age. Routledge, London (2019)
6. Premack, R.: Teachers Across America are Obsessed with Google Products—Here's how Apple and Microsoft Plan to Win Them Back. Business Insider (2018). www.businessinsider.com/google-apple-microsoft-competing-dominate-education-technology-market-2018-11. Accessed 10 Feb 2020
7. Schacter, J.: The impact of education technology on student achievement: What the most current research has to say (1999)
8. Saettler, P.: The Evolution of American Educational Technology. IAP, Bridgeport (2004)
9. Domingo, M.G., Garganté, A.B.: Exploring the use of educational technology in primary education: teachers' perception of mobile technology learning impacts and applications' use in the classroom. Comput. Hum. Behav. **56**, 21–28 (2016)
10. Ball, D.M., Levy, Y.: Emerging educational technology: assessing the factors that influence instructors' acceptance in information systems and other classrooms. J. Inf. Syst. Educ. **19**(4), 8 (2019)
11. Scherer, R., Siddiq, F., Tondeur, J.: The technology acceptance model (TAM): a meta-analytic structural equation modeling approach to explaining teachers' adoption of digital technology in education. Comput. Educ. **128**, 13–35 (2019)
12. Wilson, B.G.: Constructivist Learning Environments: Case Studies in Instructional Design, pp. 3–8. Educational Technology Publications, Englewood Cliffs (1996)

13. Bada, S.O., Olusegun, S.: Constructivism learning theory: a paradigm for teaching and learning. J. Res. Method Educ. **5**(6), 66–70 (2015)
14. Bereiter, C.: Constructivism, socioculturalism, and Popper's world 3. Educ. Res. **23**(7), 21–23 (1994)
15. Knuth, R.A., Cunningham, D.J.: Tools for constructivism. In: Designing Environments for Constructive Learning, pp. 163–188. Springer, Heidelberg (1993)
16. Dede, C.: The evolution of constructivist learning environments: immersion in distributed, virtual worlds. Educ. Technol. **35**(5), 46–52 (1995)
17. Moore, J.L., Dickson-Deane, C., Galyen, K.: e-Learning, online learning, and distance learning environments: are they the same? Internet High. Educ. **14**(2), 129–135 (2011)
18. Januszewski, A., Molenda, M. (eds.): Educational Technology: A Definition with Commentary. Routledge, Abingdon (2013)
19. Aagaard, J.: Drawn to distraction: a qualitative study of off-task use of educational technology. Comput. Educ. **87**, 90–97 (2015)
20. Tondeur, J., van Braak, J., Siddiq, F., Scherer, R.: Time for a new approach to prepare future teachers for educational technology use: its meaning and measurement. Comput. Educ. **94**, 134–150 (2016)
21. McKnight, K., O'Malley, K., Ruzic, R., Horsley, M.K., Franey, J.J., Bassett, K.: Teaching in a digital age: how educators use technology to improve student learning. J. Res. Technol. Educ. **48**(3), 194–211 (2016)
22. Zhu, C., Xu, X., Zhang, W., Chen, J., Evans, R.: How health communication via Tik Tok makes a difference: a content analysis of Tik Tok accounts run by Chinese provincial health committees. Int. J. Environ. Res. Publ. Health **17**(1), 192 (2020)
23. Anderson, K.E.: Getting acquainted with social networks and apps: it is time to talk about TikTok. Libr. Hi Tech News (2020)
24. Xu, L., Yan, X., Zhang, Z.: Research on the causes of the "Tik Tok" app becoming popular and the existing problems. J. Adv. Manage. Sci. **7**(2), 59–63 (2019)
25. Li, H., Zhang, Q.: Tik Tok app should convey correct values: taking the influence of Tik Tok on college students as an example. Theory Pract. Contemp. Educ. **5**, 19 (2019)
26. Zuo, H., Wang, T.: Analysis of Tik Tok user behavior from the perspective of popular culture. Front. Art Res. **1**(3), 1–5 (2019)
27. The Economic Times. https://economictimes.indiatimes.com/magazines/panache/tiktok-announces-edutok-programme-to-empower-first-time-internet-users-hone-their-soft-skills/articleshow/71631407.cms?from=mdr. Accessed 18 Feb 2020
28. Forbes India. http://www.forbesindia.com/article/brand-connect/tok-of-the-town/55595/1. Accessed 19 Feb 2020
29. TikTok. https://www.tiktok.com/tag/edutok. Accessed 19 Feb 2020
30. Aldunate, R., Nussbaum, M.: Teacher adoption of technology. Comput. Hum. Behav. **29**(3), 519–524 (2013)
31. John, S.P.: The integration of information technology in higher education: a study of faculty's attitude towards IT adoption in the teaching process. Contaduría y Administración **60**, 230–252 (2015)
32. Petko, D., Egger, N., Cantieni, A., Wespi, B.: Digital media adoption in schools: bottom-up, top-down, complementary or optional? Comput. Educ. **84**, 49–61 (2015)
33. Sabi, H.M., Uzoka, F.M.E., Langmia, K., Njeh, F.N.: Conceptualizing a model for adoption of cloud computing in education. Int. J. Inf. Manage. **36**(2), 183–191 (2016)
34. Mishra, P., Koehler, M.J.: Technological pedagogical content knowledge: a framework for teacher knowledge. Teach. College Rec. **108**(6), 1017–1054 (2006)
35. Borthwick, A.C., Hansen, R.: Digital literacy in teacher education: are teacher educators competent? J. Digit. Learn. Teach. Educ. **33**(2), 46–48 (2017)

36. Smith, E.E., Kahlke, R., Judd, T.: From digital natives to digital literacy: Anchoring digital practices through learning design (2018)

37. Zdravkova, K.: Reinforcing social media based learning, knowledge acquisition and learning evaluation. Proc.-Soc. Behav. Sci. **228**, 16–23 (2016)

38. Prensky, M.: Digital natives, digital immigrants. Horizon **9**(5), 45–51 (2001)

39. Underwood, J., Dillon, G.: Capturing complexity through maturity modelling. Technol. Pedagog. Educ. **13**(2), 213–225 (2004)

40. Harrison, C., Tomás, C., Crook, C.: An e-maturity analysis explains intention–behavior disjunctions in technology adoption in UK schools. Comput. Hum. Behav. **34**, 345–351 (2014)

41. Hasen, R.L.: Cheap speech and what it has done (to American Democracy). First Amend. L. Rev. **16**, 200 (2017)

42. Robinson, A., Cook, D.: "Stickiness": gauging students' attention to online learning activities. Inf. Learn. Sci. **119**(7/8), 460–468 (2018)

43. Greenhow, C., Lewin, C.: Social media and education: reconceptualizing the boundaries of formal and informal learning. Learn. Med. Technol. **41**(1), 6–30 (2016)

44. Balakrishnan, V., Gan, C.L.: Students' learning styles and their effects on the use of social media technology for learning. Telemat. Inform. **33**(3), 808–821 (2016)

45. American Psychology Society, Bradbury, N.A.: Attention span during lectures: 8 seconds, 10 minutes, or more? https://journals.physiology.org/doi/full/10.1152/advan.00109.2016. Accessed 20 Feb 2019

46. Bunce, D.M., Flens, E.A., Neiles, K.Y.: How long can students pay attention in class? A study of student attention decline using clickers. J. Chem. Educ. **87**(12), 1438–1443 (2010)

47. Bolliger, D.U., Supanakorn, S., Boggs, C.: Impact of podcasting on student motivation in the online learning environment. Comput. Educ. **55**(2), 714–722 (2010)

48. Dabbagh, N., Kitsantas, A.: Supporting self-regulation in student-centered web-based learning environments. Int. J. E-learn. **3**(1), 40–47 (2004)

Work of First-Year Students
with Terminological Units

Alena Guznova⬛, Olga Belousova⁽⊠⁾, Valery Polyakov⬛,
Leonid Mikhailiukov⬛, and Nikolay Mordovchenkov⬛

Nizhny Novgorod State Engineering and Economic University,
Knyaginino Nizhny Novgorod Region, 606340 Nizhny Novgorod, Russia
alena_guznischeva@mail.ru, ilichevalga@yandex.ru

Abstract. Each branch of science has its own terminology base that defines the concept of scientific research, which is the basis for understanding the material and forming a new scientific theory. The study is based on method of terminological analysis. While working with the material, popular scientific methods were also used: observation, analysis, synthesis, comparison and special methods: linguistic experiment, linguistic analysis of text. There is a problem of knowledge of terminology units by students of higher education institutions at the bachelor's level. The purpose of our experiment is to identify the level of determinism of professional thinking of students of the 1st year of training Agroengineering (profiles «Electrical equipment and electrical technologies» and «Technical service in agriculture». Experimental work was carried out on the basis of the Nizhny Novgorod state University of engineering and Economics (Knyaginino). The analysis of the current work program on the discipline «Speech culture and business communication» was carried out.

Keywords: Terminology · Method of terminological analysis · Observation

1 Introduction

Each branch of science has its own terminology base that defines the concept of scientific research, which is the basis for understanding the material and forming a new scientific theory. The terminology base serves as a foundation for familiarizing students with scientific theory and studying science, therefore, the process of studying terms at the initial stage of acquaintance with science is an important step in the formation of scientific knowledge among students.

A term is «a word or phrase that is the name of some concept of some field of science, technology, art, etc.» [1]. Terms are usually characterized by unambiguous application and lack of emotional coloring. J. V. Slozhenikina calls binding, unambiguity, emotional and expressive neutrality and systemicity as specific features of the term [2]. G. O. Vinokur noted the special role of the term - to name a special concept [3].

Despite the simplicity of the wording of these definitions and the conciseness of the listed features, the question of the concept of «term» and its differentiation with other concepts, for example, «terminological unit», «term system», «terminological combination», is still relevant. There are also works where the concepts «word-term» and «word-not term» appear in relation to the objectivity and subjectivity of the meaning.

© Springer Nature Switzerland AG 2021
T. Antipova (Ed.): ICIS 2020, LNNS 136, pp. 252–262, 2021.
https://doi.org/10.1007/978-3-030-49264-9_23

The purpose of this study is to interpret the concept of «terminological unit», to determine the essence of the work of first – year students with terminological units. Tasks: to determine the semantic content of the «terminological unit», to distinguish it from the concept of «term»; to determine the role of first-year students' work with terminological units; to create an algorithm for students' work with terminological units to form a holistic view of the terminological base of science.

Research methodology. The methodological basis of the research was the works of D. S. Lotte, G. O. Vinokur, B. N. Golovin, Z. I. Komarova, V. M. Leychik, as well as the works of S. G. Kazarina on typological terminology, E. I. Golovanova on cognitive terminology, D. S. Zolotukhin on approaches to the description of linguistic term systems [4] and E. V. Lopatina on the features of terminological units [5].

The research is based on the method of terminological analysis. The formation of the terminological device is the first step to obtaining new scientific knowledge, a process that includes disclosure of the essence of the concept through interpretation, discovery, values, clarification of meaning, elimination definitions, define relationships with other concepts.

When working with the material, general scientific methods are also used: observation, analysis, synthesis, comparison and special methods: linguistic experiment, linguistic analysis of the text.

2 Work of First-Year Students with Terminological Units

Terminology is the science of terms and term systems [6]. The study of special vocabulary is an integral part of professional education.

M. Clevenger claims that «our society demands men not only skilled ... but educated citizens who take an active part in industrial, labour and community affairs ...» [7, p. 18]. M. Dyer says that «professional education suffers very greatly from a lack of congruence between the actual performance of its graduates and the training programs through which they are put» [8, p. 38]. S. G. Kazarina notes the importance of studying terminology as a «tool of professional activity», focusing on the limitations of works on the typology of terminology, which is reduced to describing separate groups of words without «unity in the subjects of description, goals, methods and algorithms of work». The researcher concludes that the key to successful teaching of the language of specialty lies in the quantization of typological description of terminological systems in a concise form, which contains typological characteristics of individual units and their complexes [9, p. 15]. It is a thought that the study of terminological units as the basis of any language of the profession in the system, where all elements are mutually agreed, is productive for the development of the profession: the student should not only know the interpretation of terms, but also understand their relationship, be able to build in the system. To describe such systems S.G. Kazarin introduces the concept of «typological passport of term systems».

E. I. Golovanova speaks about moving the goal of terminological research to the internal sign nature of the term through the connection with professional knowledge, communication and professional activity, to the representation of different types of knowledge in terminological units [10]. In the study of term systems, the author proposes to adhere to the anthropological principle. In our opinion, this approach to the problem of studying terminological units is quite justified in pedagogy and teaching methods in view of personal-oriented learning, interactive learning, where subject-subject relations allow all participants to become «getters» of new knowledge, seekers and researchers of new information in the educational process.

A. Y. Bagiyan and L. B. Ezdekova, exploring the transformational potential of terminology units in the non-scientific discursive realization, note the functioning of terms as a component of copyright applications to popularize elements of social knowledge, which contributes to the formation of an objective terminological semantic network of a certain area [11]. Thus, knowledge of terminology units, ability to use them allow to determine the outlook of their medium, to determine its qualification.

The term, or terminological unit, is a unique unit of the language of science, the specificity of which, as V. V. Vinogradov noted, lies «in terms of content, in the character of meaning» [12]. The ordered set of terms constitutes a term system. P. I. Schleivis notes the presence of logical relationships in the term system in a certain period of time [13].

The fact of the need to study the term system of science, which determines the future professional activity of the student, remains undeniable. As has already been said before, modern works have little research on how to approach the study of terminological units in an integrated manner.

The meta-language function allows to describe language phenomena by means of language units, corresponding study of terminological units, their description and systematization takes place through the language. We will show work with terminological units of students studying in the direction of study 35.03.06 Agroengineering. In our opinion, work with special terms should begin primarily with language discipline to lay down principles of understanding work with elements of language for further work with terms, for finding definitions and building logical connections in the term system. To work with terms, it is necessary to master language, communicative competence. In Federal state educational standard 3 ++, in the direction of study 35.03.06 Agroengineering it is presented universal competence 4 (UC-4), which belongs to the group of universal competences «Communication» - «to be able to carry out business communication in oral and written forms in the state language of the Russian Federation and foreign language(s)» [14]. This competence is formed during the study of the discipline «Speech Culture and Business Communication» which replaced the discipline «The Russian Language and Speech Culture» under the previous approximate educational program of Federal state educational standard 3+.

As it is known, «… in the course of training it is necessary to form such competences that will make the future graduate competitive in the labor market [15]».

There is a problem of knowledge of terminology units by students of higher education institutions at the bachelor's level. Unfortunately, after passing the exam, students do not find possession of terms, since the tasks are not of an applied nature.

Students are not motivated to study and interpret, and there are no logical connections in oral and written speech.

Terminology performs cognitive and orientation functions in professional communicative behavior. The codified technique has its own logic, different from the ordinary ideas that almost everyone can have.

As you know, terminology is an integral part of the scientific and technical functional style, and the terminological development of written and oral communication indicates the level of culture of the specialist. According to E. S. Shmatova, «the richer the terminological Fund, the more developed and widely used... terminology, the more stability, efficiency and conciseness are achieved» [16]. The richer the employee's speech terminology, the higher his professionalism. The number of terms and their combination with each other may be different, depending on the topic of the statement and the level of culture of the specialist. The development of technical terminology contributes to the accurate and clear formulation of operating instructions, achieving maximum conciseness of the technical text.

For the best development of theoretical concepts, students must learn to be aware of the termination of their verbal designations. This problem was dealt with by I. N. Gorelov and K. F. Sedov, D. Slobin and J. Green [17].

As you know, one of the main ways to «adequately fix fragments of mental space» is the method of associative experiment [18]. According To K. V. Bityutskikh, «free associative experiment is one of the most common and valid ways to access images of language consciousness of carriers of a certain culture [19]». The subject is offered a word-stimulus (in our case, its role is played by a term), to which he needs to respond with the first words or phrases that come to mind. The termination of associative reactions is determined by the percentage of terms among these associates.

The purpose of our experiment is to identify the level of determinism of professional thinking of students of the 1st year of training 38.03.05 Agroengineering (profiles «Electrical equipment and electrical technologies» and «Technical service in agriculture». Experimental work was carried out on the basis of the Nizhny Novgorod state University of engineering and Economics (Knyaginino). The experiment involved 22 people.

The analysis of the current work program on the discipline «Speech culture and business communication» was carried out. The working program defined the lexical field of the experiment. Linguistic and agricultural terms were chosen as incentives. The study of this vocabulary is currently in demand among technical students who study agriculture and the culture of business communication. Determining the degree of determinism of vocabulary will help in creating a dictionary that will interpret terms, which will be a significant help in the study of disciplines and further application of knowledge in practice – in the future profession.

3 Results

We decided to organize some kind of test on students' associations. We present the results of introductory and control experiments. Its aim is to find terms among associations given to the words of linguistic and professional sphere and to compare them. It is interesting to learn where there will be more terms. It will show what speech and competencies are more developed.

Students were asked to write associations for the following words of linguistic orientation: language, speech, morphology, phonetics, speech style, norm; as well as from the professional sphere: machine, tool, soil, sowing, planting, fertilizer. They were given only 15 min. The selected stimulus terms are of key importance in the conceptual system of the subject area (see Table 1).

Table 1. Results of the introductory experiment: students' reactions on linguistic terms-incentives

№ s/p	Words - stimuli	Words-Reactions of students
1.	Language	Russian language, mouth (3), word (2), gums, conversation, taste, teeth, difficult, gums, conversation, taste, teeth dialect, speech flutter, perception, country, organ, state, Russian, mouth, country, English
2.	Speech	Word (2), communication, understanding, sentences, Mat, ligament, logic, visualization, thoughts, conversation, voice, sound, communication, letters, correct, curved, unrelated (error: incoherent), good
3.	Morphology	Science, linguistics, section, language, composition, science, linguistics structure, infinitive, spelling, word formation, sentence, rules, section of chromatics (error: grammars), word, morpheme, exam, verb
4.	Phonetics	Section of linguistics, structure, intonation, sound, letter, rules, the Russian language, pronunciation, grammar, word, sound, accent, exam, sound combinations (error: sound combinations), sound, vowel, soft hard
5.	Speech style	Literacy, communication, speech, text, beautiful dialect, taste, fashion, life, oral speech, written speech, business speech, speak, expression of thought; speech style, not all speech styles are called, artistic, the Russian language, publication, to read
6.	Norm	Measure, standard, rules, border, limit, term, pattern, prescription, morality, whole, order, law, stop, measure, rule, word, teach

Unfortunately, one student misunderstood the instruction and provided 4-5 words-reactions to 6 linguistic and 6 professional terms at once, although this point was clarified at the request of the second. In this regard, there was no reaction to the word «norm».

The word «language» caused the greatest number of reactions, although, para-doxically, it turned out to be the least terminated. And the greatest number of reactions-terms caused the word «phonetics» (Tables 2 and 3).

It is noteworthy that often the reactions were words that express the attitude to the term and their personal associations. For example, the word «language» in one student caused the association «difficult», the term «morphology» - «exam», «phonetics» - «exam», «norm» - «enough». A large number of reactions to this word turned out not to have an abstract meaning, as, for example, the word-reaction «mouth».

The word «language» caused the greatest number of reactions, although, para-doxically, it turned out to be the least terminated. And the greatest number of reactions-terms caused the word «phonetics».

Table 2. Determinism of students' associative reactions according to the introductory experiment with linguistic terms

Terms-Incentives	Language	Speech	Morphology	Phonetics	Speech style	Norm
Total associates	19	18	16	16	17	15
Total associates-terms	3	5	11	12	7	3
Percentage of associated terms	15,8%	27,8%	68,8%	75%	41,2%	20%

Reactions to the term «phonetics» were the most terminated.

Table 3. Results of the introductory experiment: students' reactions on linguistic terms-incentives

№ s/p	Words - Stimuli	Words-Reactions of students
1.	Car	Engine, wheel, petrol, money (2), speed (2), transport, wheelbarrow, robot, riding, Volga, drift, racing; Peugeot, Lexus, expensive
2.	Tool	Shovel, labor, tools, axe 2, hand, war, murder, machine gun, hoe, stick, rake, labor, dig, stick
3.	Soil	Soil, fertilizer, weeds, humus, plant, earth, grass, fertilizer, mud, earth, soil, mole, peat, vegetable garden, vegetables, potatoes, earth, earth, dirt, growing, black
4.	Sowing	Modern technologies, seed, seeds, fertilizer, machinery, field (2), planting, trees, mushrooms, cones, corn, crop, seed, flowers, plants, vegetables
5.	Landing	Plane, takeoff, arrival, airport, plant, may, plane, land, labor, tractor, field, agriculture, crop production, cottage, garden, lilies, roses, watering, sowing (error)
6.	Fertilizer	Fertilizer care, feeding, growth, protection, insects, store, soil, manure, land, vegetable garden (2), humus, black soil, bed, soil, ash, seeds, water, care

Some students were associated with the constituent components of the subjects, while others focused on the genus-species relationships. So, for the word «car», some students selected such associates as «engine», «wheel» (component parts), while others –«Peugeot», «Lexus». Someone had completely personal associations, such as «expensive», «money». One student's soil was associated with «mud», while another's planting was associated with a dacha. Unfortunately, when selecting associations, one of the students used profanity. So, to the term «car» he picked up the word-associate «car».

The most common Association for the term «tool» was the word «axe». There were students who substituted the concept of «tool» for «weapon». In this regard, they were the first to come to mind such words as: «war», «murder», «gun» , although they are studying at an agricultural University. There were also those who did not give an associate for this term and «seeding». For one of the students, the term «planting» was associated with the word «sowing», although the terms differ in subject matter.

We present the termination of associative reactions (see Table 4).

Table 4. Determinism of students' associative reactions according to the introductory experiment with linguistic terms

Terms-Incentives	Car	Tool	Soil	Seeding	Planting	Fertilizer
Total associates	15	12	16	16	17	18
Total associates-terms	6	3	7	12	11	12
Percentage of associated terms	40%	25%	43,8%	75%	64,7%	66,7%

Reactions to the term «seeding» were the most terminated.

If we talk about the results of the input experiment, it should be noted that there was a low level of possession of the ability to interpret. In this regard, students were offered tasks for the formation of the definitive part, namely the selection of definitions. For example, they were asked to replace terms from the subject area with synonyms (plant, crop, manure, seeds, cultivation). Based on these words, students had to create creative texts, and then try to visualize these concepts: draw a diagram, prepare a collage or picture of their choice. And as a final task, they had to keep a dictionary throughout the discipline «speech culture and business communication».

The results were not long in coming in the end of that semester. The students were given the same linguistic and professional terms to which they should find the associative reactions in 15 min (see Tables 5, 6, 7 and 8).

Table 5. Results of the control experiment: students' reactions on linguistic terms-incentives

№ s/p	Words - Stimuli	Words-Reactions of students
1.	Language	Letters, word (3), Cyril and Methodius (2 errors), sentences, human languages, sign languages, animal languages, computer languages, speech (4), organ, ability, gift, homeland, Russian (4), dictionary, science, Russia, literacy, human intermediary between meaning and word (2 errors: between, thought), mouth (2), teeth, long, foreign
2.	Speech	Form, communication, people, rules, word (2), language (2), conversation (3), voice (2), dialogue (2), culture, purity, nationality, speak, beautiful, direct, result of language work, correct, beautiful, literate, suitable, monologue, language, form of communication, intonation, rhythm
3.	Morphology	Section, grammar (2), signs, significant, root, ending, composition, words (2), speech, parts, language, letters, composition, science (4), part of speech (2), linguistics, teaching, syllable, parsing, grammatical meaning
4.	Phonetics	Section, study, sound (7), structure, word (2), accent, letters (2), grammar, pronunciation, science, syllables (2), language, linguistics section, structure, intonation, phone, microphone, tape recorder, accents, sound composition
5.	Speech style	History (2), system, communication (2), literature, colloquial, business, jargon, rough, language, manner, Mat, pace, culture, scientific, journalistic, colloquial, speech tools, speech, text, information (2), narrative, colloquial style, oratorical style, business style
6.	Norm	Morality (2), morality, prescription (2), term (2), measure, rule (2), standard (2), good, deviation, template, empty, relic, sample

In the control experiment, the students showed the best results, although errors were observed, such as: «Cyril and Mythodius» (with 2 errors); «human mediator between meaning and word» (with 2 errors). There were also personal associations, for example, «good» to the term «norm».

The most popular responses to the term «language» were the words «Russian» and «speech» (in 4 people each of the reactions), in second place «word» (3 students gave such reactions), followed by the word «mouth» (2 reactions).

The results of the control experiment are presented (see Table 6).

Table 6. Termination of associative reactions of students according to the control experiment with linguistic terms

Terms-Incentives	Language	Speech	Morphology	Phonetics	Speech style	Norm
Total associates	23	27	20	21	26	13
Total associates-terms	14	11	9	10	21	5
Percentage of associated terms	60,9%	40,7%	45%	47,6%	80,8%	38,5%

Reactions to the term «style of speech» were the most terminated.

Table 7. Results of the control experiment: students' reactions on professional terms-incentives

№ s/p	Words - Stimuli	Words-reactions of students
1.	Car	Machine gears, man, labor, mechanism (2), auto, movement, robot, motor, tractor, car, Volga, reaction, machinery (2), carburetor, work, engine, speed (4), transmission, riding (2), transport, make, type, steering wheel, wheels, gasoline
2.	Tool	Labor (2), shovel (4), knife, weapon, aid, device, land (2), hands, work, axe, tools, processing, rake, seeds, plow, cultivator
3.	Soil	Earth (7), looseness, turf, field (2), plant (3), soil (3), seedlings, bed, roots, moisture, fertilizer (2), nature, land, grass, insects, furrow, black soil
4.	Sowing	Method, cultivation, microorganisms, substances, field (2), crop, soil, garden (2), labor, hands, hoe, watering can, arable land, season, combine, seeds (3), grain (3), ear, flour, plants
5.	Landing	Tolerance, tightness, gap, difference, garden, garden, flight, plane, vegetable (3), fruit, fertilizer (2) seeds (2), beds, watering (2), land (3), seedlings, shovel, combine, field, trees, machine
6.	Fertilizer	Substance (2), nutrition (2), fertility (2), soil, land (2), manure (2), plants (2), garden, Supplement, bed, help, care, result, top dressing, growth, processing, chemistry, crop, humus

The most popular reaction to the term «car» was the word «speed». In connection with the awareness of students about the modern development of technology, the word «robot» appeared among the associates.

As a result of the experiment, the most popular Association for the term «tool» is the word «shovel», and for the term «soil» - «land», for «sowing» - «grain», for «planting» 2 words at once: «vegetable» and «land», and in relation to fertilizer, opinions were divided: «substance», «nutrition», «fertility», «land», «manure», «plants». There are students who understand the relationship between the components of the same field of knowledge. So, the term «planting» was associated with another term from the experiment – «fertilizer». Unfortunately, the term «landing» was chosen by one of the students for associations that are not related to agriculture (flight, plane).

The results of the experiment are presented (see Table 8).

Table 8. Termination of associative reactions of students according to the control experiment with professional terms

Terms-Incentives	Car	Tool	Soil	Seeding	Planting	Fertilizer
Total associates	26	16	17	20	22	19
Total associates-terms	16	6	10	10	12	11
Percentage of associated terms	61,5%	37,5%	58,8%	50%	54,5%	57,9%

Reactions to the term «car» were the most terminated.

4   Conclusion

The purpose of this study, i.e. to interpret the concept of «terminological unit», to determine the essence of the work of first-year students with terminological units, was gained. We didn't only describe the notion, but also analyzed it in practice during the experience.

In general, the percentage of termination of associative reactions has increased. Students learned to select more words as responses to the proposed terms. They began to remember the material passed in the classroom. In a word, the experiment was a success, but we observed the low level of the skill to interpret the terms.

If we compare the percentage of terms among the associations of input and control experiments in linguistic and professional sphere, we can conclude that during the first test it was 41,4% in the linguistic and 52,5% in the professional one. As for resulting test, in the linguistic one it is 52,3% and in the professional – 53,4%, almost the same. It shows that the speech in the both spheres are developed identically. Of course, the results may be different in some other group and course, but they suit to our purpose of the research.

To improve the level of usage of terminological units we suggest students to match definitions and terms, to use synonyms to terms, to make creative texts on the basis of 5 terms, to visualize the terms by means of schemes, pictures and collages, analyze in the linguistic way the text from the student's books, to observe the reactions on the terms of the students during the study. The students must understand the ties between the terminological units, they should work with vocabulary and make their ones in their notebooks.

We would like to continue working on this theme and to find new methodology for the result to be higher. To teach students to use clusters of terms for better understanding and memorizing them. We want to describe the results of this experiment in the future research.

References

1. Lopatin, V.V., Lopatina, L.E.: Russian explanatory dictionary. https://dic.academic.ru/dic.nsf/ruwiki/16180. Accessed 22 Feb 2020
2. Slozhenikina, V.: The Term: live as life (why the term can and should have options). http://www.zpu-journal.ru/e-zpu/2010/5/Slozhenikina. Accessed 22 Feb 2020
3. Vinokur, G.O.: Some phenomena of word formation in Russian technical terminology. In: Proceedings of Moscow Institute of Philosophy, Literature and History (MIFLI), pp. 3–54. MIFLI, M. (1939)
4. Zolotukhin, D.S.: Basic approaches to the description of linguistic terms. https://cyberleninka.ru/article/n/osnovnye-podhody-k-opisaniyu-lingvisticheskih-terminosistem. Accessed 21 Feb 2020
5. Lopatina, E.V.: Features of Terminological Units. https://cyberleninka.ru/article/n/osobennosti-terminologicheskih-edinits. Accessed 22 Feb 2020
6. Komarova, Z.I.: Semantic Structure of the Special Word and its Lexicographic Description. Ural University Publishing House, Ekaterinburg (1991)
7. Clevenger, M.: Where «tip» is tops. In: American Education, pp. 16–19. The Office, Columbus (1974)
8. Dyer, M.: Competency-based teacher education. In: American Education, pp. 38–39. The Office, Columbus (1974)
9. Kazarina, S. G.: Typological terminology as a differentiated linguistic discipline. https://cyberleninka.ru/article/n/tipologicheskoe-terminovedenie-kak-differentsirovannaya-lingvisticheskaya-distsiplina. Accessed 23 Feb 2020
10. Golovanova, E.I.: Cognitive terminology: problems, tools, directions and prospects of development. https://cyberleninka.ru/article/n/kognitivnoe-terminovedenie-problematika-instrumentariy-napravleniya-i-perspektivy-razvitiya. Accessed 23 Feb 2020
11. Bagiyan, A.Y., Ezdekova, L.B.: Transformational potential of terminological units in non-scientific discursive implementation: linguocognitive analysis of determology of elements of term system of basic and critical technologies. Sci. Dialogue 7, 9–29 (2018)
12. Vinogradov, V.V.: Terminology and a norm. Higher School, M. (1972)
13. Shleivis, P.I.: Linguistically relevant characteristics of terminological units. Questions Theor. Appl. Linguist. 4(2), 21–27 (2016)
14. Federal State Educational Standard of Higher Education – Bachelor's Degree in the Direction of Study 35.03.06 Agro-Engineering. https://www.timacad.ru/sveden/files/350306_2017.pdf. Accessed 23 Feb 2020
15. Belousova, O.A., Polyakov, V.M., Mikhailiukov, L.V.: Methods diagnosing terminological culture of students studying in agricultural institutions. Mod. pedagog. Educ. 4, 109–112 (2019)

16. Shmatova, E.S.: Legal terminology: problems of interpretation and desire for unification. http://confer.hses-online.ru/pdf/1-12.pdf. Accessed 23 Feb 2020
17. Petrova, E.A.: Terminality as a basis for developing skills of professionally oriented communication. In: Actual Problems of Linguistics and Methods of Teaching Foreign Languages 2014, OLA, Omsk, pp. 46–51 (2014)
18. Rusova, N.Y.: Text. Culture. Education: Scientific and Methodological Guide. NSPU Publishing House, Nizhny Novgorod (2009)
19. Bitiutskikh, K.V.: Reflection of the Corporate Culture of the Military in the Linguistic Consciousness of the Cadets. Publishing House of Chelyabinsk State University, Chelyabinsk (2015)

Accounting Student Training Trends at Russian Universities in Digital Age

Olga Efimova$^{(\boxtimes)}$ ⓘ and Olga Rozhnova$^{(\boxtimes)}$ ⓘ

Department of Accounting, Analysis and Audit,
Financial University under the Government of the Russian Federation,
Moscow 125993, Russia
{Oefimova, ORozhnova}@fa.ru

Abstract. This paper reports the data collected from interviews with over 300 participants of the accounting educational process and presents the evidence that students, university lecturers, employers are waiting for changes in educational process and its adaptation to digitalization. Using the survey data, we provide insights into why and how accounting students of different educational levels as well as university professors and employers consider contemporary challenges connected with digital transformation. The research found out that the vision of the subject from the various interested parties is different. Summarizing the survey results, we suggest the approach to adapt and prepare accounting students to the requirements of digital economy.

Keywords: The accounting educational process · Digital technologies in accounting · Accounting students' expectations · Educators and employers

1 Introduction

The issues of the changing roles of an accounting profession have been emphasized by plenty of researches in many countries. The digital transformation of economy has prioritized the issues for adaptation of an accounting profession to working under new circumstances.

The cause-and-effect relationships exist between the benefits and threats of the digital economy and specific issues in accounting profession and education A. Lawson et al [13]. New technologies are changing the character of the traditional accounting activities and create the mismatch between accountant competencies and the requirements of digitization as Pincus [15], Harper and Dunn [8] describe.

Recent corporate technological and accounting scandals have proved that such skills as ethics and responsibility are getting more and more important for accounting profession. Ali [1], Volkova [20] argue that with the help of digital technologies it becomes possible to more accurately and systematically assess uncertainties and risks and integrate them into the overall system of the company strategic management. Numerous research evidence that at present mainstream investors are increasingly interested in useful data and information concerned with ESG aspects Amel-Zadeh and Serafeim [2], Barker and Eccles [4]. To fulfill these expectations professional accountant needs to collect, integrate and disclosure both financial and non-financial

© Springer Nature Switzerland AG 2021
T. Antipova (Ed.): ICIS 2020, LNNS 136, pp. 263–274, 2021.
https://doi.org/10.1007/978-3-030-49264-9_24

data taking into account their relevance for stakeholders Efimova and Rozhnova [6], Seele [16], Wiek [21].

Numerous researches and essays on employment in a digitized economy provide evidence that changes in the demand for the professionals determine the new requirements for their preparation Dolphin et al. [5]; Apostolou et al [3]; Hoffman, 2017 [9], Lawson et al [13]; Wyness and Dalton [22]. Moreover, the results of vast studies provided by Pincus et al. [15], Kamordzhanova and Solonenko [10] indicate that development of professional knowledge alone is not sufficient to fill the gap between emerging challenges and requirements and established practice of accounting preparation. To investigate and asses these new requirements it is necessary to get the full vision of the preferences, expectations and needs of all accounting educational process participants.

2 Literature Review

As far as businesses are being transformed by the impact of digital technologies the role of accountancy and finance professionals is getting to change too. New ways of doing business, shaped by technology and shifting regulatory environments, mean that accountants in business and practice are facing challenges and exciting opportunities.

Seele [16], Hoffman [9] suppose that the new technologies will not only change the way the financial information providing and aggregating but will create an entirely new business environment, allowing greater flexibility to professional accountants.

According to Kuznetsov, Kuter, Gurskaya [12], Lawson [13], Wiek [21] digital technologies lead to creation of new professional hybrids when accountants and finance professionals develop from a service function to a business-critical service, central to strategic decision making. Kuter et al. [11], Tanaka and Muyako [19]; Gomaa [7] argue that professional accountants need to be aware of the whole process of business value creation and to identify the possible actions and their consequences. Harper and Dunn [8]; Stoner [18] suppose that accounting graduates must be able to evaluate the reliability and relevance of preparing and collecting the information from different sources as well as using data analysis as a tool for developing meaningful, action-oriented business decisions. In continuation of the above Lawson et al. [13] suggest that accounting and finance graduates need the skills for providing enhanced reporting of risk exposures, for reporting information to support decisions on deploying capital to grow the business profitably, for supporting the long-term value creation in their enterprises, and for communicating the ways in which accounting can promote the successfull business strategy.

Tanaka and Muyako [19] noticed that colleges and universities now are starting to explore the potential value that digital technologies could deliver to their organizations. Universities actively use software to solve organizational and administrative issues, to implement automated training, general-purpose programs, specialized information retrieval systems. McKnight [14], Shaaronin [17] emphasize that the technologies are being used in the traditional learning process in order to improve the quality of education and involve a huge number of learners that are not limited to physical space through online training.

In the same time Lawson [13] had outlined the lag in the content and quality of accounting students' preparation from the real needs. Pincus [15] argued that technology advances have not significantly changed either what we teach (curriculum) or how we teach (pedagogy). Automation rapidly increases the skills gap for jobs in accounting and finance. McKnight [14] noticed that successful digital conversion for classrooms, districts, and states is not determined by the technology, but by how technology enables teaching and learning.

Numerous studies have discussed accounting graduates' employability skills (Lawson et al [13], Stoner [18], Wiek et al [21], Shaaronin [17]). They have investigated IT skills and knowledge that were most relevant to accounting students in providing competent and professional services. According to the research results Information Technology skills are increasingly becoming important for accounting education.

Harper and Dunn [8] based on International Accounting Accreditation Standard 'Information Technology Skills and Knowledge for Accounting Graduates' and PwC white paper—Data Driven: What Students Need to Succeed in a Rapidly Changing Business World‖ emphasize accounting graduates' competencies in data analytics, data management and other business information technologies.

As the studies (Stoner [18], McKnight et al [14], L.Wyness and F. Dalton [22]) show all these issues create the need to strengthen IT knowledge as well as the formation of integrated thinking, allowing accounting graduates to understand and identify key value drivers and associated risks and opportunities to avoid the prominent problem of information asymmetry.

3   Materials and Methods

In order to conduct this study, the opinions of following stakeholders of the accounting educational process were investigated: university students (all levels of education), university professors and employers as well as their attitude to the new requirements for accounting profession connected with digital transformation of economy. To enhance our understanding the problem we have developed questionnaires for various groups of university students. Bachelors, masters, postgraduates, university professors and employers were involved in the survey. Students of 5 Moscow universities, employers and professors from 11 universities in Moscow, Saint Petersburg, Novosibirsk, Chelyabinsk, Nizhny Novgorod, Voronezh and Kazan took part in the survey. The questioning was conducted at scientific and practical conferences on modern education, digitalization, and accounting in 2017–2018. The results of the study were processed at the end of 2018. Since then, there have been changes in curricula and processes aimed at enhancing digitalization. To date, certain positive changes can be noted in relation to digitalization among accounting students as well as lecturers and employers.

For analysis students of accounting and financial undergraduates as well as postgraduate programs were selected. In total, 348 questionnaires were processed and analyzed. Further, we analyzed opinions of the Russian university professors and lectures teaching accounting and analytical disciplines (46 questionnaires processed).

In addition, we analyzed the vacancies and requirements of employers presented on the website of the Institute of Professional Accountants of Russia (IPAR) in order to assess the degree of compliance of students' expectations for future professional activities and the needs of employers[1].

For research purposes we used the results of discussions at scientific conferences (both for professionals and students). Finally, we analyzed the results of surveys provided by leading consulting companies (Big Four), as well as international professional organizations (IFAC, ACCA, ICAEW, PwC).

As the study showed the most students receive information about practical application of digital technologies in professional sphere from the Internet. Figure 1 allows to see a full picture of the estimates obtained.

The survey results show a clearly insufficient role of universities in preparing accounting students to work with digital technologies in professional sphere. It seems obvious that universities and colleges should be aware of this issue and urgently make appropriate adjustments to their curricula and the accounting disciplines teaching approaches.

As the study demonstrates (Fig. 2), accounting students of all levels (undergraduate, graduate and postgraduate) in their overwhelming part do not feel familiar with digital technologies in professional sphere—as most of them assess their level of knowledge as medium and low.

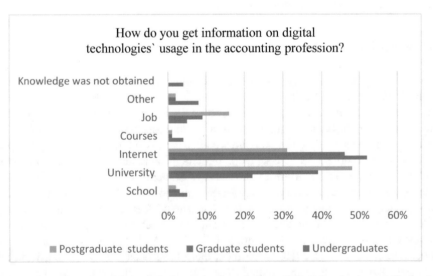

Fig. 1. Main recourses and channels for getting knowledge and digital technology skills for the accounting profession.

---

[1] https://www.ipbmr.org. Mode of access: http://www.ipbmr.ru/.

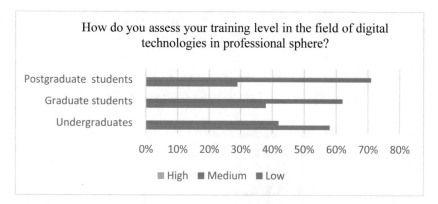

Fig. 2. Assessment by students of their knowledge in the field of digital technologies.

At the same time the majority of students (60%) believe that they need skills of information technologies for a successful career, and 40% suppose that such knowledge would not be superfluous. No one doubts that professional accountants need such skills. However, most of the students see the demand for this knowledge as distant perspective. According to the Fig. 3 more than 50% of students suppose that such skills will be in demand in 2 to 3 years or even later.

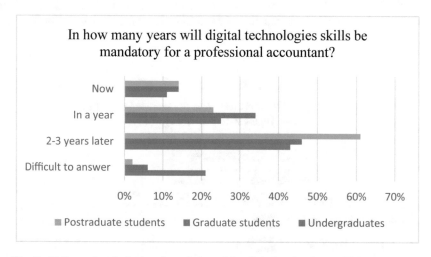

Fig. 3. Estimated period when knowledge of the digital technology will be in demand.

It is important to note that students were asked to choose one of the possible areas. This explains the relatively low percentage of digital technology importance for corporate and financial reporting in the opinion of interviewed students. The possibility of choosing several professional areas would significantly increase their share of the priority ones.

In the professional area students mainly see the need in using digital technologies in financial analysis, audit and financial consulting (Fig. 4).

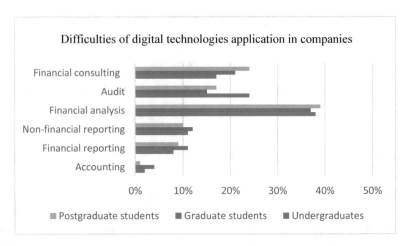

Fig. 4. Professional activity areas in which knowledge of technology is particularly vital.

Students were asked to assess the main barriers to widespread use of digital technologies in the accounting practice of the companies which they work for. Figure 5 allows to see the results of the analysis.

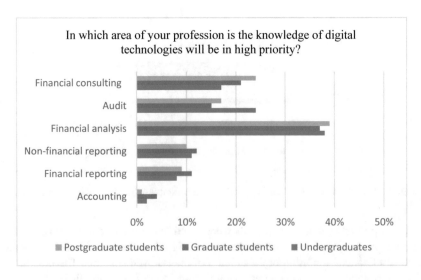

Fig. 5. The main barriers of digitalization.

According to Fig. 5 as the main reasons for the lack of active digitalization in practice, students pointed out three interrelated causes including lack of technological support, lack of specialists and financing.

In order to improve their digital skills, the students offer:

- studying the discipline "Digital technologies in accounting" (as well as in analysis, audit, finance);
- using the technologies in the preparation of practical tasks in various disciplines;
- increasing training hours for information technology related disciplines;
- developing analytical projects in which it is necessary to use structured and unstructured data; cases that expand the horizons of students;
- using digital technology in the period of training practice that is mandatory for students of Russian universities.

At the next stage of the study we analyzed the attitude of university professors and lecturers to the need for digital transformation of accounting and analytical disciplines. Various studies have outlined that the views of academics on the subject-matter taught in universities differ from views of practitioners I.M. Ali et al. [17]. In this case the main aspects of the study included the following points: 1) the need to integrate digital technologies and the disciplines taught; 2) whether teachers themselves use such technologies, and if not, are they planning to do it; 3) in which areas of their professional activities university professors and lecturers use digital technology.

The analysis made it possible to draw the following conclusions. The absolute majority of professors and lecturers do not support direct integration of digital technologies and the disciplines taught (over 70%), believing that the main focus of the teaching process should be concentrated on the conceptual foundations and methodological aspects of accounting disciplines and their practical application. More than 60% of university lecturers responded that they do not possess digital technologies at a high level, but they are planning or are already improving their skills in this field. Interviewed lecturers (who operate digital technologies) noted that they use these technologies primarily for scientific and research purposes, as well as for preparing and developing guidelines.

The final stage of the survey was to study the requirements of employers for the professional skills and abilities that university graduates should possess. According to provided analysis of the statistical data and available vacancies of recruitment agencies, as well as focus group surveys, the following was investigated: the employer's vision on the competencies that the employee should have; the employer's attitude to training at different levels of the professional education system; the student's views on the requirements of the employers and these requirements themselves. As we could conclude the ongoing changes in the sphere of up-to-date business are accompanied by changes in the field of human resource management. They include the transformation from narrow specialization and limited liability to broad professional responsibility; from a planned career to a flexible choice of professional development path; from the manager responsibility for the personnel development to the employees` responsibility for their own career.

4 Discussion and Results

Although data collected through survey can suffer from several issues (response bias, selection bias, attribution bias), the survey offers a way to collect information and provide insights in questions that cannot be answered at that point by published data.

Our key findings fall in the following main categories.

First, we found out the main sources used by students of accounting professions to obtain information about digital technologies and their practical application for professional purposes. The majority of the students indicated Internet resources. This fact seems paradoxical because at the same time, the curriculum for university students at all levels (bachelors, masters, and postgraduate students) contains disciplines related to information technologies. However, this can be explained by the fact that IT disciplines are given without reliable linkage with the professional subject areas and teaching approach does not adequately take into account the specifics and needs of their use in accounting, analytical and financial areas.

The second set of results relates to how the students themselves assessed their level of digital technology. Most of the students acknowledged that their level of technological training is not high. It seems understandable that the percentage of the students that consider their level as insufficient varies depending on their level of education, decreasing from 58% for bachelors to 28% for postgraduates. This assessment suggests that accounting students are not confident in their digital skills, which can create difficulties for them in building a career and offering their services in the labor market.

The third set of results relates to when, according to interviewed students, the knowledge and skills of digital technologies will be relevant in their professional field. A relatively small percentage of accounting students believe that they need skills in digital technologies at present to perform their current professional duties. At the same time, most students believe that they will need such skills in 2–3 years.

The fourth set of results relates with the clarification of professional areas for which the application of digital technology is of primary importance. Most of the students surveyed indicated financial analysis and consulting. On the one hand, this seems to be reasonable, but, on the other hand, it makes one think about the need to draw students' attention to the importance of using digital technology for the purposes of professional judgments.

The fifth set of results focuses on the main barriers of digital technology application in practice. As such barriers, the students noted above all, the lack of technical support and the lack of appropriate specialists to enable the digital transformation of accounting, analysis and audit tasks. It also makes the necessary changes to the existing system of training accounting specialists.

The sixth set of results relates to studying the students' vision of how to solve the existing difficulties of digital transformation in the accounting profession. The students proposals include the improvement of the educational process by increasing the number of hours for the disciplines related to digital technologies, involving practitioners, and the development of case studies and research projects for students allowing them to practice their skills in using digital technologies under the supervision of professionals.

The seventh set of results relates to studying the university professors' vision of the subject. The survey showed that many issues stay debatable. Many professors surveyed believe that students should receive knowledge and skills of digital technologies by studying relevant disciplines, and then apply this knowledge in practice. At the same time the majority of professors consider the integration of accounting and analytical disciplines with the necessary knowledge and skills of digitalization as important option for graduates. Not all lecturers are ready for this transformation due to many factors, including lack motivation, poor organization of the educational process, conservatism. The survey showed that for many lecturers the future of the accounting profession remains controversial, but the majority of respondents positively assess the imFinally, we consider employers requirements for university graduates. Generally, the employers wish younger applicants pass through the university that gives them relevant knowledge, digital literacy, communication skills, competence in resolving conflict situations, imparts certain ethical principles and other soft skills. Special knowledge that business also needs may be given by the employers themselves or under their own control. Companies are now looking for certain competencies (among which the digital skills are mandatory) and talents. It is more efficient and faster for them to form a position for a talented graduate in a company, therefore recruiters constantly analyze the market and look for talentspact of digitalization on the profession.

5   Recommendations

Based on provided analysis we suppose that the basic disciplines related to digital technologies should be studied at undergraduate degree programs (such disciplines include XBRL technologies, blockchain, basic skills in working with Big Data). Besides, in our opinion, universities and colleges should provide more on-line programs and courses on digital technologies and their practical implementation. Graduates are needed to be advanced-level users of digital technologies.

Masters are supposed not only to be advanced users of digital technologies but also be able to find non-standard applications of these technologies to non-standard situations, as well as to search challenging issues related to the use of digital technologies. Master programs should be oriented on the multifunctionality of the specialists trained, their ability to demonstrate creative thinking in different situations using the set of knowledge and skills across a range of disciplines. MAs need to have formed integrated awareness of business ecosystem and the ways of making managerial decisions. Profound training, virtual reality simulators allow modeling both economic and non-economic processes in the virtual environment. As a result of such activity students get opportunity to model their actions in professional sphere.

Postgraduates and PhD students should be taught to find the ways of accounting methodology development, its concepts according to existing and anticipated digital opportunities as well as to define necessities in new digital technologies. The main outcome of postgraduate education should be the formation of integrated thinking, which will allow postgraduates to consider the strategic goals of the organization and apply information and analytical support to the whole system. So the main result of postgraduate programs should include the forming of integrated thinking that will allow graduates to consider the strategic organization goals and the information and analytical support of these goals as the whole system.

Achieving consistency between work functions, accounting competencies and digitalization requires continuous training for university teachers. The university's role in this regard is to organize the process of improving the qualifications of academics, taking into account the requirements of employers and the development of modern digital technologies. University should have a tutor on applied digital education of accountants. It is necessary to constantly determine the link between content of the training programs and the requirements of employers (current and expected). The accounting and analytical programs should provide practical training of using digital technologies for each discipline (with the involvement of relevant professionals, if necessary).

It seems extremely useful to incorporate in educational process such kind of activity as teem-working projects based on the principle of collective learning via on-line platforms. Students can work in mini-teams on the accounting, analytical and financial tasks provided by the on-line platform, alternately changing the roles of an educator and a student. Thus, by sharing knowledge, students improve their professional competencies, increase their rating for the correctness of performing tasks and observing deadlines. While being trained students obtain the necessary level of digital knowledge in accounting as well as teem working skills.

Finally, as being underlined, digital technologies allow to form open on-line platforms where academics, students, employers, professional communities and other interested parties can share experience and learning resources, forming educational hub of knowledge and receiving timely response from employers.

6 Conclusion

Our study demonstrates that there is number of challenges for accounting profession and, accordingly, accounting education caused by digitalization. The study investigates the key stakeholders of the accounting educational process: university students (of all levels of education), academics and employers as well as their attitude to the new requirements for accounting profession connected with digital transformation of economy. As the result of conducted study, we can conclude that all analyzed participants of the accounting education process are aware of the changes inevitability caused by digitalization. The provided analysis of expectations and demands among interested parties allows us to propose the following ways of mitigating the most important, in our opinion, considered issues concerning accounting education in digital economy.

Training accountants, corresponding to the reliable level of their mobility in digital economy, requires students to achieve such competencies as digital technology knowledge at the user level within the accounting professional activities (blockchain, XBRL, big data technologies and others); the ability to formalize the accounting and analytical tasks for IT specialists; the skills to conduct accounting activities remotely, to organize remote work of subordinates; the ability to organize the process of self-learning and subordinates/team training, increasing their knowledge due to international professional experience and digital technologies.

The results obtained in the survey point out several directions for further research. Taking into account that the proposed recommendations were based on a study with some limitations (the number of respondents and questions asked were not very significant; investigated groups were not divided into various sub-groups (for example, working students; non-working students; students who have good knowledge and skills of digital technologies, etc.), the study was conducted for a limited period of time and it would be advisable to follow how the views of the respondents change over time. For further research, the authors intend to investigate the differences between groups using statistical coefficients (e.g. gender differences, status of students vs. lecturer, place of residence).the authors believe that the research direction will be supported and developed by different specialists in other scientific fields.

References

1. Ali, I.M., Kamarudin, K., Suriani, N.A., Nur Zulaikha Saad, N.Z., Afandi, Z.A.M.: Perception of employers and educators in accounting education. J. Proc. Econ. Finan. **35**, 54–63 (2016)
2. Amel-Zadeh, A., Serafeim, G.: Why and How Investors Use ESG Information: Evidence from a Global Survey. Harvard Business School (2017). http://nrs.harvard.edu/urn-3:HUL. InstRepos:dash.current.terms-of-use
3. Apostolou, B., Dorminey, J.W., Hassell, J.M., Rebele, J.E.: Analysis of trends in the accounting education literature (1997–2016). J. Account. Educ. **41**, 1–14 (2017)
4. Barker, R., Eccles, R.G.: Should FASB and IASB be responsible for setting standards for nonfinancial information? Green Paper. University of Oxford, Sand Business School (2018)
5. Dolphin, T. (ed.): Technology, globalisation and the future of work in Europe: essays on employment in a digitized economy, IPPR (2015). http://www.ippr.org/publications/technology-globalisation-and-the-futureof-work-in-euro
6. Efimova, O., Rozhnova, O.: The corporate reporting development in the digital economy. In: Antipova, T., Rocha, A. (eds.) Digital Science. DSIC18 2018. Advances in Intelligent Systems and Computing, vol. 850, pp. 150–156. Springer, Cham (2019)
7. Gomaa, M.I., Markelevich, A., Shaw, L.: Introducing XBRL through a financial statement analysis project. J. Account. Educ. **29**, 153–173 (2011)
8. Harper, C., Dunn, C.: Building better accounting curricula (2018). https://sfmagazine.com/post-entry/august-2018-building-better-accounting-curricula/
9. Hoffman, C.: Accounting and Auditing in the Digital Age (2017). http://xbrlsite.azurewebsites.net/2017/Library/AccountingAndAuditingInTheDigitalAge.pdf
10. Kamordzhanova, N.A., Solonenko, A.A.: Trends in the development of the accounting profession in the world of unstable economy. Audit J. **1–2**, 120–133 (2017)
11. Kuter, M., Gurskaya, M., Andreenkova, A., Bagdasaryan, R.: The early practices of financial statements formation in medieval Italy. Account. Historians J. **44–2**, 17–25 (2017)
12. Kuznetsov, A.V., Kuter, M., Gurskaya, M.: The formation of accounting education in Russia. In: 5th International Conference on Accounting, Auditing, and Taxation (ICAAT 2016). https://doi.org/10.2991/icaat-16.2016.27
13. Lawson, A., Blocher, E.J., Brewer, P.C., Cokins, G., Sorensen, J.E., Stout, D.E., Wouters, M.J.: Focusing accounting curricula on students' long-run careers: recommendations for an integrated competency-based framework for accounting education. Issues Account. Educ. **29** (2), 295–317 (2014)

14. McKnight, K., O'Malley, K., Ruzic, R., Horsley, M.K., Franey, J.J., Bassett, K.: Teaching in a digital age: how educators use technology to improve student learning. J. Res. Technol. Educ. **48**(3), 194–211 (2016)
15. Pincus, K.V., Stout, D.E., Sorensen, J.E., Stocks, K.D., Lawson, R.A.: Forces for change in higher education and implications for the accounting academy. J. Account. Educ. **40**, 1–18 (2017)
16. Seele, P.: Digitally Unified Reporting: how XBRL-based real-time transparency helps in combining integrated sustainability reporting and performance control. J. Cleaner Prod. **136**, 66–77 (2016)
17. Sharonin, Y.: Digital technologies in higher and professional education: from personally oriented smart-didactics to blockchain for targeted specialist training. J. Modern Prob. Sci. Educ. **1** (2019). http://science-education.ru/ru/article/view?id=28507. Accessed 10 Feb 2019
18. Stoner, G.: Accounting students' IT application skills over a 10-year period. Account. Educ. **18**(1), 7–31 (2009)
19. Tanaka, S., Muyako, S.: Information technology knowledge and skills accounting graduates need. Int. J. Bus. Soc. Sci. **6**(8) (2015). http://www.ijbssnet.com/journals/Vol_6_No_8_August_2015/5.pdf
20. Volkova, O.N.: About the future of accounting profession and academic discipline. Audit Statements. (5–6), 31–42 (2017). Available in Russian
21. Wiek, A., Withycombe, L., Redman, C.L.: Key competencies in sustainability: a reference framework for academic program development. Sustain. Sci. **6**, 203–218 (2011). https://doi.org/10.1007/s11625-011-0132-6
22. Wyness, L., Dalton, F.: The value of problem-based learning in learning for sustainability: undergraduate accounting student perspectives. J. Account. Educ. **45**, 1–19 (2018)

Digital Engineering

Using Metrics in the Throughput Analysis and Synthesis of Undirected Graphs

Victor A. Rusakov$^{(\boxtimes)}$ (ID)

National Research Nuclear University MEPhI
(Moscow Engineering Physics Institute), Moscow 115409, Russia
VARusakov@mephi.ru

Abstract. The usual representations of the communication environment are graphs with certain properties. Like reliability, throughput is one of the most important characteristics of such graphs. Metric tasks often arise when simplifying complex and practically important problems on graphs. A traditional metric, such as the usual shortest paths, forms the basis of the traditional throughput index. In this case, the metric is used to obtain the distribution of multi-colour flows in graphs more complex than trees. To achieve better results than when using ordinary shortest paths, one can use the Euclidian metric. If one starts with the Kleinrock formula for the average packet delay, then the Euclidian (quadratic) metric allows one to practically refuse multiple distributions over the shortest paths with variable edge *lengths* in the cut saturation procedures. The same Euclidian metric describes the distribution of the flow of any colour in an arbitrary graph as the best approximation to the ideal distribution in a complete graph in the sense of quadratic deviation. Such independence of the result from the Kleinrock formula demonstrates the effectiveness of linear metric models in the throughput analysis and synthesis of graphs. The Euclidian metric also allows you to introduce the throughput index of an arbitrary graph into these tasks in the form of an abstract measure. Therefore in such tasks one can completely disregard the distribution of the flows. Theoretical results are illustrated by an example of graph synthesis.

Keywords: Shortest paths · Minimal cuts · The Moore-Penrose pseudo inverse of the incidence matrix · The Euclidian metric on graphs · Distribution of multi-coloured flows · Packet switching computer network

1 Introduction

Graphs are widely used as a model of objects in the surrounding world. A Packet-switched computer network is one of the types of such objects. There is a clear correspondence between the communication nodes and graph vertices and also between the trunk communication channels and the graph edges in a model.

From now on a graph will be understood as a finite connected undirected graph without loops and multiple edges having k vertices and h edges.

Information flows enter the network nodes from the outside and load the channels during their movement through intermediate nodes to the destination nodes. The latter transmit flows outside the network.

© Springer Nature Switzerland AG 2021
T. Antipova (Ed.): ICIS 2020, LNNS 136, pp. 277–287, 2021.
https://doi.org/10.1007/978-3-030-49264-9_25

There are flows of different k colours in the graph each of which is determined by the destination vertex. This is because for any intermediate vertex, the flow should move further only depending on the destination vertex, but not on how it got to this intermediate vertex [1]. So each vertex t is a receiver of the flow of colour t and also it's a source of the flows having colours $1, ..., t - 1, t + 1, ..., k$ [2]. When simultaneously viewing two or more flows of *one and the same* colour, for some edges the flows can go in opposite directions. In this case the magnitudes of flows having opposite signs are subtracted from one another. But there is no subtraction for the flows having *different* colours in such a situation. These flows are considered to each simply belong to its direction of the edge [3].

A flow is a sequence of packets in the packet-switched computer network. In the real world it takes random amounts of time for any packet to go through the network from its origin node to the destination one. The simplest probabilistic models are often used to get the average amount of such time estimated for the whole network, not for the single packet or for some pair or subset of nodes. A number of supplementary assumptions and terms are incorporated into these models. Some of these additions are quite unrealistic, but in practice the simplest models give not bad results due to the hodge-podge of flows in the network. By this heuristic way the average packet delay τ during transmission over the network was obtained, $\tau = \gamma^{-1} \sum_i \lambda_i \tau_i$, where γ is the total intensity of the packet arrival in the network, λ_i is the same value for the i-th trunk communication channel, and τ_i is the average delay of the packet having length μ^{-1} on this channel with a capacity σ_i, $\tau_i = (\mu\sigma_i - \lambda_i)^{-1}$ [4].

Large amounts of accurate data are very rare [5, 6]. The situation is just the same for the network's draft stage. At this stage, they also tend to obtain throughput estimates under the condition that equal opportunities are provided to all pairs of nodes [7, 8]. That's why the matrix of the so-called uniform traffic is often used. Such a matrix has all equal entries except the zero ones on the main diagonal. These entries are time-average intensities of the input flows. Such entries are equal to 1 in a normalized traffic matrix.

The τ expression can be used as an objective function of the problem of finding the distribution of a multi-colour flow for a given traffic matrix and a set of other constraints. The distribution found should minimize the value of the average delay τ. A dual problem statement is also used. The distribution found should maximize the value of the $f = \gamma\mu^{-1}$ total flow through the network for a given set of constraints with $\tau \leq \tau_{max}$ among them. The traffic matrix may vary proportionately when searching for the distribution.

2 Flow Distribution and Graph Metrics

There is a huge number of graphs even for modest k and h. So the problem of finding the distribution of a multi-colour flow should be solved many times when synthesizing networks that have the best throughput index. But the combination of many factors practically makes pointless the search for even one *exact* solution to the problem of the distribution of a multi-colour flow. Among these factors are the heuristic nature and non-linearity of the τ expression, the low certainty of many of the original data, their possible non-stationarity, and the large calculations necessary to get even a sole *exact* solution.

As a result, additional assumptions and methods are used to improve the situation. Expanding τ_i into a power series and assuming that the flows in the network are small enough, we immediately get $\tau \approx \sum_i (\mu\sigma_i)^{-1} \lambda_i \gamma^{-1} = \sum_i (\mu\sigma_i)^{-1} f_i f^{-1}$, where $f_i = {}_i\mu^{-1}$, $f = \gamma\mu^{-1}$. In essence, this means the *linearization* of the τ expression of the objective function of the flow distribution task. Along with this, the network flow is *broken down* into a set of flows for the pairs s,t of the vertices $\forall s \neq t$. Further, suppose that only one pair of vertices, u,v, forwards the flow $f = f_{uv}$, and the quantity $(\mu\sigma_i)^{-1}$ is considered the *length* of the i-th edge – as in the problem of finding the usual shortest paths, $i = 1,\ldots,h$. Then, in order to minimize τ, the entire flow of f_{uv} must be directed along the u,v-shortest path. The same applies to all other pairs of vertices. The *re-composition* of the obtained edge flows, taking into account the flow's colour, gives the final distribution. Changing the traffic matrix proportionately, one maintains the inequality $\tau \leq \tau_{\max}$. The paradox of the distribution along the shortest paths lies in the fact that it, as an idea, arises with small edge loads, but practically used with much larger ones when $\tau = \tau_{\max}$ and some edges are close to saturation.

The following consideration may appear at this point. If it is important for us to unload the edges most loaded with the flow from u to v, then why not distribute it so that $\min_{\{f(u,v)\}} \max_i |f_i(u,v)|$ is reached on it? Here $\{f(u,v)\}$ denotes the set of the *edge* flows. Each element of this set forwards the flow $f = f_{uv}$ from u to v. And $|f_i(u, v)| = f_i$ is the i-th non-negative component of such an *edge* flow. Of course, when dealing with heuristics, one cannot affirm anything with absolute certainty. However, you should consider Hamming's aphorism [5], which states: "Do not try too hard to optimize the small pieces of a tightly interrelated system because it will cost you more than you gained when you put the parts together". We also note that the criterion for the distribution of the flow f_{uv} described in this paragraph is an example of the hypothetical use of a cubic metric on a graph [9].

To work around the problem, methods for refining the distribution along the shortest paths tend to direct the flows for the u,v-pairs of vertices along the paths of greater length, bypassing heavily loaded edges. An example is the "cut saturation" procedure [2, 10]. It reuses the distribution along the shortest paths during the construction of the final solution. In this case, the *lengths* of the edges of the graph vary from one construction step to another, depending on their load.

One can almost completely avoid such multiple steps by using a property of *quadratic* approximations. Such approximations are used to minimize errors, when (oppose to the case with linear laws) significant deviations are more undesirable than small ones. Here, the obvious simplest interpretation is a proportional increase in the *lengths* of the edges of the graph with increasing fractions of the flow $f = f_{uv}$ along them [11]. So the objective function of the flow distribution task will look as follows $\sum_i [(\mu\sigma_i)^{-1}(f_i f_{uv}^{-1})] f_i f_{uv}^{-1} = \sum_i (\mu\sigma_i)^{-1}(f_i f_{uv}^{-1})^2$ for the pair of vertices u,v, $\forall u \neq v$. Inside the square brackets of the left side of the last equality is the *length* of the edge with the number i. For normalized flow ($f_{uv} = 1$) the objective function of the flow distribution task will look as follows $\sum_i (\mu\sigma_i)^{-1} f_i^2$, where f_i is the dimensionless fraction of the flow of a magnitude of 1 directed from one vertex of the pair to another. This fraction goes along the edge with number i.

To obtain a description of such a distribution, we use an explicit representation of the matrix G^+, where G is the incidence matrix for a weighted graph (the weighted incidence matrix) and «$+$» symbolizes the Moore-Penrose pseudo inverse. Then the component of this distribution for an arbitrary edge (i,j) of the graph will look like $p_{ij}(z_{ui} - z_{uj} + z_{vj} - z_{vi})$ [11, 12]. In the general case, there is a factor $f_{uv} = f(u,v) \neq 1$. Thus, loading an arbitrary edge (i,j) of the graph with the flow of the colour t will look as follows $\sum_{r=1}^{k-1} f(s_r, t) p_{ij}(z_{s,i} - z_{s,j} + z_{tj} - z_{ti})$, where $f(s_r, t)$ is the s_r, t-th entry of the traffic matrix. The formation of such a load is described in the Introduction, and the matrices $P = (p_{ij})$ and $Z = (z_{ij})$ are described in [11, 12].

Lemma 1 [11]. For commonly used uniform traffic, the distribution of the flows using the Euclidian metric has the following remarkable property. Namely, the full loads of both directions of the network's arbitrary channel (i,j) are the same.

Proof. Let uniform traffic be normalized, i.e., for all non-diagonal entries $f(s_r, t) = 1$. Then for any colour t $\sum_{r=1}^{k-1} f(s_r, t) p_{ij}(z_{s,i} - z_{s,j} + z_{tj} - z_{ti}) = \sum_{r=1}^{k-1} p_{ij}(z_{s,i} - z_{s,j} + z_{tj} - z_{ti}) = p_{ij} \sum_{r=1}^{k-1} (z_{s,i} - z_{s,j} + z_{tj} - z_{ti}) = p_{ij} \sum_{r=1}^{k} (z_{s,i} - z_{s,j} + z_{tj} - z_{ti}) = p_{ij}(\sum_{r=1}^{k} z_{s,i} - \sum_{r=1}^{k} z_{s,j}) + p_{ij} \sum_{r=1}^{k} (z_{tj} - z_{ti}) = kp_{ij}(z_{tj} - z_{ti}) = -kp_{ij}(z_{ti} - z_{tj})$.

Here we have taken advantage of the fact that for $s_r = t$ the flow of the colour t along any edge (i,j) is equal to zero. The well-known [13] property of the matrix $Z = (z_{mn})$ is also used, namely: $\sum_n z_{mn} = 1, \forall m$.

For each colour t_a from the set of receivers $\{t_a\}$ let $z_{t_a j} - z_{t_a i} \geq 0$, that is, for all these colours, nonzero flows go from i to j. For all other colours t_b from the set of receivers $\{t_b\}$ $z_{t_b j} - z_{t_b i} < 0$, that is, such flows go from j to i. The total cardinality of both sets is k. Denote the total load of the direction from i to j as $l_{ij} = \sum_{t_a} kp_{ij}(z_{t_a j} - z_{t_a i})$ and the total load of the direction from j to i as $l_{ji} = -\sum_{t_b} kp_{ij}(z_{t_b j} - z_{t_b i})$.

Further suppose that $l_{ij} \neq l_{ji}$. Then $l_{ij} - l_{ji} = \sum_{t_a} kp_{ij}(z_{t_a j} - z_{t_a i}) + \sum_{t_b} kp_{ij}(z_{t_b j} - z_{t_b i}) = kp_{ij}[\sum_{t_a} (z_{t_a j} - z_{t_a i}) + \sum_{t_b} (z_{t_b j} - z_{t_b i})] = kp_{ij} \sum_t (z_{tj} - z_{ti}) \neq 0$. But for the *symmetric* [12] matrix Z, the equality $\sum_m z_{mn} = 1, \forall n$, holds [13]. The contradiction obtained proves that $l_{ij} = l_{ji}$.   □

In practice, the selection of channel capacities must be made from a small finite set [7]. In many procedures either these capacities are assigned before the routing is performed, or a routing is first performed and capacities then assigned [10]. If the second typical approach is used, then the quadratic objective distribution function of the normalized flow described above will look like $(\mu\sigma)^{-1} \sum_i f_i^2$. In other words, the flow $f_{uv} = 1$ of any pair of u,v vertices must be distributed along the edges so that the usual sum of squares of its components is minimal. To obtain a description of such a distribution, we use the explicit representation of the matrix A^+ [9, 11, 14, 15]. Then the component of this distribution for an arbitrary edge (i,j) of the graph will look like $p(z_{ui} - z_{uj} + z_{vj} - z_{vi})$. The loading of this edge by the flow of the colour t can be obtained as in Lemma 1. Thanks to the equal capacities of all edges, the indices of the factor p_{ij} can be omitted, so this load is $-kp(z_{ti} - z_{tj})$. Matrices P and Z are described in [9, 11, 14, 15].

Let X and Y be vector spaces with dimensions h and k correspondingly over the field of real numbers. The arbitrarily introduced *orientations* of the edges in the absence of limits to their capacity turns the incidence matrix A of an undirected graph having $\pm$ 1 as non-zero elements in each column into a useful instrument for the *linear transformation* $Ax = y$ of the edge flows $x \in X$ into the vertex flows $y \in Y$. From now on the incidence matrix A will be understood to be such a matrix.

Let $\{x_t \in X \mid Ax_t = y_t\}$ where $y_t = y(t)$ has an entry t equal to $1-k$, and the remaining entries are equal to $+1$. Any of the vectors $x_t = x(t)$ represent the transfer of the flow of 1 of the colour t from s_r, $r = 1, \ldots, (k-1)$, to t along the edges of the graph. Then, without additional designations one should search for the extreme norm along all such x_t.

Let C designate the incidence matrix of a complete graph. The indices «c» and «a» together with $x(t)$ are used to indicate the following: $x_c(t) = C^+ y(t) = C^+ y_t$ and $x_a(t) = A^+ y(t) = A^+ y_t$. Also, the symbol «$'$» further means matrix transposition.

3 The Euclidian Metric and the Best Quadratic Approximation of the Flow

Preservation of the essence of the problem of throughput in linear models when one has obtained them from more complex models means the following: the possibility of constructing throughput estimates (the ability to analyze throughput) exists without any explicit use of the τ expression described above. The material of this section is devoted to these constructions within the framework of such linear models [11].

The basis of throughput analysis is the distribution of a multi-colour flow along the edges of an arbitrary connected graph. The task of finding such a distribution remains difficult even after the simplifying actions with the colour components of the flow described in the beginning of Sect. 2. Requirements for search methods and the description of distributions increase many times due to the exceptional vastness of the search area during graph *synthesis*.

Consider a complete graph with the same k number of vertices. It has a valuable symmetry property. This property will be used in conjunction with other typical conditions for throughput analysis and graph synthesis. Such conditions include the same capacities of the edges and (normalized) uniform traffic.

For these conditions, the *obvious* distribution of the flow of the colour t, $t = 1, \ldots, k$, is one that delivers the flow along a path of length 1 directly from any source vertex to a single receiver vertex t. As with reliability analysis [15], the symmetry of the complete graph and the uniformity of traffic do not allow, in accordance with the *Principle of Non-Sufficient Reason* [16], to send a nonzero flow between any source vertices. Denote such a flow from s_r, $r = 1, \ldots, (k-1)$, to t in a complete graph as $f_c(t)$, Fig. 1. The flow directions are unambiguous. So the *signs* of these $f_c(t)$ components and the arbitrary *orientations* of the edges are in obvious conformity.

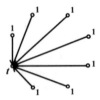

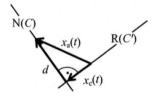

Fig. 1. The flow $f_c(t)$ constructed for complete graph ($k = 8$)

Fig. 2. The spaces and vectors of Lemma 3 for $y_0 = y(t) = y_t$

Lemma 2 [11, 15]. The vector $f_c(t)$ is proportional to $C^+ y_t : f_c(t) = cC^+ y_t = ck^{-1} C' y_t$.

Proof. Using Lemma 2 [11, 15] one immediately get $c = 1$. $\square$

An arbitrary graph differs from a complete one with the same number of vertices – some edges are absent. It simply means you can't send a flow between vertices that are not connected by an edge. So it is clear that the flow $C^+ y_t = k^{-1} C' y_t$ may be absent in an arbitrary graph for an arbitrary colour t among $\{x_t \in X | A x_t = y_t\}$. However, one can always search for the *best approximation* in a certain sense to $k^{-1} C' y_t$, that way we can reveal the absence of some edges by means of such a flow.

Other things being equal, the throughput of a complete graph is maximal thanks to the fact that the flow distribution like $f_c(t)$ for any colour in a complete graph is the best one. It seems natural to call such a distribution *ideal* due to the absence of intermediate vertices between any source and receiver for a flow of any colour t. Therefore, the *magnitude* of the *smallest* deviation of the approximate flow in an arbitrary graph from $k^{-1} C' y_t$ can be used as the throughput measure of this graph for the flow of the colour t. The description of the throughput for the flow of the colour t in a complete graph in the form of a vector $k^{-1} C' y_t$ is the natural reference point here.

To determine the meaning of the term "best approximation", we introduce the scalar product into X in the usual way, and by the square of the length of the vector x we mean (x, x).

Here and below let N(H) and R(H') mean the H kernel and the H' image of an arbitrary matrix H, respectively.

Insert zero columns into the matrix A at the places of the missing edges in an arbitrary graph. Matrix A^+ will have zero rows with the same numbers. Let $\forall y_0 \in R(A)$. Clearly, $y_0 \in R(C)$.

Lemma 3 [11, 15]. $(C^+ y_0, A^+ y_0 - C^+ y_0) = 0$. Proof. [11, 15]. $\square$

Figure 2 illustrates Lemma 3. The vectors $d = A^+ y_0 - C^+ y_0$ and $C^+ y_0$ for $y_0 = y(t) = y_t$ are the legs of a right-angled triangle with the hypotenuse $A^+ y_0$ for the same y_0.

Let the number of vertices k and $y_0 = y(t) = y_t$ be fixed. Then, according to Lemma 2, the vector $x_c(t)$ is also constant. Then, according to the Pythagorean theorem, the Euclidian lengths of vectors d and $x_a(t) = A^+ y_t$ are minimal *simultaneously* for an arbitrary graph with the incidence matrix A. According to a well-known [17, 18] property of the Moore-Penrose pseudo inverse, the Euclidian norm of the vector $x_a(t) = A^+ y_t$ will be minimal among $\{x(t) \in X | A x(t) = y_t\}$. In this case you can always use the length of the vector $x_a(t)$ instead of the length of vector d as a measure of the graph's throughput for the flow of the colour t. The vector $x_a(t) = A^+ y_t = pA'Zy_t$, so the sought measure of throughput is $((x_a(t), x_a(t)))^{1/2} = k(p(z_{tt} - k^{-1}))^{1/2}$ for this colour.

4 Multi-colour Flow and Synthesis Using the Euclidian Metric

Throughput, along with the reliability and cost of resources, is the most important factor in the graph synthesis. The Euclidian metric allows one to efficiently solve the main task of throughput analysis, namely, the task of the distribution of a multi-colour flow [11, 12]. Therefore, it can be used in standard synthesis procedures and their modifications in the same way as usual shortest paths are used. At the same time, its quadratic nature and analytical capabilities provide the advantages noted in Sect. 2.

However, the possibilities of the Euclidian metric in synthesis are not limited to this. An abstract measure of the throughput of an arbitrary graph has been described above for a single flow of any colour. This measure allows the composition to be used to introduce such a measure also for a multi-color flow.

For a Euclidian, that is quadratic, metric, the next way of composition seems natural. Namely, the use of a weighted sum of indices for flows of different colours. In this case the t-colour flow index may be used as the weight of itself. Note that with this criterion, the worst distribution of a flow of any colour will automatically be highlighted with the highest weight [11, 12]. Thus, if the common factor k^2 is discarded, then the required sum is $p \sum_t (z_{tt} - k^{-1}) = p(\text{Sp}Z - 1)$.

The objective function is accompanied by constraints in synthesis. Here the limitation is the number of h edges allowed for use in the synthesized graph. This is a simple form of the resources constraint.

There are other constraints. The synthesis of graphs with throughput as an objective function is no less complicated than, say, the synthesis of reliable graphs. In the latter, reliability *indicators* of graph *elements* should be taken into account. Their significant changes entail a change in the objective function of the synthesis task where there are limited available resources [15].

Similarly, the need to take into account traffic non-stationarity cancels the desire to recklessly improve the flow distribution of the graph to be synthesized when traffic intensity is averaged over time. [19–21].

Faced with multicriteria synthesis, a set of the non-interior (Pareto efficient) solutions or scalarizing of the objective functions are traditionally used. Both approaches are very difficult to put into practice. In our case, in the analysis and synthesis of reliable graphs [15] and here working on throughput, a single metric basis is used. In both cases, the Euclidian metric allows us to formulate a uniform criterion of quadratic proximity to a complete graph. The condition of low probability of failure of the edges can be reduced to a set of simple constraints [15].

As a result, the task formulation of a high throughput graph's synthesis taking into account the non-stationarity of the (normalized) uniform traffic and the case of a small edges failure probability may look as follows. For a given $3 < k$, $k \leq h$ and

$|d_i - d_j| \le 1 \, \forall i, j$ find a graph with a minimum value of $p(\mathrm{SpZ} - 1)$ [15]. The search can be based, for example, on the sequence of adding and deleting edges when starting from some initial graph. The rank 1 perturbations used in this case do not require large computations for the correction of the inverse matrices [15, 22, 23].

5   Synthesis Example

Here are two graphs. One of them is given; the other is the result of synthesis (see, for example, [24]), the conditions of which are described above in Sect. 4. For both graphs, $k = 23$, $h = 28$, and all edges have the same capacity. All non-diagonal entries of the matrix of the averaged normalized uniform traffic are equal to 1.

The given graph is a well-known version of one of the stages of the ARPA network [10, 25, 26]. In Fig. 3, for each edge, the values of the full normalized load of each of its two directions are given. The flow distribution was determined using the Euclidian metric, as described above. According to Lemma 1, both directions of any edge are loaded equally by flows of different colours if the traffic is uniform.

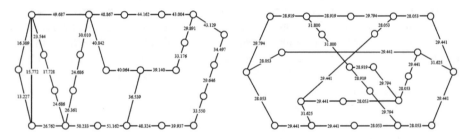

Fig. 3. The given graph   **Fig. 4.** Synthesized graph

Note the significant unevenness of these full loads of *different* edges. Under the described throughput analysis conditions, this unevenness is a characteristic of this graph. A proportional change in the traffic matrix leads to saturation of the edges with the highest full loads 51.162, 50.233, 49.687, 48.867, 48,324. At the same time, edges with the smallest full loads 13.227, 15.772, 16.309, 17.728 remain substantially underloaded.

Figure 4 shows the synthesized graph and the loading of its edges. Full loads of different edges are substantially aligned. Largest full loads: 31.800, 31.625, ... Least full loads: 28.053, 28.919, ...

This convergence of full loads means a significant increase in the efficiency of using limited resources. And a simultaneous decrease in their largest values means an increase of graph throughput in terms of Delay/Flow. Recall that the abstract measure of quadratic proximity to the complete graph was used as the synthesis's object function, and the distribution as such of the multi-colour flow was not applied at all.

6   Results

1. A simplified flows distribution task is used for throughput analysis in practice. The simplification consists in linearizing the objective function and splitting the multi-colour flow into flows between subsets of graph vertices. At the final stage, the recomposition of the multi-colour flow is done. The paradoxical nature of using ordinary shortest paths for flow distribution is shown.
2. We propose using the Euclidian (quadratic) metric on graphs instead of repeatedly applying the usual shortest paths in procedures such as cut saturation.
3. The distribution of the flow of any colour in an arbitrary graph may be presented as the best quadratic approximation to the ideal distribution in the complete graph if the Euclidian metric is used.
4. Such an abstract measure of quadratic proximity to a complete graph makes it possible to totally ignore the distribution of flows in the synthesis process.
5. The presence of this measure means that the Euclidian metric provides a metric basis for the analysis and synthesis of graphs according to a single criterion for throughput and reliability.
6. The Euclidian metric makes for easily verifiable additional conditions during the synthesis of a graph with throughput as the objective function. This allows you to limit the effect of traffic non-stationarity, and also take into account the reliability parameters of the elements of the graph.

7   Conclusion

The usual shortest path is used as the metric basis for the heuristic routing procedures. The paradoxical nature of the use of these paths is shown. The use of the Euclidian metric in throughput analysis instead of repeatedly applying the usual shortest paths is substantiated. Justification for this is not explicitly based on the well-known expressions from the simplest heuristic models of delay/flow throughput. This verifies the high effectiveness of linear metric models in the analysis of the throughput of graphs and their synthesis. The graph's throughput index based on the Euclidian metric has been introduced and justified. Such an abstract measure of quadratic proximity to a complete graph makes it possible to totally ignore the distribution of flows in the synthesis process. This abstract measure based on the Euclidian metric also allows you to take into account the real traffic's non-stationarity and the reliability parameters of graph elements. The advantages of the Euclidian metric are illustrated by the example of a graph synthesis with throughput as the objective function.

References

1. Davies, D., Barber, D.: Communication networks for computers. Wiley, Hoboken (1973)
2. Chou, W., Frank, H.: Routing strategies for computer network design. In: Proceedings of the Symposium on Computer Communication Networks and Teletraffic, Polytechnic Institute of Brooklyn, pp. 301–309 (1972)
3. Hu, T.C.: Integer Programming and Network Flows. Addison-Wesley, London (1970)
4. Kleinrock, L.: Analytic and simulation methods in computer network design. In: Proceedings of the AFIPS Conference SJCC, vol. 36, pp. 569–579 (1970)
5. Hamming, R.W.: Numerical Methods for Scientists and Engineers. McGraw-Hill, New York (1962)
6. Voevodin, V.V.: Computational Foundations of Linear Algebra. Nauka, Moscow (1977)
7. Frank, H., Kahn, R.E., Kleinrock, L.: Computer communication network design – experience with theory and practice. In: Proceedings of the AFIPS Conference SJCC, pp. 255–270 (1972)
8. Frank, H., Frisch, I.T., Chou, W.: Topological considerations in the design of the ARPA computer network. In: Proceedings of the AFIPS Conference SJCC, pp. 581–587 (1970)
9. Rusakov, V.A.: On Markov chains and some matrices and metrics for undirected graphs. In: Antipova, T. (ed.) ICIS 2019. Lecture Notes in Networks and Systems, vol. 78, pp. 340–348. Springer, Cham (2020). https://doi.org/10.1007/978-3-030-22493-6_30
10. Frank, H., Chou, W.: Topological Optimization of Computer Networks. Proc. IEEE **60**(11), 1385–1397 (1972)
11. Rusakov, V.A.: Analysis and Synthesis of Computer Network Structures. Part 1. Analysis. Moscow Engineering Phys. Inst. Report: VNTI Center No Б796153, Moscow (1979)
12. Rusakov, V.A.: On the Moore-Penrose pseudo inverse of the incidence matrix for weighted undirected graph. Procedia Comput. Sci. **169C**, 147–151 (2020). https://doi.org/10.1016/j.procs.2020.02.126
13. Kemeny, J., Snell, J.: Finite Markov Chains. University series in undergraduate mathematics. Van Nostrand, Princeton NJ (1960)
14. Rusakov, V.A.: Matrices, shortest paths, minimal cuts and Euclidian metric for undirected graphs. Procedia Comput. Sci. **145**, 444–447 (2018). https://doi.org/10.1016/j.procs.2018.11.104
15. Rusakov, V.A.: Using metrics in the analysis and synthesis of reliable graphs. In: Samsonovich, A.V. (ed.) BICA 2019. Advances in Intelligent Systems and Computing, vol. 948, pp. 438–448. Springer, Cham (2020). https://doi.org/10.1007/978-3-030-25719-4_58
16. Polya, G.: Mathematical Discovery. Wiley, New York (1962, 1965, 1981)
17. Albert, A.E.: Regression and the Moore-Penrose Pseudoinverse. Academic Press, New York (1972)
18. Beklemishev, D.V.: Additional Chapters of Linear Algebra. Nauka, Moscow (1983)
19. Rusakov, V.A.: Synthesis of computer network structures and the problem of small certainty of initial values. USSR AS's Scientific Council on Cybernetics. In: Proceedings of the 5th All-Union School-Seminar on Computing Networks, vol. 1, pp. 112–116. VINITI, Moscow-Vladivostok (1980)
20. Rusakov, V.A.: Non-stationary flows in computer networks: a linear basis for their study. USSR AS's Scientific Council on Cybernetics. In: Proceedings of the 7th All-Union School-Seminar on Computing Networks, vol. 1, pp. 118–124. VINITI, Moscow-Erevan (1983)
21. Rusakov, V.A.: On the regularity of the displacement of the mean estimate for the throughput with non-stationary traffic. USSR AS's Scientific Council on Cybernetics. In: Proceedings of the 9th All-Union School-Seminar on Computing Networks, vol. 1.2, pp. 48–52. VINITI, Moscow-Pushchino (1984)

22. Rusakov, V.A.: Implementation of the methodology for analysis and synthesis of computer network structures using Markov chains. In: Engineering-Mathematical Methods in Physics and Cybernetics, issue 7, pp. 41–45. Atomizdat, Moscow (1978)
23. Rusakov, V.A.: Reconstruction of the Euclidian metric of an undirected graph by metrics of components. In: Natural and Technical Sciences, vol. 2(52), pp. 22-24. Sputnik +, Moscow (2011)
24. Rusakov, V.A.: A technique for analyzing and synthesizing the structures of computer networks using Markov chains. In: Computer Networks and Data Transmission Systems, pp. 62–68. Znaniye, Moscow (1977)
25. Frank, H., Chou, W.: Network properties of the ARPA computer network. Networks **4**, 213–239 (1974)
26. Van Slyke, R., Frank, H., Chou, W.: Avoiding simulation in simulating computer communication networks. In: Proceedings of the AFIPS Conference, 4–8 June, pp. 165–169. NCCE (1973)

Machine Learning

Link-Sign Prediction in Signed Directed Networks from No Link Perspective

Quang-Vinh Dang$^{(\boxtimes)}$ (iD)

Industrial University of Ho Chi Minh City, Ho Chi Minh City, Vietnam
dangquangvinh@iuh.edu.vn

Abstract. Predicting future sign of connections in a network is an important task for online systems such as social networks, e-commerce and other services. Several research studies have been presented since the early of this century to predict either the existence of a link in the future or the property of the link. In this study we present a new approach that combine both families by using machine learning techniques. Instead of focusing on the established links, we follow a new research approach that focusing on no-link relationship. We aim to understand the move between two states of no-link and link. We evaluate our methods in popular real-world signed networks datasets. We believe that the new approach by understanding the no-link relation has a lot of potential improvement in the future.

Keywords: Signed network · Machine learning · Link prediction

1 Introduction

Many real-world systems can be modelled by a signed-directed graph [11], such as connection graph between users on social networks like Facebook, citation graphs in scientific publication, recommendation graph in e-commerce, or the activities like view/like/dislike graphs on YouTube.

In order to improve the quality of user experience in these systems and increase the traffic usage hence the revenue of the systems, the service providers usually provide a recommendation service to suggest users what service to use next. The recommendation engine should be able to understand the users to recommend the service that the user want to use or the item that the user want to buy. Besides the well-known collaborative filtering technique, another approach to look at the problem is to consider the recommendation as the **link-attribute** prediction problem [11], i.e. we would like to predict the property of the link between two nodes. For instance, in e-commerce systems like Amazon, if we consider buyers and items are nodes and connections between them are the rating scores the buyers give to the items, link-attribute prediction shares a same goal with collaborative filtering.

In this study, we limit to the problem of **link-sign prediction**, i.e. we consider only two possible attributes of a link which are *positive* or *negative* [19]. We can think of positive links as a node (user on Facebook, buyer on Amazon, viewer on YouTube) *like* another node, and negative links as *dislike* relation. If we can predict accurately the signs of the links, we can improve significantly the quality of the recommender systems [10].

T. Antipova (Ed.): ICIS 2020, LNNS 136, pp. 291–300, 2021.
https://doi.org/10.1007/978-3-030-49264-9_26

The problem of link-sign prediction has attracted a lot of studies in the last two decades. However, the problem is divided into *link-prediction* and *sign-prediction* separately. In this study we consider unifying the two problems into one single framework.

2   Related Works

The problem of link-sign prediction is usually divided into the link-prediction problem [23] and the sign-prediction problem [19]. The link-predictor aims to predict the existence of the link in the future, given the non-existence of this link at present. The sign-predictor aims to predict the sign of a link in the future, given that the existence of the link is confirmed. The difference is visualized in Fig. 1.

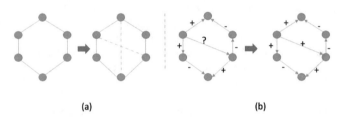

(a)   (b)

Fig. 1.   (a) Link-prediction; and (b) Sign-prediction

2.1   Link-Prediction Problem

Link-prediction can be roughly divided into proximity-based and learning-based methods [22].

Proximity-based methods mostly rely on homophily theory [26] which states that similar users tend to connect to each other's. The proximity metrics can be nodal proximity-based metrics [38] or structural proximity-based metrics [23]. The nodal-based proximity methods use the personal profile information of the nodes to design the proximity metric. For instance, the authors of [37] proposed to use demographics information such as age, education and occupation. In [36] the authors also used geographic information in physical space to estimate the distances between two nodes in cyber-space. Several studies employed user activities on social networks such as keywords, hashtags or semantic of the content that the users share or view on social network to calculate their similarity [40].

Among different metrics designed in different studies, cosine similarity is mostly used as the proximity metric [22]. Other popular distance metrics have been used in literature are KL-divergence [31] or Jaccard's coefficient [30].

Structural-based methods do not rely on the individual information of the nodes but based on the topological information of the graph [23]. An early attempt is credited to [1] that defines the Adamic-Adar measure. The metric is inspired by the Erdos-Renyi model [2]. Both theoretical and empirical studies showed that the likelihood of a link depends on the size of direct neighborhood of a node [23]. Other well-known structural-based measures that rely on neighborhood information are SimRank [16], Sorensen Index or HP/HD Index [29].

Another kind of structural-based proximity is to rely not on the neighborhood information but on the connection between two nodes. The core idea is to calculate the distance between two nodes as the surrogate estimation of how close they are to each other, hence how likely that they will connect. One of very early approaches belongs to [17]. The core idea of Katz index [17] is to combine all connected paths between two nodes with associated weights as their corresponding length. The authors of [43] updated the Katz index to use local-path information. Other studies [35] modified the PageRank algorithm for the link prediction problem. Recently, [39] added the interaction among paths, i.e. the authors consider the interaction between nodes that belong to different paths into consideration, to achieve a better prediction. [4] explored a very new and interesting aspect of the network, which is the uncertainty of the network topology, i.e. when we are not 100% sure if there is a link or no-link between two nodes.

On the other hand, learning-based link-prediction methods build machine learning models to predict the probability that a link will be established in the future. In order to build the machine learning models, features needed to be extracted. The features might be like the similarity metrics described above, i.e. neighbor-based features such as common friends [3] or path-based features such as the shortest path between two nodes [24]. The most common algorithms are logistic regression [27] or SVM [41].

2.2   Sign-Prediction Problem

In contrast to link-prediction problem where the problem is formed as predicting whether a link will be established in the future or not, the sign-prediction problem tries to predict the attribute of the link, here positive or negative sign, given the existence of the link in the future [13].

Early attempts in sign-prediction used rule-based systems, such as "friends of friends are friends" and "friends of enemies are enemies" for prediction [13]. In order to speed up the calculation, the prediction is performed as matrix multiplication. The following studies turned the problem into a low-rank matrix factorization problem [25] as in the recommender systems.

Starting with the work of [19], *structural balance theory* and *social status theory* are used to interpret the structure of the social graphs hence to predict the sign of the connections. We visualized two theories in Fig. 2 and 3 [6, 11]. Several studies have followed the work of [19] such as [5] by extending the cycle length or [15] by integrating the social theories into matrix factorization techniques. In the same direction, [18] focused on micro-structure of three users with bidirectional links and their similarities, for the sign prediction.

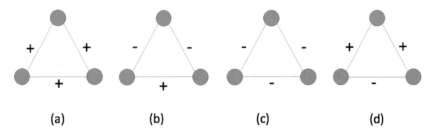

(a)   **(b)**   **(c)**   **(d)**

Fig. 2. Visualization of structural balance theory [19]. *Structural balance theory* considers the triads *(a)* and *(b)* are balanced, while *(c)* and *(d)* are not. The *weak balance theory* considers *(c)* as balanced as well, regardless the direction.

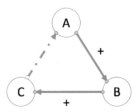

Fig. 3. The sign of the dash line from *C* to *A* is inferred by their social status. Because + + *A* → *B* and *B* → *C*, therefore hence the *social status theory* states that *B* > *A* and *C* > *B*, so *C* > *A*, leads to the prediction that the sign of line from *C* to *A* is negative.

The authors [32] argued that, (i) even structural balance theory and social status theory played an important role in existing research studies on link sign prediction, they are not scalable up to the size of the modern networks, and (ii) a fully observed network is usually not available in practice. They develop Bayesian node features based on partially observed networks and used a logistic regression classifier for link sign prediction. However, the obtained performance is weak compared to other recent studies.

Graph embedding techniques attracted a lot of attention recently [12]. Many graph embedding research studies are inspired by word embedding techniques [28] as they share a similar goal. The authors of [11] combined word embedding technique, random walk and Long-Short Term Memory (LSTM) networks to utilize the time information of the network for sign-prediction.

As pretrained models are more and more popular in natural language processing community and achieve a lot of success, transfer learning has been used in sign-prediction as well. This direction has a lot of potential usage as it does not require labelled data [42].

Most existing works in sign-prediction assumes the previous task which is link-prediction has been done, so they assume the existence of the link and focus only on the sign of the link.

3   Understanding No-Link

As we discussed in Sect. 2, the work of link-sign prediction is usually divided into two steps: link-prediction and sign-prediction [20]. Given that the performance of state-of-the-art link-prediction algorithms are far from perfect [22], it is not realistic to expect a high-accuracy prediction in sign-prediction if we consider the task link-sign prediction as an entire task.

Given that most of a network should be no-link [11], it is important to understand the no-link relationship to achieve a higher-accuracy prediction. Hence, we focus on the *move* from no-link into link between two nodes. We collected features based on sociology theories and other studies [20] such as Balance Theory, Status Theory, Reciprocity or Frequent Subgraph [19]. We also derived features automatically through some graph embedding techniques such as DeepWalk [28], LINE [33] and node2vec [12]. We try to capture the semantic of the network just before a link is established. Our approach shares some similarities with the work of [21] as we combine both explicit and implicit features, but different from [21] we treated different links with different weights. Furthermore, in [21] the authors simply add two explicit and latent scores then solving an optimization problem, so they consider two sets of features as the same weight without any learning process.

Regarding the no-link relations, we consider the time when a node decides to make a link to a particular other node but no other nodes. At this very point of time we consider that, it means the first node decides to not make a link to other nodes. We collected the same features but with label of no-link.

Instead of treating the problem as the traditional classification problem [11, 19] we convert the problem into the ranking problem. We use a state-of-the art ranking algorithm implemented in XGBRanker version 1.0.0[1] which has just been released at the time of writing. In our opinion, this approach makes more sense as what we need to optimize in practice, such as in an e-commerce system, is the ranking metric, i.e. which item a user will most likely like, rather than a normal classification problem [20]. Indeed, we can use the output of any classification problem as the ranking score but actually they are optimized for different purposes.

4   Experiments

4.1   Datasets

We used well-known signed network datasets that have been used widely in the literature: Epinions, Slashdot and Wikipedia RfA [11, 19]. They are realworld signed networks collected from e-commerce website (Epinions), sharing news website (Slashdot) or crowd-sourcing encyclopedia (Wikipedia). In these websites, users can form explicitly trust (positive)/distrust (negative) links to other users. The connections are provided with the established time that allow us to understand the history of the network.

---

[1] https://xgboost.readthedocs.io.

We display some basic statistics of the dataset in Table 1 [11].

Table 1. Basic statistics of datasets. WCC stands for *weakly connected component.*

Epinions		Slashdot wikipedia	
# of nodes	119 217	82 140	7 118
# of edges	841 200	549 202	103 747
fraction of edges	$6e-5$	$8e-5$	$2e-3$
+ edges (%)	85.0	77.4	78.8
− edges (%)	15.0	22.6	21.2
largest WCC (%)	99.1	100	100
average # of directed connection	590	327	418
# of triads	13 375 407 1 508 105		790 532
fraction of triads	$1.35e-10$ $5.46e-11$		$4.25e^{-9}$

The authors of [21] studied the change from no-link to link between two nodes given the they have common neighbors. The analysis is presented in Fig. 4 for Epinions dataset. It is agreed between several recent studies [20, 21] that the process of moving from no-link to link is really important.

For each dataset, we divide the train/validation/test set according to the ratio of 60:20:20 by temporal order, i.e. we use 60% of the links that are established first in each network as the training set, 20% of the links that are established next for the validation set, and 20% of the links that are established last for testing purpose.

In order to avoid overfitting, we apply early stopping technique. Other hyperparameters are optimized by using Bayesian optimization.

4.2   Evaluation Metric

As the matter of fact, the network in general is extremely balanced: most of the potential links do not exist. As we present in Table 1, there is around only less than 0.1% of the potential links in the network, i.e. a connection that might be exist between any two nodes, is established in a network. It is understandable because usually a typical person can only make a few connections, and even if the degree of a node is thousands it is still a very small number compared to billions of nodes in modern networks like Facebook.

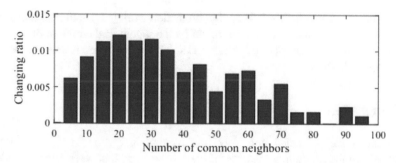

Fig. 4. Changing ratio given the common neighbors between two nodes [21]. The X-axis represents the number of common neighbors in 15 months and the Y-axis represents the proportion of two nodes that formed a link.

Some popular metrics which are sensitive to the imbalanced dataset such as accuracy score or even F-score are not usable to evaluate link prediction problem [9]. AUC is used more widely [11] but the AUC is designed only for a binary classification problem. In this paper we will use a more generalized AUC metric designed by [14] which has been adapted in multi-class classification problem [8, 9].

The generalized AUC is chosen because it is based on the same principle of measuring the ranking order as in typical AUC. Like typical AUC, the generalized AUC ranges from 0.0 to 1.0, higher is better.

4.3   Experimental Results

We compare our method with several other baseline methods, such as multi-class classification [11]. We note that it is our method is not comparable to existing works because it focuses on both link-prediction and sign-prediction at the same time.

Here, we consider the problem of predicting at a given point of time, given that there is no link between node A and node B, which action will be performed by node A: (i) establish a positive link, or (ii) establish a negative link, or (iii) do not establish a link.

The result is presented in Table 2. We notice that in general the ranking approach achieve a better performance as the AUC by its nature is optimized by a ranking method.

Table 2. Generalized AUC of our method in different datasets.

Methods	Epinions	Slashdot	Wikipedia
Classification	0.72	0.74	0.77
Ranking	0.74	0.75	0.79

Like other link-prediction [19] or sign-prediction work [11], both methods perform best in Wikipedia dataset. The reason might be the topological structure of Wikipedia that is much more dense than other networks, but we let it be an open question for further research works.

5 Conclusions

In this study we presented our work on link-sign prediction in signed networks. Together with the preliminary work of [20], they are very first working in considering link-prediction and sign-prediction into one problem. By combining the problems, we eliminate the accumulated error over multiple prediction steps. We note that the approach presented in this paper is a topological-based approach [34], i.e. we only rely on the features that can be derived from the topology of the network and we do not use any personal information of node such as gender or activity history.

In the future research we plan to incorporate more temporal features as described in [11]. We believe that by understanding the movement from no-link to a link in a network we will gain more insights of the link-sign prediction problem. In other words, we would like to know what bring a node to a very precise point of time before he or she decide to form a link or not, not only just before a connection is made. Another interesting research direction is to predict the time when a link is made rather than considering this time as a given input. On the other hand, we would like to incorporate trust calculation to the network as the links as the interaction between nodes [7].

References

1. Adamic, L.A., Adar, E.: Friends and neighbors on the web. Soc. Netw. **25**(3), 211–230 (2003)
2. Baraba´si, A.L., et al.: Network Science. Cambridge University Press, Cambridge (2016)
3. Benchettara, N., Kanawati, R., Rouveirol, C.: A supervised machine learning link prediction approach for academic collaboration recommendation. In: RecSys, pp. 253–256. ACM (2010)
4. Chen, X., Guo, J., Pan, X., Zhang, C.: Link prediction in signed networks based on connection degree. J. AIHC **10**, 1747–1757 (2019)
5. Chiang, K., Natarajan, N., Tewari, A., Dhillon, I.S.: Exploiting longer cycles for link prediction in signed networks. In: CIKM, pp. 1157–1162. ACM (2011)
6. Dang, Q.V.: Trust assessment in large-scale collaborative systems. Ph.D. thesis, University of Lorraine, France (2018)
7. Dang, Q.V., Ignat, C.L.: Computational trust model for repeated trust games. In: Trustcom/BigDataSE/ISPA, pp. 34–41. IEEE (2016)
8. Dang, Q.V., Ignat, C.L.: Measuring quality of collaboratively edited documents: the case of wikipedia. In: CIC, pp. 266–275. IEEE Computer Society (2016)
9. Dang, Q.V., Ignat, C.L.: Quality assessment of wikipedia articles without feature engineering. In: JCDL, pp. 27–30. ACM (2016)
10. Dang, Q.V., Ignat, C.L.: dTrust: a simple deep learning approach for social recommendation. In: CIC, pp. 209–218. IEEE (2017)

11. Dang, Q., Ignat, C.: Link-sign prediction in dynamic signed directed networks. In: CIC (2018)
12. Grover, A., Leskovec, J.: node2vec: scalable feature learning for networks. In: KDD, pp. 855–864. ACM (2016)
13. Guha, R.V., Kumar, R., Raghavan, P., Tomkins, A.: Propagation of trust and distrust. In: WWW (2004)
14. Hand, D.J., Till, R.J.: A simple generalisation of the area under the ROC curve for multiple class classification problems. Mach. Learn. 45(2), 171–186 (2001)
15. Hsieh, C., Chiang, K., Dhillon, I.S.: Low rank modeling of signed networks. In: KDD, pp. 507–515. ACM (2012)
16. Jeh, G., Widom, J.: Simrank: a measure of structural-context similarity. In: KDD, pp. 538–543. ACM (2002)
17. Katz, L.: A new status index derived from sociometric analysis. Psychometrika 18, 39–43 (1953)
18. Khodadadi, A., Jalili, M.: Sign prediction in social networks based on tendency rate of equivalent micro-structures. Neurocomputing 257, 175–184 (2017)
19. Leskovec, J., Huttenlocher, D.P., Kleinberg, J.M.: Signed networks in social media. In: CHI, pp. 1361–1370. ACM (2010)
20. Li, X.: Towards practical link prediction approaches in signed social networks. In: UMAP (2018)
21. Li, X., Fang, H., Zhang, J.: FILE: a novel framework for predicting social status in signed networks. In: AAAI, pp. 330–337. AAAI Press (2018)
22. Li, Z.L., Fang, X., Sheng, O.R.L.: A survey of link recommendation for social networks: methods, theoretical foundations, and future research directions. ACM Trans. Manag. Inf. Syst. 9, 1–26 (2018)
23. Liben-Nowell, D., Kleinberg, J.M.: The link-prediction problem for social networks. JASIST 58, 1019–1031 (2007)
24. Lichtenwalter, R.N., Lussier, J.T., Chawla, N.V.: New perspectives and methods in link prediction. In: KDD, pp. 243–252. ACM (2010)
25. Ma, H., Yang, H., Lyu, M.R., King, I.: Sorec: social recommendation using probabilistic matrix factorization. In: CIKM, pp. 931–940. ACM (2008)
26. McPherson, M., Smith-Lovin, L., Cook, J.M.: Birds of a feather: homophily in social networks. Ann. Rev. Sociol. 27(1), 415–444 (2001)
27. O'Madadhain, J., Hutchins, J., Smyth, P.: Prediction and ranking algorithms forevent-based network data. ACM SIGKDD Explor. Newsl. 7, 23–30 (2005)
28. Perozzi, B., Al-Rfou, R., Skiena, S.: Deepwalk: online learning of social representations. In: KDD, pp. 701–710. ACM (2014)
29. Ravasz, E., Somera, A.L., Mongru, D.A., Oltvai, Z.N., Baraba'si, A.L.: Hierarchical organization of modularity in metabolic networks. Science 297, 1551–1555 (2002)
30. Scellato, S., Noulas, A., Mascolo, C.: Exploiting place features in link prediction on location-based social networks. In: KDD, pp. 1046–1054. ACM (2011)
31. Shen, D., Sun, J., Yang, Q., Chen, Z.: Latent friend mining from blog data. In: ICDM, pp. 552–561. IEEE Computer Society (2006)
32. Song, D., Meyer, D.A.: Link sign prediction and ranking in signed directed social networks. Soc. Netw. Anal. Min. 5(1), 1–14 (2015)
33. Tang, J., Qu, M., Wang, M., Zhang, M., Yan, J., Mei, Q.: LINE: large-scale information network embedding. In: WWW, pp. 1067–1077. ACM (2015)
34. Tang, J., Chang, Y., Aggarwal, C., Liu, H.: A survey of signed network mining in social media. ACM Comput. Surv. 49, 1–37 (2016)

35. Tong, H., Faloutsos, C., Pan, J.Y.: Fast random walk with restart and its applications. In: ICDM, pp. 613–622. IEEE (2006)
36. Wang, D., Pedreschi, D., Song, C., Giannotti, F., Baraba´si, A.: Human mobility, social ties, and link prediction. In: KDD, pp. 1100–1108. ACM (2011)
37. Xu, Y., Rockmore, D.N.: Feature selection for link prediction. In: PIKM, pp. 25–32. ACM (2012)
38. Yang, Y., Chawla, N.V., Sun, Y., Han, J.: Predicting links in multi-relational and heterogeneous networks. In: ICDM, pp. 755–764. IEEE Computer Society (2012)
39. Yao, Y., Zhang, R., Yang, F., Tang, J., Yuan, Y., Hu, R.: Link prediction in complex networks based on the interactions among paths. Phys. A **510**, 52–67 (2018)
40. Yuan, G., Murukannaiah, P.K., Zhang, Z., Singh, M.P.: Exploiting sentiment homophily for link prediction. In: RecSys, pp. 17–24. ACM (2014)
41. Yuan, W., He, K., Guan, D., Zhou, L., Li, C.: Graph kernel based link prediction for signed social networks. Inf. Fusion **46**, 1–10 (2019)
42. Yuan, W., Pang, J., Guan, D., Tian, Y., Al-Dhelaan, A., Al-Dhelaan, M.: Sign prediction on unlabeled social networks using branch and bound optimized transfer learning. Complexity **2019**, 4906903:1–4906903:11 (2019)
43. Zhou, T., Lü, L., Zhang, Y.C.: Predicting missing links via local information. Eur. Phys. J. B **71**(4), 623–630 (2009)

Smart Cities in Digital Age

Smart Mobility: A Systematic Literature Review of Mobility Assistants to Support Multi-modal Transportation Situations in Smart Cities

Nelson Pacheco Rocha[1]([✉]) [iD], Ana Dias[2] [iD], Gonçalo Santinha[3] [iD],
Mário Rodrigues[4] [iD], Alexandra Queirós[5] [iD],
and Carlos Rodrigues[3] [iD]

[1] Institute of Electronics and Informatics Engineering of Aveiro,
Department of Medical Sciences, University of Aveiro, Aveiro, Portugal
npr@ua.pt
[2] GOVCOPP - Governance, Competitiveness and Public Policies,
Department of Economics, Industrial Engineering, Management and Tourism,
University of Aveiro, Aveiro, Portugal
anadias@ua.pt
[3] GOVCOPP - Governance, Competitiveness and Public Policies,
Department of Social, Political and Territorial Sciences, University of Aveiro,
Aveiro, Portugal
{g.santinha,cjose}@ua.pt
[4] Institute of Electronics and Informatics Engineering of Aveiro,
Águeda School of Technology and Management, University of Aveiro,
Aveiro, Portugal
mjfr@ua.pt
[5] Institute of Electronics and Informatics Engineering of Aveiro,
Health Sciences School, University of Aveiro, Aveiro, Portugal
alexandra@ua.pt

Abstract. The study reported by this article aimed to identify mobility assistants designed to support transit situation in multi-modal transportation networks using smart cities' infrastructures. Therefore, a systematic review was performed based on a search of the literature. A total of 16 articles were included in the systematic review and all of them aimed to contribute for the development of mobility assistants to support multi-modal transportation situations in smart cities, either proposing algorithms to optimize routes planning or presenting specific applications.

Keywords: Smart cities · Smart mobility · Mobility assistants · Systematic review

1 Introduction

Urban mobility is related to the movement of individuals, both obligatory (e.g., home-to-work trips) and voluntary (e.g., leisure), with the goal of accessing desired destinations. It depends not only of the available transportation, but also of other

© Springer Nature Switzerland AG 2021
T. Antipova (Ed.): ICIS 2020, LNNS 136, pp. 303–312, 2021.
https://doi.org/10.1007/978-3-030-49264-9_27

characteristics, such as locations of both activities and households, physiological, intellectual and socioeconomic needs of the individuals, purpose of the movement, movement length or travel time distribution [1, 2].

According to the current concepts of multi-modal urban mobility, public transportation should be combined with other motorized (including small size electric vehicles) and non-motorized modes, as well as with new forms of vehicle ownership (e.g., car-sharing or ridesharing) [3].

Within the smart city paradigm, smart mobility [4] is often seen as supported by information technologies able to adequately orchestrate services designed to improve urban movements [2]. Moreover, information services such as personal mobility assistants can help to surpass difficulties of the travelers facing multi-modal transit situations [3].

The study reported by this article aimed to determine the current state of mobility assistants to support multi-modal transportation situations using smart cities' infrastructures. This is useful to inform smart cities' stakeholders about state-of-the-art solutions and researchers about gaps of the current research.

2 Methods

The study was informed by the following research questions:

- RQ1: What are the current research trends related to mobility assistants using smart cities' infrastructures to support multi-modal transportation situations?
- RQ2: What types of smart cities' data are being used?
- RQ3: What are the maturity levels of the solutions being reported?

Boolean queries were prepared to include all the articles that have in their titles, abstract or keywords one of the following expressions: 'Smart City', 'Smartcity', 'Smart-city', 'Smart Cities', 'Smartcities' or 'Smart-cities'. The resources considered to be searched were two general databases, Web of Science and Scopus, and one specific technological database, IEEE Xplore. The literature search was concluded on January 2020.

As inclusion criteria, the authors aimed to include all the articles that report the development of mobility assistants to support multi-modal transportation situations explicitly using smart cities' infrastructures.

Considering the exclusion criteria, the authors aimed to exclude all the articles not published in English, without abstract or without access to full text. Furthermore, the authors also aimed to exclude all the articles that report overviews, reviews, or solutions that do not explicitly require smart cities' infrastructures, as well as, article reporting studies not relevant for the specific objective of this systematic review.

After the removal of duplicates and articles without abstract, the selection of the remainder articles according to the outline inclusion and exclusion criteria was performed in the following steps: (i) the authors assessed all abstracts for relevance and those clearly outside the scope of this systematic review were removed; (ii) the abstracts of the retrieved articles were screened against inclusion and exclusion criteria to exclude non relevant articles; (iii) the full texts of the eligible articles were retrieved and screened for inclusion; and (iv) the full text of the included articles were analyzed and classified. In all these four steps the articles were analyzed by at least two authors and any disagreement was discussed and resolved by consensus.

3 Results

A total of 14953 articles were retrieved from the initial search on Web of Science, Scopus and IEEE Xplore.

The initial step of the screening phase yielded 14797 articles by removing the duplicates (144 articles) or the articles without abstracts (12 articles).

Based on abstracts, 4776 articles were removed since they were not published in English, or they are overviews or reviews, editorials, prefaces, and announcements of special issues, workshops or books. Moreover, 9802 articles were removed because they do not target the development of mobility assistants to support multi-modal transportation situations using smart cities' infrastructures.

Finally, the full texts of the remaining 219 articles were screened and 203 articles were excluded because they do not meet the inclusion criteria. Therefore, 16 articles were considered eligible for this systematic review.

From those 16 articles, 12 were published in conference proceedings [5–16], three were published in scientific journals [17–19] and one was published as a book chapter [20]. All the included articles intend to contribute for the development of mobility assistants to support multi-modal transportation situations using smart cities' infrastructures. However, four of them propose algorithms that can be used by mobility assistants [6, 11, 12, 16], while the remainder focus on the development of specific applications [5, 7–10, 13–15, 17–20]. The aims of the 16 included articles are presented in Table 1.

3.1 Algorithms

The algorithm proposed by article [16] is useful to find optimal routes in a multimodal transportation network composed of a set of mono-modal networks (e.g., tramway, train or metro), road traffic networks and available parks in a city. The global multi-modal network is abstracted into sub-graphs. Contrary to traditional approaches where for the optimal path resolution the multimodal network is considered as a whole, the proposed solution consists on making intermediate calculations for each sub-graph before considering the whole network [16]. Results of the algorithm simulations under MATLAB show that when the global graph is abstracted into sub-graphs, the problem is simplified and consequently the execution time is reduced [16].

Article [6] proposes a method for planning optimal routes considering both the distance from origin to the destination and the passengers' density of public transports. This essentially requires finding shortest paths for both minimum distance and minimum total density. For that, a two-layer graph model was used. Consequently, the optimal solution is obtained by a two steps process: (i) first, a shortest path is found using conventional methods in the distance graph; then (ii) the solution vector that is

Table 1. Goals of the retrieved studies.

#	Aim	Year
[5]	To present a mobility assistant that handles end-to-end itineraries that may involve multiple green, shared and public transportation	2013
[6]	To propose a method for planning optimal routes considering both the distance from origin to the destination and the passengers' density of public transportation	2013
[7]	To propose a mobility assistant based on a stochastic time-dependent model for public transportation networks by leveraging a set of historical travel smart card data	2015
[8]	To present a mobility assistant focused on car-based multimodality, where the users always start the trips with their private vehicle but can also use public transportation to reach the destinations	2016
[17]	To present a crowd sensing-based mobility assistant in which the updates of transports schedule information relies on automatic stop event detection of public transportation vehicles using mobile sensing	2015
[9]	To present a mobility assistant that takes into account uncertainty (e.g., delays in time of arrivals, impossibility to board, or walking speed)	2015
[10]	To present a mobility assistant that leverages on comprehensive urban data (e.g., traffic network data or real-time traffic speed data), aiming to provide accurate and effective recommendations	2015
[18]	To present a mobility assistant able to support micro-navigation and to provide crowd-aware route recommendation	2016
[20]	To propose an uncertainty-aware mobility assistant to advise on how to use a given transportation system	2016
[11]	To propose a two steps algorithm to provide tourists with safe and efficient itineraries considering mobility policies	2017
[19]	To present a mobility assistant to support micro-navigation of travelers that are unfamiliar with multimodal transportation	2017
[12]	To present algorithms that incorporate dynamic transit schedule data while balancing the availability of bikes among the bike stations	2018
[13]	To present a mobility assistant that provides a multimodal route solution combining public transportation with carpooling	2018
[14]	To provide a mobility assistant that grounds on an approach for deliberative agents using mental attitudes, in order to overcome the information overload and proactively help travelers	2018
[15]	To present a mobility assistant that exploits an ecosystem of devices of a smart city	2018
[16]	To propose an approach of assistance taking advantages of multimodal urban transportation means, which is based on an abstraction of a city multimodal graph	2019

found in the first step is used to extract which lines are supposed to be used and at which stop the traveler must transfer in order to travel in vehicles with least number of passengers.

In turn, article [11] also propose a two steps algorithm, but for a different aim: to provide tourists with safe and efficient itineraries and at same time promoting sustainable mobility in the city by considering mobility policies. The proposed algorithm contains two phases: (i) transformation of the initial problem to an equivalent arc orienteering problem, where the scores and time costs are associated solely to the routes; and (ii) finding a near optimal solution to the transformed problem [11].

To evaluate the efficiency and accuracy of the proposed algorithm, the authors created a real-life dataset related to the city of Barcelona contained 800 test instances with different graph topologies. By evaluating the algorithm using this data set, the authors concluded that it finds solutions with accuracy requiring few interactions, which makes it suitable for mobile devices [11].

Article [12] is also focused in a dual optimization problem. The goal is to maximize the number of bike stations that are balanced and to optimize the route planning process by incorporating dynamic real-time data about schedule delays. A modified random-walk method is used to estimate the bike station's net flow and identify the stations that are unbalanced and for which it is required to either pick-up bike from or drop-off bikes to. In turn, for the route planning the proposed method [12] searches the shortest path from the origin to the destination using the standard cost metrics included in the Open Trip Planner framework.

Two heterogeneous sources of real world data from the city of Warsaw were used to evaluate the proposed algorithm and the authors concluded that the approach might allow the travelers to take advantage of the large number of bike stations located nearby transit stops rather than walking [12].

3.2 Applications

Considering the articles related to studies aiming to develop mobility assistants, most of them present proof of concepts [5, 7–10, 14, 15, 17, 19, 20] and only two report applications that were evaluated by end users [13, 18].

The purpose of [14] was to argue how a mobility assistant can overcome great amount of information. The authors proposed an agent based on the Belief, Desire and Intention model [22] and, according to them, the agent beliefs are retrieved from external services as well as from the mobile device the assistant is executed on, the intentions (i.e., selected plans from the plan database) are also located in the external services and the mobile device and the desires are encoded in the saved travel routes. However, the authors only present the approach and intended as future work to develop and implement the concept [14].

The study reported in article [5] aimed to develop a mobility assistant to make practically feasible the balancing of efficiency of time, energy, pollution and cost. The mobility assistant is composed by a smartphone application and a set of web services to gather and interpret relevant sources of data including transports status (e.g., train timetable, train delay, underground load, road traffic and deviation from standard travel time) and data retrieved from social networks, that is handled as a text message (e.g., for the detection of traffic jams or road bumps) [5].

In turn, the aim of the mobility assistant proposed by article [8] is to generate different ranked lists of possible multimodal routes that include also parking spaces.

The solution is intended to exploit heterogeneous data sources about the road infrastructure, mainly from third party data providers, such as digital map of the road networks and parking data [8].

Article [17] focuses on the communication framework using Extensible Messaging and Presence Protocol to support the development of a crowd assisted mobility assistant that publishes events resulting from the aggregation and interpretation of real-time public transportation data, including data collected by sensors that automatically detect halt events of public transportation vehicles at the stops.

Considering that popular route planners (e.g., Google Maps) have major drawbacks (e.g., the query results are the same no matter whether the departure time falls in peak or off-peak hours), article [7] reports a study aiming the definition of stochastic time-dependent models that take both travel time and waiting time into account and optimize both the speediness and reliability of routes. The resulting mobility assistant might leverage both dynamic (e.g., historical travel smart card data) and static (e.g., bus travel time and waiting time) data sources to recommend routes adapted to traffic situations.

The mobility assistant present by article [9] intends to consider uncertainties related to the expected arrival time of the different modes of transportation available in a city. Therefore, it is supported by a platform able to handle large volumes of heterogeneous urban data (e.g., public transportation network or real-time public transportation data) to provide dynamic updates [9]. By using open data about public transportation of the city of Rome, the authors concluded that their approach improves the planning of the routes and reduces average expected travel time.

The mobility assistant presented by article [20] is also supported by dynamic transportation network updates. For that, the authors defined a knowledge base, the network snapshot, with all the information available about a multimodal transportation network (e.g., real-time updates on the estimated times of arrival or GPS data collected from public transportation vehicles).

Similarly, articles [10] and [15] also argue that the aggregation of comprehensive urban data (e.g., real-time speed data) can optimize the recommendations to travelers.

The aim of the proposed mobility assistant by article [10] is to generate inter-modal trips and compute scores for a final decision. For that purpose, data from different sources, including road map data, public transportation data, traffic data and weather data are collected and aggregated [10].

In turn, the mobility assistant presented in article [15] also collects data from different sources, namely real-time traffic data, parking space vacancy data, public transportation data or car sharing data. Some of these data are foreseen to be collected by an ecosystem of Internet of Things (IoT) devices (e.g., traffic sensors that can sense vehicles passing by or installed at open-access parking lots) [15].

The objective of the study reported by article [19] is to support individuals who are unfamiliar with the transfer situation. This can be achieved by the development of a mobility assistant for positioning and navigation considering both the macro and micro (i.e., fine-grained contextual guidance) levels. The authors considered several requirements, including smooth transit (i.e., providing real-time data and indoor navigation guidance for travelers to smoothly transit at stations where they need to change) and indoor modelling (e.g., typical routes from one underground line to another or time-dependent patterns). Furthermore, the authors identified that Wi-Fi and RFID are suitable localization techniques for the use case scenario under consideration.

Article [18] presents a mobility assistant that uses the IoT paradigm (e.g., real-time interaction of travelers' smartphone devices with public transportation vehicle to sense the presence of on board passengers) aiming to improve the experience of public transportation usage by providing micro-navigation and crowd-aware route recommendation.

The mobility assistant has been integrated into the municipal bus infrastructure in Madrid, Spain, since 2013, and is available to the public as a free smartphone application that was downloaded by 750 users since it was first released. The mobility assistant was improved over two consecutive user trials aiming a technical test of the system effectiveness and the assessment of the user experience [18]: (i) the first study was devised to collect quantitative feedback from a broad set of users of the mobility assistant by integrating a short questionnaire into the application; and (ii) the goal of the second study involving ten participants was to analyze the attitudes and feelings that the participants developed during the use of the application. Several participants expressed positive experiences with regard to improved information accessibility [18].

The mobility assistant presented by article [13] aims to combine public transportation with carpooling services. The authors propose a set of layers to represent different transports mode and interlayer which represent the travelling time (by foot) required to transit from a transport mode to another. The Dijkstra algorithm was considered in combination with the graph representing the network. Furthermore, to match travelers to cars during the ride matching process, the authors considered that the car travels from origin to destination along a pre-determined optimal route, and travelers need to travel to the closest possible stop.

The authors conducted an evaluation process of the mobility assistant focused on technical and functional aspects, including its usability, accessibility, ease of use, robustness as well as functionality. For that questionnaire were requested to be filled in by the 71 test users and the mobility assistant was on average assessed positively [13].

4   Discussion and Conclusion

Considering the current trends of research related to mobility assistants using smart cities' infrastructures to support multi-modal transportation situations (i.e., the first research question), four of the retrieved articles report algorithms to determine optimal multimodal routes, while the remainder 12 articles report the development of mobility assistants.

In terms of algorithms, although the general goals are the same, the studies present different specific aims and, consequently, various approaches are envisaged: (i) in article [16] the global multimodal network is abstracted into sub-graphs related to either a particular mode of transport or the road traffic network of the city; (ii) article [6] proposes a method for planning routes considering both the distance from origin to the destination and the passengers density of public transports; (iii) article [11] proposes a method to provide tourists with safe and efficient itineraries considering mobility policies; and (iv) article [12] aims to maximize the number of bike stations that are balanced and to optimize the route planning process.

Considering the articles reporting the development of mobility assistants, a general concern is to aggregate heterogeneous sources of data to provide the users with optimal routes considering not only the distance between the origin and the destination but also other parameters such as the availability of alternative transport modes [13] or the crowd density of public transports [18].

In terms of data sources, in addition to static data sources (e.g., public transportation data [5, 7, 9, 10, 15], digital maps of the roads network [8, 10], existing parks in the city [8]) the mobility assistants use dynamic data sources, including real time road traffic [10, 15], real time public transportation data (e.g., deviation from standard travel time or underground load) [5, 7, 9, 17, 20], car sharing data [20], weather data [10] and social networks data (e.g., for the detection of traffic jams or road bumps) [5]. Furthermore, one article reports the use of travel smart card data [7], one article reports the use of GPS data [20] and other articles report the application of different devices such as mobile phones and other IoT devices to collect dynamic data (e.g., to detect halt events of public transportation vehicles [17], to monitor traffic, park space vacancy [15] or the positioning of the travelers [19]).

Regarding the proposed solutions maturity level (i.e., the third research question), it should be mentioned that most of the articles report prototypes that were developed to demonstrate the feasibility of the concepts. Indeed, only two articles (i.e., 12.5% of the included articles) report prototypes that were assessed by real users [13, 18]. Moreover, looking in detail for the assessments being performed, there is a lack of robust methodological approaches since ad-hoc questionnaires were used, the design of the studies and the measured outcomes were poorly described and, in general, the number of participants was small. These results are in line with the results of other studies analyzing smart cities' applications (e.g., [23, 24]).

Although the review selection and the data extraction of this systematic review were rigorous, it should be acknowledged that this study has limitations, namely the dependency on the keywords and the selected databases. Nevertheless, after this systematic review is possible to state that, in the case of mobility assistants, relevant arguments were made regarding the importance of smart cities' infrastructures. Furthermore, this systematic review evidences the lack of robust solutions.

Acknowledgements. This work was financially supported by National Funds through FCT – Fundação para a Ciência e a Tecnologia, I.P., under the project UI IEETA: UID/CEC/00127/ 2019.

References

1. Vidović, K., Šoštarić, M., Budimir, D.: An overview of indicators and indices used for urban mobility assessment. Promet-Traffic Transp. **31**(6), 703–714 (2019)
2. Benevolo, C., Dameri, R.P., D'Auria, B.: Smart mobility in smart city. In: Empowering Organizations, pp. 13–28. Springer, Cham (2016)

3. All, S.P., Klug, K.: Key to Innovation Integrated Solution Multimodal Personal Mobility. European Commission, Brussels (2013)
4. Papa, E., Lauwers, D.: Smart mobility: opportunity or threat to innovate places and cities. In: 20th International Conference on Urban Planning and Regional Development in the Information Society (REAL CORP 2015), pp. 543–550 (2015)
5. Motta, G., Sacco, D., Belloni, A., You, L.: A system for green personal integrated mobility: a research in progress. In: Proceedings of 2013 IEEE International Conference on Service Operations and Logistics, and Informatics, pp. 1–6. IEEE. (2013)
6. Nasibov, E., Berberler, M.E., Diker, A.C., Atilgan, C.: Optimal journey planning depending on distance and passenger density parameters. In: 7th International Conference on Application of Information and Communication Technologies, pp. 1–4. IEEE (2013)
7. Ni, P., Vo, H.T., Dahlmeier, D., Cai, W., Ivanchev, J., Aydt, H.: Depart: dynamic route planning in stochastic time-dependent public transit networks. In: 18th International Conference on Intelligent Transportation Systems, pp. 1672–1677. IEEE (2015)
8. Di Martino, S., Rossi, S.: An architecture for a mobility recommender system in smart cities. Procedia Comput. Sci. **98**, 425–430 (2016)
9. Berlingerio, M., Bicer, V., Botea, A., Braghin, S., Lopes, N., Guidotti, R., Pratesi, F.: Mobility mining for journey planning in Rome. In: Joint European Conference on Machine Learning and Knowledge Discovery in Databases, pp. 222–226. Springer, Cham (2015)
10. Yu, L., Shao, D., Wu, H.: Next generation of journey planner in a smart city. In: International Conference on Data Mining Workshop (ICDMW), pp. 422–429. IEEE (2015)
11. Mrazovic, P., Larriba-Pey, J.L., Matskin, M.: Improving mobility in smart cities with intelligent tourist trip planning. In: 41st Annual Computer Software and Applications Conference (COMPSAC), vol. 1, pp. 897–907. IEEE (2017)
12. Tomaras, D., Kalogeraki, V., Liebig, T., Gunopulos, D.: Crowd-based ecofriendly trip planning. In: 19th IEEE International Conference on Mobile Data Management (MDM), pp. 24–33. IEEE (2018)
13. Dimokas, N., Kalogirou, K., Spanidis, P., Kehagias, D.: A mobile application for multimodal trip planning. In: 9th International Conference on Information, Intelligence, Systems and Applications (IISA), pp. 1–8. IEEE (2018)
14. Kuster, C., Masuch, N., Sivrikaya, F.: Toward an interactive mobility assistant for multi-modal transport in smart cities. In: International Conference on Service-Oriented Computing, pp. 321–327. Springer, Cham (2017)
15. Lai, C., Boi, F., Buschettu, A., Caboni, R.: SmartMobility, an application for multiple integrated transportation services in a smart city. In: Proceedings of the 15th International Conference on Web Information Systems and Technologies (WEBIST 2019), pp. 58–66 (2019)
16. El Moufid, M., Nadir, Y., Boukhdir, K., Benhadou, S., Medromi, H.: A distributed approach based on transition graph for resolving multimodal urban transportation problem. Int. J. Adv. Comput. Sci. Appl. **10**(9), 449–454 (2019)
17. Farkas, K., Fehér, G., Benczúr, A., Sidló, C.I.: Crowdsensing based public transport information service in smart cities. Infocommun. J. **6**(4), 13–20 (2014)
18. Handte, M., Foell, S., Wagner, S., Kortuem, G., Marrón, P.J.: An internet-of-things enabled connected navigation system for urban bus riders. IEEE Internet Things J. **3**(5), 735–744 (2016)
19. Retscher, G., Obex, F.: A cooperative positioning service for multi-modal public transit situations. J. Navig. **71**(2), 371–388 (2018)
20. Botea, A., Berlingerio, M., Braghin, S., Bouillet, E., Calabrese, F., Chen, B., Gkoufas, Y., Nair, R., Nonner, T., Laumanns, M.: Docit: an integrated system for risk-averse multimodal journey advising. In: Smart Cities and Homes, pp. 345–359. Morgan Kaufmann (2016)

21. OpenTripPlanner: An open source multi-modal trip planner. https://github.com/opentrip planner/OpenTripPlanner. Accessed 31 Jan 2020
22. Georgeff, M., Pell, B., Pollack, M., Tambe, M., Wooldridge, M.: The belief-desire-intention model of agency. In: International Workshop on Agent Theories, Architectures, and Languages, pp. 1–10. Springer, Heidelberg (1998)
23. Rocha, N., Dias, A., Santinha, G., Rodrigues, M., Queirós, A., Rodrigues, C.: Smart cities and public health: a systematic review. Procedia Comput. Sci. **164**, 516–523 (2019)
24. Rocha, N., Dias, A., Santinha, G., Rodrigues, M., Queirós, A., Rodrigues, C.: A systematic review of smart cities' applications to support active ageing. Procedia Comput. Sci. **160**, 306–313 (2019)

An Overview of Aspects of Autonomous Vehicles' Development in Digital Era

Nadezhda Rozhkova[1] ⓘ, Darya Rozhkova[2(✉)] ⓘ,
and Uliana Blinova[2] ⓘ

[1] State University of Management, Moscow 109542, Russian Federation
[2] Financial University under the Government of the Russian Federation,
Moscow 125993, Russian Federation
rodasha@mail.ru

Abstract. This study focuses on the future of transport automatization. Engineering achievements such as creation of dedicated tracks on motorways and in urban areas in many countries of the world, global competition for new innovative markets, first victims of autonomous driving and other events make us turn to this topic. The relevance is associated with its promise in the field of technology development and improving the quality of life. The purpose of this article is to study the potential problem and prospects of autonomous transport, as well as the state of legal regulation. Conclusions are made about main advantages of car full automation: the safety, economic, environment and social gain, efficiency and convenience, mobility. By eliminating the human driver, the problem of legal regulation of autonomous vehicles arises. Technology of car automatization, disruptive by nature, represents a necessity to change existing legal norms of civil, criminal, administrative law etc. We have identified a major problems and directions of legislative initiatives for Russian Federation.

Keywords: Digital technologies · Autonomous vehicle · Autonomous driving · Driver assistance system · Self-driving cars

1 Introduction

Nowadays economies are changing greatly due to development of emerging markets, introduction of policies, new technologies and changing behavior of customers in relation to property ownership. Digitization changes industries, boosts their development without exception. An automotive industry is also subject to change. Connectivity, diverse mobility, autonomous vehicle and electro-cars will be driven by these factors. Most experts are sure that all four these trends will boost the development of each other, and the industry will be changed dramatically [1]. Even though many researches are conducted in the automotive sphere, there is no clear vision for the future.

Appearance of electro-cars and autonomous vehicles (AV) will be first in the high-income areas with high number of potential users, strict policies on carbon dioxide emissions, with all necessary infrastructure and where the cost of the technology will be less than income from it. If the cost represents high proportion of income, it will not be

© Springer Nature Switzerland AG 2021
T. Antipova (Ed.): ICIS 2020, LNNS 136, pp. 313–324, 2021.
https://doi.org/10.1007/978-3-030-49264-9_28

efficient to adopt until this technology becomes cheaper. Most experts are sure that an autonomous vehicle will be electric. Thus, promotion of self-driving cars can be considered from two main aspects: autonomous driving as a new technology and electrification. In the case of electric cars battery has a significant proportion of a car cost and it greatly influences commercial use of electric vehicles which in future have potential to become autonomous. However, the price of the battery has declining trend and after few years it can become available for large number of users. Fully autonomous vehicles are the most industry-changing technology; however, its development and cost of acquisition will not make it publicly available in the next few years. Nevertheless, partially autonomous cars can start to penetrate the market and play a crucial role in changing of customer's behavior, regulations, infrastructure and business models for autonomous vehicle era.

Today we constantly meet stories in the scientific literature and a press about autonomous vehicle that can leave drivers without work. The machine, which does not need a man behind the wheel or even a wheel itself, is supposed to become as familiar as once elevators without elevator operators or long-distance calls without the help of telephone operators. According to research J'son and Partners «The world market for self-driving cars in 2020–2035», the sales of AV will rise from 330 thousand cars in 2017 to 30.4 million cars per year by 2035. In the report it is underlined that «mass use of self-driving cars might face, in some countries, serious legal obstacles from regulators» [2]. Moreover, legal aspects and legal challenges of AV are presented in recent works of Beiker (an importance of a partnership between academia, industry, and government to establish policies for autonomous driving) [3], Brady (Australian law and automated cars) [4]. Thus, autonomous vehicle is an area for which legislative regulation is still a serious challenge.

Following the research path, the present paper addresses the following set of research questions:

- main advantages of AV adoption,
- main regulation that influences a probability of AV adoption,
- some practical steps in Russian legislation.

The paper proceeds as follows. Section 1 reviews theoretical basics of autonomous vehicles' development. Section 2 provides the experience of autonomous vehicles regulation, current stage and problems. In Sect. 3 we made an attempt to evaluate legal aspects and further changes in Russian legislation.

The scientific novelty of the article is to identify the potential of autonomous transport and establish its legal status.

2 Theoretical Basics of Autonomous Vehicles' Development

2.1 Autonomous Vehicles: Types and Main Advantages

In an age of rapid technological development, many of fantastic ideas can be translated into reality. One of these is an autonomous vehicle (self-driving vehicle, driverless car). The automotive industry is changing the direction of its development towards driving automation. Already in our days, many have heard about this project and it can be assumed that this innovation will come into our lives and will radically change the understanding of the word «transportation». For many years, the idea of unmanned vehicles has been trying to bring to life (starting from 70s of the XX century in Japan). Today this area continues to evolve, active development is underway around the world. However, over the past 10 years a lot of impressive work has been done.

Six levels of driving automation span from no automation to full automation have been developed in SAE international's J3016:

«0» - no automation – complete lack of automatic control of the car;

«1» - driver assistance – technologies for automatic vehicle control (cruise control, automatic parking system) may be used, but the driver should be required to intervene in the process of driving in case of unforeseen circumstances, the human driver perform all remaining aspects of the dynamic driving task;

«2» - partial automation – the driver "assists" in the automatic vehicle control system which perform steering and acceleration/deceleration;

«3» - conditional automation – the driver may not control the vehicle on sections of the road with predictable traffic, but has the obligation to be prepared to take a control;

«4» - high automation – level similar to previous but does not require driver attention;

«5» - full automation – the only duties of the driver are starting the vehicle and setting the coordinates of the destination [5].

This standard has been adopted in UK, Australia and USA. Most cars in our time are in the «1» and «2» levels of driving automation. The human driver performs all aspects of the dynamic driving task but cruise control, automatic parking is used. Large companies (Google, Toyota, Tesla, Volvo) strive to achieve complete automation of car.

In scientific research, the term «automated vehicle» applies to levels 4 and 5.

The technologies of car automatization show its high development prospect and therefore continues to evolve. During our investigation, many positive aspects of the project were discovered. One of the most important is an increase of traffic capacity. The whole process of selecting and calculating the route should be automated, which will allow determination the fastest route with high accuracy.

The main concept of a driving automation is the automated control of the vehicle, and many problems associated with the human factor will disappear, in particular, road traffic accidents. There are three types of accidents: the fault of the driver, the fault of the pedestrian and the fault of the poor technical condition of the vehicle. Thus, autonomous cars can significantly reduce the number of accidents and incidents that occur due to driver errors. The opinion of experts and accident statistics together prove the safety of autonomous transport.

An AV can affect the reduction in the cost of goods due to the fact that the cost of its transportation will decrease. The absence of drivers means that there is no need to spend money on salaries. The time of delivery and the amount of spent fuel resources will decrease. This indicator has a positive effect on the amount of spending.

Another important point when using AV is environmental friendliness. Google experts predict that autonomous vehicles can reduce the number of cars used by 90%. According to statistics, the average American car stays in the parking 95% of the time. Carlo Ratti, director of the MIT Senseable City Laboratory, gives an even more optimistic assessment: each autonomous vehicle can diminish from 10 to 30 private vehicles. Because of this, the emission of harmful substances into the atmosphere will be greatly reduced, and the emission of carbon dioxide will also be greatly decreased. The problem of the car and the environment has been of great concern to ecologists in recent decades, but it is with the advent of AV the situation can radically change. Plus, the vacant multi-million parking spaces can be used for landscaping many cities.

AVs will promote services of carsharing and ride-sharing and make private ownership less attractive. Demand for these services will attract new players to the industry. It will increase need for additional regulations. To attract participants tax benefits can be increased or government subsidies introduced. Carsharing and ride-sharing will also mean lower car sales and it will have an impact on automobile manufacturers.

To sum up, we consider that the main advantages of car full automation are:

1. Safety.
2. Economic, environment and social gain.
3. Efficiency and convenience.
4. Mobility.

Despite all benefits which appear with autonomous driving, concerns arise about future of taxi and bus driver. If the cars become fully autonomous, they will lose their jobs.

These people will have an incentive to be against spread of AVs. In this case it is possible for government to soften the situation by creating additional working places and employing drivers there. If such people get support from state, there will be much more support from customer side. Furthermore, liabilities and safety details are unknown, and a lack of personal information safety may be under question without introduction of the new privacy standards.

To eliminate uncertainties on the interaction of transportation system government should make research and implement new policies on autonomous vehicle, privacy, liability and security.

2.2 Existing Regulation that Influences AVs Adoption

Kyoto Protocol is an existing document that can influence AV's adoption. It targets on reducing emissions of greenhouse gases. The maximum amount of emissions (measured as the equivalent in carbon dioxide) that a party may emit over a commitment period in order to comply with its emissions target is known as a Party's assigned amount. The individual targets for Annex I Parties are listed in the Kyoto Protocol's Annex B [6].

Different countries have various policies on carbon dioxide emissions, cities have different level of development of charging stations, customers in some regions are readier to accept innovations than other citizens and it all creates conditions for electro-mobiles creation and market penetration. These vehicles are becoming commercially available according to demand for them and cost associated with the technology.

There is different need for regulation of technological advanced according to the state of its development and implementation. Different policies can be introduced by governments to soften the consequences of innovations. Small technological advances or regular changes in legislation will not cause big change in existing laws. Autonomous vehicle is a major development in automotive sphere, and it will need governmental interventions.

For example, government can raise awareness in society, improve trust of the population, especially in the sphere of personal data protection and storage as it depends on the technology used. It can also show its interest in the technology by creating stimulus for further development. For instance, this can be done by purchasing the technology, organizing research and development funding, loans, grants, providing necessary for development, implementation and operation infrastructure, carrying liability risks on the technology, creating support schemes or deciding to stop step by step use technology.

The acceptance of citizens will be the most successful in large cities of developed countries, which have high parking fees or free parking for electric or hybrid cars, strict CO_2 restrictions, electricity discounts, tax credits and subsidies, low dependence on fossil fuel and so on. In small cities and the countryside traditional fuel vehicles will remain preferable for a longer time due to lack of necessary charging infrastructure and low battery capacity. As batteries are becoming cheaper and have more capacity the difference between urban and rural infrastructure will become less noticeable and electrical vehicles will be able to take priority over fuel cars.

In some countries, for instance, United States fundamental steps for preparation of robocars implementation are being taken. The new standards for licensing and testing are being developed not at the national level, but on state level. It may create contradictions between states. There is also legislation on tax deductions, tax credits and environmental pollution: The Clean Air Act (42 U.S.C. § 7401) is a United States federal law designed to control air pollution on a national level [7]. For major sources, Section 112 requires that EPA establish emission standards that require the maximum degree of reduction in emissions of hazardous air pollutants; American Taxpayer Relief Act of 2012 [8]. The Federal Internal Revenue Service (IRS) tax credit is for new EV purchased for use in the U.S. The size of the tax credit depends on the size of the vehicle and its battery capacity.

Each state of the United States has its own legislation for autonomous vehicles' regulation. Over the past few years, 33 States have either signed laws to regulate AVs or announced initiatives to allow self-driving vehicles appear on public roads.

In EU the following legislation influencing adoption of electric and autonomous vehicles exist. EU Emissions Trading System (EU ETS). By 2021, phased in from 2020, the fleet average to be achieved by all new cars is 95 g of CO_2 per kilometer. From 2019, the penalty is €95 from the first gram of exceedance onwards. The cars Regulation gives manufacturers additional incentives to produce vehicles with extremely low emissions (below 50 g/km) by super credits. Each low-emitting car is counted as 1 from 2016 to 2019 and vehicles in 2020 [9].

The Council of the European Union in the Netherlands approved the testing of automated cars on the roads in 2015 and updated the legislation in 2017, according to which tests without drivers are now allowed [10]. In 2015, the Swedish government for the first time considered the issue of testing self-driving vehicles and concluded that the tests can be carried out on Swedish roads. In July 2017, it signed the relevant law. In 2014, the French government announced its intention to allow testing of autonomous cars in 2015. 2,000 km of roads are open for testing throughout the country. In 2013, the UK government allowed the testing of the vehicles on public roads. In 2018, the government announced its desire to put autonomous cars into use by 2021. Germany has approved a law that allows companies to start testing AVs on public roads. Drivers are allowed to take their hands off the steering wheel and do simple tasks during trips like looking at the phone.

In 2013, the EU authorities created the public-private partnership SPARC-a project entirely dedicated to the financing and development of robotics. In February 2017, the European Parliament adopted a resolution "Norms of civil law on robotics". The document, which consists of more than a hundred items, is devoted to various aspects and problems of robotics and artificial intelligence.

Certainly, the market for unmanned vehicles has an investment attractiveness; research and testing conducted by various companies is presented in Table 1.

Table 1. The volume of investment of some companies in the development of autonomous vehicles

Company	Country	Investments amount
Kamaz	Russia	300 billion rubles
Cognitive Technologies	Russia	750 million rubles
Hyundai	South Korea	22 billion dollars
Google	USA	1.1 billion dollars
GM и Uber	USA	1.5 billion dollars
Yandex	Russia	Not disclosed
Tesla	USA	Not disclosed

Russian companies invest in development and have only prototypes, but there are already potential unmanned vehicle industry leaders who actively test their work models (prototypes) on the roads (Google, General Motors).

3   Autonomous Vehicles Regulation

3.1   Current Stage of Autonomous Vehicles and Electronic Vehicles Development

The main success factor of electric vehicles in Norway is the large subsidies and benefits for the purchase of these vehicles. To encourage sales, the Norwegian government has abolished import duties on electric motor vehicles as well as taxes on their registration and sale. The owners drive free of charge on toll roads, use ferries, have parking discounts and use bus lanes in city centers. It is planned that by 2025 only cars with zero emissions of harmful substances will be sold in Norway.

The Swedish government plans to build a network of roads that allow EV's owners to charge electric cars on the road without stopping. It was reported in April 2018 by newspaper The Guardian. Innovative roads will be equipped with a "charged section" in which a contact rail is laid like in metro. It can charge the car through the hard wire which is in contact with the rail.

China seriously intends to compete with other countries in the self-driving segment. In December 2017, Beijing became the first Chinese city to allow the testing of autonomous vehicles. China has extended tax benefits for electric vehicles. The grace period has now been extended to 2020 for electric vehicles, hybrids, plug-in hybrids and fuel cell vehicles. The extension of benefits should motivate manufacturers to expand the range of electric vehicles on the Chinese market. Also, the reason was the coming in 2019 tightening of environmental requirements for vehicles sold in China. The exact date of the ban is discussed in the government.

Germany has repeatedly declared its readiness to fully switch to electric vehicles in the next 10 years, even though there are still many obstacles to achieve this goal. Earlier, the country even sent a resolution to the European Union asking for a ban on the sale of new cars with gasoline and diesel engines since 2030. However, the automotive industry in Germany faced an unexpected difficulty: the world simply does not have enough raw materials to produce batteries.

The Minister of economy of Taiwan said in the autumn of 2017 that the region plans in the coming years to completely switch to electric vehicles. The provincial authorities decided to take the example of France and the UK, which in 2040 will impose a ban on the sale of cars with internal combustion engine. But, as the Minister explained, Taiwan is going to switch to electric transport not only for the sake of improving the environment, but also to stimulate the economy.

In the UK, the sale of cars on diesel and gasoline will be banned from 2040. The newspaper Times wrote about it. Subsequently, the official announcement was made by the British Minister of environment Michael Gove. Deliveries of hybrid cars that have both an electric motor and a diesel or gasoline engine will also be banned.

In Russian Federation the State Duma began preparing proposals on legal regulation of robotics and the use of artificial intelligence. It is planned to create a legal framework for the development of new technologies.

3.2 The Problem of Legal Regulation of Autonomous Vehicles

The problem of legal regulation of autonomous vehicles can negate all its advantages. In connection with the need for legal regulation, it is necessary to rethink the legal aspects of:

- traffic law,
- insurance law,
- administrative law,
- civil law,
- criminal law.

It is necessary to determine the administrative offenses and criminal offenses associated with the use of autonomous vehicles. Also, it is required to determine the subjects of civil liability for causing property damage through the fault of unmanned vehicles. All these problems remain unresolved and require the development of a new legislative framework that will regulate this side.

If you consider world experience, especially developed countries, then the legislation of most countries currently prohibits the use of fully autonomous cars on public roads. Separate legislative initiatives regarding autonomous transport have been adopted in the EU countries. So, the EU directive 347/2012 and 351/2012 of January 1, 2016 prescribes to equip all trucks registered or used within the EU with special systems that ensure road safety. Nevertheless, the Convention on Road Traffic (the Vienna Convention on Road Traffic) requires drivers to constantly monitor the vehicle, which makes it difficult to test autonomous vehicles on public roads and launch a mass production. The EU wanted to amend the convention to enable the development of technology. To date some amendments have been done. Still it is prescribed that every driver shall at all times be able to control his vehicle. But the term «vehicle systems which influence the way vehicles are driven» has been appeared.

Today regulation in the field of autonomous vehicles has both successful and unsuccessful examples. In 2016 the USA National Highway Traffic Safety Administration (NHTSA) officially equalized the human driver and the computer that drives Google's unmanned vehicle. This solution greatly simplifies the development of autonomous vehicles. It can be argued that the United States in solving this problem has moved beyond the rest of the world. There are separate laws passed in some states (for example, Florida, Nevada, Michigan) that regulate the testing and movement of unmanned transport systems. Arizona lawmakers in the US last years tried to initiate legislation regulating the use of autonomous vehicles on state roads. But they could not solve the problem of who should be responsible for the accident that occurred with the participation of the "system driver": the owner of the car, the company that developed the technology, or the automaker who made and sold the car.

In California, in December 2015, authorities announced preliminary rules for autonomous vehicles. The fixing of these rules in the form of law is necessary before one of the vehicles can be sold to consumers. They require drivers, if necessary, to be ready to take complete control of their car. The rules proposed by the California Department of Automobiles suggest that drivers remain responsible for complying with traffic regulations, whether they are driving or not. Next step has been done in December 2019, when autonomous, light-duty trucks were allowed to use for commercial purposes on public roads in California.

Unfortunately, today Russia is less than other countries in the development of legislation in the field of the use of autonomous vehicles. Therefore, the discussion initiated in our country is very important for the implementation of the roadmaps, as well as the program for creating domestic autonomous vehicles in general.

At a meeting of the Presidium of the Council under the President of the Russian Federation, Strategy for the innovative development of transport in Russia was approved. One of the goals are to develop the market for partially and completely autonomous vehicles with a share of domestic manufacturers of 60% by 2035. The development of this technology is planned to be carried out at the expense of the state funds with a gradual decrease in favor of private investors.

4 Legal Aspects and Further Changes

In Russian Federation fundamentals of civil legal relations in the sphere of automobile transportations of passengers are stated in three documents:

- Civil code of the Russian Federation (part 2, section IV, Ch. 40);
- Federal law of 08.11.2007 № 259-FZ "Charter of road transport and urban land electric transport". This Federal law regulates relations arising in the provision of services by road and urban land electric transport, which are part of the transport system of the Russian Federation. This Federal law defines the General conditions of carriage of passengers and baggage, goods, respectively, buses, trams, trolley-buses, cars, trucks.
- Resolution of the Government of the Russian Federation of 14.02.2009 №112 (amended on 26.11.2013) «On approval of Rules of transportations of passengers and luggage by road and urban ground electrical transport» .

The absence in Russian legislation of the very concept of autonomous vehicles one of the main problems hindering the widespread use of autonomous cars in Russia.

Two main blocks of questions remain unresolved at both the international and Russian levels, one of which is related to who is responsible if accident (while there are two working versions - the software owner and the car owner), and the other - with personal data. To drive on the roads, an autonomous vehicle must collect a huge amount of information about the traffic situation and the purpose of the trip, and then transfer this information to the management company. However, there are very clear requirements related to data depersonalization.

In Russia there is a draft Federal Law No. 710083-7 «On the pilot operation of innovative vehicles and amendments to certain legislative acts of the Russian Federation». The subject of regulation of the document is relations that arise «in connection with the pilot operation of innovative vehicles, including highly automated vehicles, on public roads» (article 1). It is necessary to pay attention to four concepts: trial operation, innovative vehicles, highly automated vehicles, public roads. Autonomous vehicles in the text are called «highly automated vehicle» . There is also the concept of «innovative vehicles» , in which new design solutions are applied that qualitatively change the basic operational indicators, and which cannot be evaluated in accordance with the current standardization documents. However, this draft of the Federal Law is still under discussion and its concepts has not been launched. Key directions of legal aspects' improvement and two main blocks of questions remain unresolved in Russian legislation are presented in Fig. 1.

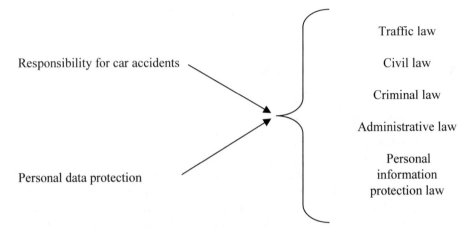

Fig. 1. Key directions of legal aspects' improvement

To our understanding, the first direction for improvement of legal aspects is the traffic law. It is currently provided that «a driver» – a person driving a vehicle, a drover, driving packs, riding animals or a herd along the road. A driver is equated with a driver training». It is considered that «a vehicle» is a device designed to transport people, goods or equipment installed on it on the roads.

As we can see, it is necessary to define autonomous vehicles. Currently, term is not universally recognized and normatively shown in the international law. Nevertheless, the term "highly automated vehicle" takes precedence over the term "unmanned vehicle". The often-used term "unmanned", "driverless" is less appropriate, since it emphasizes the absence of a driver (pilot) in the vehicle, and this cannot always be realized with the current level of technological development, and the term "unmanned" does not take into account the presence of intermediate levels of automation.

Of course, many changes to be made in terms of delimiting the roads intended for AV, as well as road signs. For example, in the future, familiar traffic signs may disappear. The car will read the tags using machine vision, as well as «communicate» with other drones using Vehicle-2-Vehicle (V2V) technology. It is not necessary for passengers to see, hear and know all this.

The second area for improvement is a civil law. The Commission on the legislative regulation of autonomous vehicles can make the following changes to the Civil Code of the Russian Federation: in art. 1079 of the Civil Code of the Russian Federation introduce a wording equating autonomous vehicles with the concepts of «vehicle» for the further possibility of applying the rules of art. 1079.

According to the amendments to art. 1079 of the Civil Code of the Russian Federation, the obligation to reimburse rests with the legal entity or citizen who owns a source of increased danger on the basis of ownership. As well it will reflect the fact that property liability for damage caused by autonomous vehicles as a source of increased danger should occur both when they are purposefully used and when their harmful properties are manifested spontaneously (for example, in case of damage caused by spontaneous movement of autonomous vehicles).

The third area of improvement is a criminal law. The driver is responsible for driving an autonomous vehicle in accordance with article 264 of the Criminal Code of the Russian Federation.

Russian lawmakers proposed to allow the use of autonomous vehicles on public roads if there is a driver in his cabin, who, if necessary, can switch the control process to manual mode. That means that to the question who should be responsible for the accident involving the «automotive driver» : the owner of the car, the company that developed the artificial intelligence system, or the car manufacturer – the answer is «the driver».

We emphasize that, in our opinion, this is a key issue, and at the initial stages of car automation it seems reasonable. In the future, however, the question arises of the fault of the driver, in case of incorrect control system settings. Never before manufacturers faced the need to produce cars with protection against the fact of an accident or from mistakes made by other drivers. In other words, if you get into an accident involving another driver, the manufacturer of your car will most likely have nothing to do with it. In the case of AV, questions will arise regarding the correctness of the «decisions» made by the automotive systems based on the algorithms in it.

The supplier of the car and its artificial intelligence, as a rule, is a legal entity, which entails amendments to the Civil code of the Russian Federation and the code of Administrative Offenses of the Russian Federation. However, the Criminal code of the Russian Federation does not dictate the liability aspects of legal entities. Since driving a vehicle is a potentially dangerous type of activity that can cause mild, moderate and severe harm to the health of citizens, it will be necessary to create a fundamentally new branch of criminal law in the world legal practice in cooperation with human rights defenders and representatives of development companies.

It makes sense for Russian lawmakers to adopt the experience of the United States and take the «soft» path: act through recommendations in Subjects of the Russian Federation, so that hypotheses are tested and work solutions are identified. This approach also has drawbacks: an autonomous car that works, for example, according to the standards adopted in the Kaluga Region, it may be outlawed in Moscow.

5   Conclusion

In modern times, when there is so much attention attracted to development of artificial intelligence and «self-driving cars» a lot of questions arise on its regulation. Autonomous vehicle being at the same time electric provide a lot of benefits for economies, but these technologies at the same time have significant drawbacks. These drawbacks may be softened only with intervention of state. Automobile producers and high-tech companies continue to develop improvements of self-driving cars. Many countries have already started to prepare legal background for the technology. Some states of USA have signed different legal acts on the autonomous vehicle regulation. Most developed countries in Europe allowed AVs to appear on public roads. However, there are still various issues that should be put under legal framework and decrease the public anxiety.

It should be noted that the introduction of innovative technologies in any area of our lives entails changes, including in the legal field. In our opinion, it is not necessary to establish bans, since the development of technology is an exceptionally right step in a successful future. It is necessary that the legal regulation of this sphere be advanced and that, with reasonable regulation and protection of the interests of subjects of various legal relations, it would not deter, but contribute to the development of advanced technologies in the interests and under the control of the human mind.

References

1. Bryant W.S.: How governments can promote automated driving 25–41 (2016)
2. Mosquet, X., et al.: Revolution in the driver's seat: the road to autonomous vehicles. Boston Consult. Group **11** (2015)
3. http://json.tv/en/ict_telecom_analytics_view/the-world-market-for-self-driving-cars-in-2020–2035
4. Beiker, S.A.: Legal aspects of autonomous driving. Santa Clara L. Rev. **52**, 1145 (2012)
5. Brady, M.A.: Is Australian law adaptable to automated vehicles? Griffith J. Law Hum. Dignity **6**(3), 35–71 (2019)
6. Automated driving levels of driving automation are defined in new SAE International standard J3016. SAE International TM. https://web.archive.org/web/20161120142825, http://www.sae.org/misc/pdfs/automated_driving.pdf
7. Kyoto Protocol: Kyoto Protocol's Annex B. United Nations Climate Change. https://unfccc. int/process/the-kyoto-protocol
8. The Clean Air Act (42 U.S.C. § 7401) of USA of 3rd June (2014). https://www.epa.gov/ sites/production/files/2014-06/documents/kcbx-nov-20140603.pdf
9. American Taxpayer Relief Act of USA (2012). https://www.legalmatch.com/law-library/ article/american-taxpayer-relief-act-of-2012.html
10. European Commission. EU Emissions Trading System (EU ETS). https://ec.europa.eu/ clima/policies/ets_en

Health Policy and Management

Comparing Paper Versus Digital Registries When Introducing a Web-Based Platform Registration for Social Care Services

Ana Isabel Martins[1,2] ⓘ, Hilma Caravau[1,3] ⓘ, Ana Filipa Rosa[1,3] ⓘ,
Ana Filipa Almeida[1,3], and Nelson Pacheco Rocha[1,3(✉)] ⓘ

[1] Institute of Electronics and Informatics Engineering of Aveiro,
University of Aveiro, Aveiro, Portugal
npr@ua.pt
[2] Department of Electronics, Telecommunications and Informatics,
University of Aveiro, Aveiro, Portugal
[3] Medical Sciences Department, University of Aveiro, Aveiro, Portugal

Abstract. The study presented in this article aimed to analyze the shifting process from a typical paper registration to a registration performed by a web-based platform in two different services for older adults of a social care institution. A total of 16,960 registers related to a nursing home and a home care service were analyzed. With the web-based platform, the rate of inconsistencies declined 21.08% for the nursing home and 24.89% for the home care service.

Keywords: Older adults care services · Social care institutions · Care management systems · Comparative study · Web-based platform

1 Introduction

The use of digital technologies in health and social care sectors improves the quality and efficiency of services, and consequently the quality of care delivered. Although the digital change is happening subtly in the social care sector, it promises good outcomes for the patients' quality of life. Studies that compare typical approaches with technology-based solutions might contribute to support decision-making by organizations and their administrators. It is also relevant to assess whether the performance of the solutions is similar in different types of services considering their specificities.

The study presented in this article aimed to analyse the shifting process from a paper registration to a web-based platform, the Ankira® platform [1], both in nursing home and home care service, and was informed by the following research questions:

- RQ1: Does the adoption of a digital solution for the registration of daily social care procedures improve the quality of the registrations?
- RQ2: Does the adoption of a digital solution represent an added value for patients and caregivers?

© Springer Nature Switzerland AG 2021
T. Antipova (Ed.): ICIS 2020, LNNS 136, pp. 327–336, 2021.
https://doi.org/10.1007/978-3-030-49264-9_29

Besides this introduction, in the following sections the authors present the related work, details of the web-based platform, the methods, and the reached results. Moreover, a discussion of the results will be included, and a conclusion will be drawn.

2 Related Work

Despite the efforts of the individuals and their informal caregivers to maintain the autonomy and independence, and to remain in their natural environments as long as possible [2, 3], with the ageing process arise several needs and problems that often require specialized social care services. Therefore, due to the demographic ageing, the social care context is changing [4]. All over the world, social care services are supporting an increasing number of older adults presenting a broad range of complex needs, while facing reduced or static funding [4–6].

Similarly to the healthcare sector, where the development of information services for healthcare providers, patients and informal caregivers, both in hospital context or at home environment, have received great attention to improve quality, efficiency and patient experience, the social care sector also requires a large scale digital change [4, 7, 8]. Concerning the potential of digital solutions, there are proposals of the use of information technologies in the social care sector, in order to operate as a mediator between different actors of the social care provision [8–10]. The articulation between services and caregivers is important to guarantee patients' quality of life.

There is a general impression that the health and social care sectors have difficulties to manage the changes related to the introduction of digital technologies [8]. Attempts to introduce digital technologies can fail for a number of reasons, including [4, 8–16]: resistance towards digital adoption; concerns about privacy, safety and security, autonomy, non-maleficence and beneficence; perceived usability and support; the pressure related to serve an increasing number of individuals with scare resources, that does not allow the introduction of new and disruptive working processes; the difficulties in finding the resources needed to implement digital technology, both hardware and software; lack of adequate solutions namely in terms of interoperability; difficulties in making the cultural changes needed; and difficulties in having in-house experience required to lead and implementing the change.

There are studies that have evaluated the effectiveness and efficiency related to the introduction of digital technologies in the health care sector [8, 17, 18]. In turn, in the social care sector is difficult to find distinct digital change examples [8, 19]. However, there is the need to advocate the use of digital solutions to enable the digital change of the social care sector [7]. For that, empirical evidence and case studies can be a helpful way of conveying positive and negative learning from technological implementations to support social care services.

3   Web-Based Platform

The selected web-based platform, the Ankira® platform [1], was launched in mid-2013 and supports managing administrative, care, social, and clinical information of elderly patients from social care services such as nursing homes, day care centres, and home care services. It has also been adapted by other care services such as facilities for the mentally disabled.

Regarding a patients' dossier, information is organized in five main areas:

- Basic Activities of Daily Living, which includes bathing and personal hygiene, feeding, mobility, toilet and dressing.
- Instrumental Activities of Daily Living, including cleaning of the home and the clothes, shopping, or accompanying to the outside.
- Clinical support, which comprises medical appointments, medication or treatments (e.g., from nurses, physical therapists or speech therapists).
- Psychosocial support, including psychological assessments and social evaluations.
- Group activities, such as intellectual or sports activities.

In each of these areas, the care team may perform regular evaluations of the patients, for instance regarding their functional dependences or health condition, and plan the care. Caregivers may then register daily records of the activities performed (e.g., records such as baths given, administered medication, vital signs measurement, or the participation in a group activity), either using a web application or a mobile application.

All these records provide caregivers with information to both evaluate the evolution of the patient and the level of accomplishment of goals set in its individual plan. Managers may also use the registered information to assess service-wide indicators and obtain metrics to help make strategic decisions.

4   Methods

In this article, the authors report an experimental study that aimed to analyse the introduction of a web-based platform in a Portuguese social care institution. The shifting process from registers in paper format to a technological platform was assessed in two different services for older adults, namely a nursing home and a home care service. The objective was to verify which one of the approaches may represent an added value for the quality of the registrations, services, technical team and consequently for the patients.

To assess and compare the registration quality between both approaches (i.e., registers in paper format versus registers with the web-based platform) in a nursing home and a home care service, this study was performed in two phases and the respective data collection occurred between September and December 2019.

The research team randomly selected and collected paper and digital registers related to different areas. For the nursing home (i.e., phase 1) the registers studied were: hygiene, alimentation, drug administration, elimination, and positioning. For the home care service (i.e., phase 2), the registers studied were: hygiene, alimentation, drug administration and home hygiene. Then the quality of the information registered was analysed and the identified inconsistencies were accounted. As inconsistencies we considered, missing information, filling errors, information recorded in wrong fields or duplicate information.

The hygiene register includes information about patients' hygiene plan. Thus, the caregiver must register each procedure related to the hygiene, including, hair wash, nail care, skin hydration, intimate hygiene, among others. In paper format, this register includes checking the boxes of accomplished procedures, registering the execution time and signing the name of the person responsible for the care execution. In the web-based platform, the hygiene register includes two different registries: bath register and hygiene care register. The bath register, defined according to the patients' established plan, includes the type of bath and, when necessary, additional information in an observations field. In turn, the hygiene care register refers to hygiene care such as oral or partial hygiene. In the observations field it is also possible to include extra information.

Considering the alimentation register in paper format, there is a monthly sheet per patient in which the caregiver responsible for the feeding should sign the corresponding box (e.g., breakfast, morning supplement, lunch, snack, dinner or supper). There is also a field to register if dietary supplements were consumed, as well as observations. In the web-based platform the caregiver must access to the patient card that contains the daily meals that should be delivered according to the defined plan (usual meal and/or supplement) and check the corresponding one. Additional information can also be registered in an observations field.

Concerning the drug administration register, it is based on the medication plan defined by the nursing team. Care registries may be used by caregivers to monitor and evaluate the support that is provided and to follow the evolution of their patients. In paper format, the registration sheets include checking boxes for the caregiver to sign. There is one box per meal (e.g., fast, breakfast, lunch, snack, dinner or supper). There is also a daily observation field. In the web-based platform each patient has a medication plan card, with the number of doses planned for the day and the respective medication. After the login and the selection of the respective menu, the caregiver must choose the period of taking the medication (e.g., lunch) and save the data. There is also a field for observations that should be completed whenever necessary.

The elimination register, only applied in nursing home context, compiles the information about the type and frequency of patient elimination as well as the material used (e.g., adult diapers). It can be seen both as a health care register, allowing to understand if the patient has any abnormal functioning by analysing defecation and urination frequency, and also as a stock control of elimination items, as diapers. In paper format, the caregiver registries the change of diapers as well as the size of the diapers being used and checks the box with information about the type of elimination (i.e., faeces and/or urine). The register also requires the signature of the responsible for the procedure execution. In the web-based platform, to perform an elimination register it is necessary to select the period of the day (e.g., morning, afternoon or evening). By default, elimination items planned for the selected time/period are marked. The caregiver must indicate the type of elimination (i.e., faeces and/or urine) and any changes in faeces and/or urine. If the patient goes to the toilet, it is also possible to do this register, indicating date and time in the 'Go to toilet' option.

Concerning the positioning register, the aim is to register the number of times that the caregiver positions patients who are unable to do it themselves. As in elimination, the positioning register is only performed in the nursing home context. In paper format, the caregiver must write the abbreviation of the positioning (e.g., right lateral recumbent is register as RLR). The caregiver responsible for the procedure execution must also sign his name on the register. In the web-based platform, to register positioning the caregiver selects the type of positioning or selects the position in which the patient was placed. It is also possible to register additional information in an observations field.

The home hygiene register, only applied in the home care service, include the services of tidying and cleaning patients' home, in areas of patient exclusive use (e.g., bedroom or bathroom). The responsible for performing the defined task, in line with the planned, must check the corresponding box and sign his name on the paper register. In the web-based platform, it is only need to login and select the respective scheduled task (e.g., changing the sheets).

Following, will be presented the results of the comparative analysis. It should be noted that the number of fields in each one of the six registers type can be different, in paper and in the digital format. For example, in paper it is always necessary to have a field destined to the caregiver signature and in some cases to indicate the time of the procedure execution. The web-based platform overcomes this situation since it automatically records the identity and time of entry with the login action, and thus the registries are associated with this information.

5 Results

A total of 16,960 registries were analysed, 13,630 nursing home registries and 3,330 home care service registries. Of those, 7,839 were in paper format, and 9,121 were in digital format using the web-based platform. Table 1 presents the total of registries per service and format and Table 2 presents the type of registries.

Table 1. Registries per service and format.

	Paper format	Web platform	Total
Nursing home	6,390	7,240	1,3630
Home care	1,449	1,881	3,330
Total	7,839	9,121	**16,960**

Regarding the quality of the information registered in the nursing home, in the hygiene paper registries, a percentage of 23.41% of inconsistencies was verified. In the web-based platform, the inconsistencies percentage decreased to 1.64%. The alimentation register had a clear improvement with the introduction of the web-based platform, dropping from an inconsistencies rate of 48.68% to 18.66% of inconsistencies in

Table 2. Registries per service, type and format.

Register	n	Service	Paper format	Web platform	Total
Hygiene	3,943	Nursing home	1,350	1,830	3,180
		Home Care	414	349	763
Alimentation	4,179	Nursing home	1,736	1,736	3,472
		Home Care	311	396	707
Drug Administration	2,095	Nursing home	392	1,359	1,751
		Home Care	344	–	344
Elimination	2,493	Nursing home	1,302	1,191	2,493
Positioning	2,734	Nursing home	1,610	1,124	2,734
Home Hygiene	1,516	Home Care	380	1,136	1,516

digital format. Concerning the drug administration register, 33.67% of inconsistencies were stated in paper format and 22.78% in the web-based platform. The elimination register had 25.81% of inconsistencies in the paper format and 4.48% in the web-based platform registrations. Finally, in the paper format, the positioning register had 61.91% of inconsistencies while in digital format the percentage of inconsistencies was 36.91%. Figure 1 presents the percentage of inconsistencies by type of register resulting from comparing the paper registries to digital registries of the nursing home.

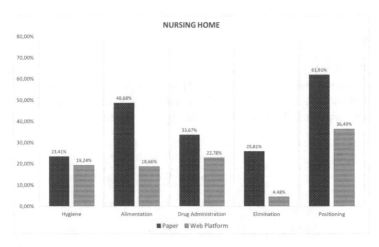

Fig. 1. Inconsistencies per type of register in the nursing home (paper versus web-based platform).

In what is concerned to the quality of the information registered in the home care service, the percentage of inconsistencies that was verified in the hygiene register was 2.41% in the paper format and 0.81% in the web-based platform. In tun, for the alimentation register, 10.67% of inconsistencies were stated in paper format and 3.86%

in digital format. Moreover, the home hygiene register had 31.25% of inconsistencies in the paper format and 2.56% in the web-based platform registrations. Figure 2 presents the percentage of inconsistencies by type of register resulting from comparing the paper registries to digital registries in the home care service.

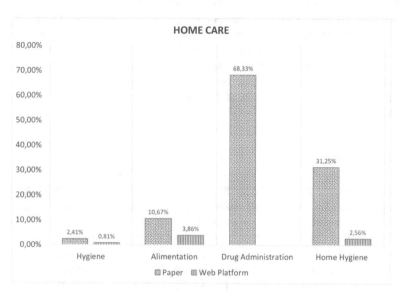

Fig. 2. Inconsistencies per type of register in the home care (paper versus web-based platform).

Table 3. Errors rate per service/type and improvement rate (%).

	Nursing Home		Home Care	
	Paper	Web Platform	Paper	Web Platform
Hygiene	23.41	19.24	2.41	0.81
Alimentation	48.68	18.66	10.67	3.86
Drug Administration	33.67	22.78	68.33	–
Elimination	25.81	4.48	–	–
Positioning	61.91	36.49	–	–
Home Hygiene	–	–	31.25	2.56
Weighted mean	41.10	20.00	27.40	2.50
Improvement rate	21.08		24.89	

Considering the institution where this study took place, for the home service the drug administration was still being recorder in paper format. Since, this clinic information demands a lot of attention from all the team, the institution decided to make the transition to digital format in phases. For that reason, there were no digital registers about drug administration in the home care service.

All types of records that can be compared between the analysed services, suffered a decrease in the percentage of inconsistencies with the introduction of the digital system for record management. Table 3 presents the error rates per service for the paper and web-based platform format in both nursing home and home care service, as well as the weighted mean of inconsistencies and the improvement rate, calculated by subtracting the weighted average of inconsistencies on paper for the weighted average of inconsistencies after the adoption of the web-based platform. The weighted mean of inconsistencies in the nursing home in paper format was 41.10%, while in the digital format was 20.00%. In its turn, the home care service had a weighted mean of 27.40% in paper format and 2.50% in digital format. With the web-based platform adoption, the rate of inconsistencies declined 21.08% in nursing home and 24.89% in home care service.

6   Discussion

This study explored the shifting process from a typical paper registration to a registration performed by a web-based platform in a social care institution for older adults. This was a two-phase study that focused in two different services that have very specific operating dynamics, a nursing home and a home care service.

Comparing the number of inconsistencies in the paper records in the two services (before the adoption of the web-based platform), the nursing home has more problems in terms of the information reliability than the home care service. This may happen because of the nature of the service itself, and the way records are made. In the home care service, there is less rotation of the staff teams and the records are made at the patients' home, right after the care is provided, leading to fewer filling inconsistencies. On the other hand, at the nursing home, the registrations are made at the end of the shift and for several patients at the same time, which leads to greater number of errors in registrations. Despite the different error rates in paper format, both services benefited significantly from the adoption of a web-based platform for registers due to a reduction of the inconsistencies rate (21.08% for the nursing home and 24.89% for the home care service).

Regarding the nursing home, the decreasing of inconsistencies rate is a promising result. Even thought, there are some constraints that difficult the registration process, such as the lack of mobile equipment available for that propose. At the present time, there is only one computer available for the staff to complete the registrations, and they normally do it at the end of the shift which may cause inconsistencies in the registrations, such as loss of information, filling errors and unreliable registers.

Although the home care service had a reduced rate of inconsistencies before de adoption of the web-based platform (28.15%), there was an evident decrease in inconsistencies rate to residual values (2.56%). The implementation of a digital system has shown benefits even in services that are already well-structured, as is the case of the home care service. Drug administration is one of the areas where quality records are critical, and the results of the paper format registration show that in the case of the home care service there is a large rate of inconsistencies (68.33%). As already mentioned, these registrations are still being made on paper, but will be introduced in the web-based platform in the future, and therefore have a potential for improvement. This is essential due to the impact that an error in the distribution of medication to a patient can cause.

The shifting process seemed to be well accepted by the staff, as the web-based platform adoption facilitates and makes many registrations faster and allows some data to be filled in automatically. For example, the authentication feature (i.e., login) avoids the need for signature and registration time as this is done automatically (contrary to what happens in paper format).

The ecological footprint of a digital system to register the care provided is minimal when compared with a traditional approach of paper registration. The amount of paper used in this type of social care services is immense, as most records are daily made and per patient over years of institutionalization and require a large physical space to store them, besides being a challenge to retrieve a past register that has already been archived.

7   Conclusion

The use of digital solutions in the social care sector is still a poorly implemented practice, as there are many constrains to overcome in the digitization process. This is a sector that has a great potential in terms of the adoption of technological solutions, as today it is still very traditional and paper oriented. Bottom line, the ones who will most benefit from these improvements, both in terms of quality of service and quality of life, will be the patients.

This study represents a contribution to the digitization of the social care sector, as it analyses the shifting process from a typical paper registration to a registration performed through a web-based platform in two different services for older adults of a social care institution. The results are unequivocal regarding the benefits of adopting information technologies for the registration of daily care procedures delivered to older adults. Both the nursing home and the home care service significantly decreased inconsistencies rates even in procedures whose registers had low inconsistencies rate in paper format. All types of registers analysed (i.e., personal hygiene, alimentation, drug administration, elimination, positioning and home hygiene) were improved with the adoption of the web-based platform. Despite the difficulty that normally exists in changing procedures in the social care sector, there was a good acceptance by the staff, as they saw their work facilitated since the act of registering daily procedures became easier and faster.

Further work is needed to study this shifting process, namely, to analyse the transition between approaches, and to evaluate what the involved participants perceive as difficulties and improvements, and how this process could be smoother, without implications for the care delivery dynamics.

Acknowledgements. This work was financially supported by National Funds through FCT – Fundação para a Ciência e a Tecnologia, I.P., under the project UI IEETA: UID/CEC/00127/2019.

References

1. Loureiro, N., Fernandes, M., Alvarelhão, J., Ferreira, A., Caravau, H., Martins, A.I., Cerqueira, M., Queirós, A.: A web-based platform for quality management of elderly care: usability evaluation of Ankira®. Procedia Comput. Sci. **64**, 666–673 (2015)
2. Bedaf, S., et al.: Which activities threaten independent living of elderly when becoming problematic: inspiration for meaningful service robot functionality. Disabil. Rehabil. Assist. Technol. **9**(6), 445–452 (2014)
3. Wiles, J.L., Leibing, A., Guberman, N., Reeve, J., Allen, R.E.S.: The meaning of 'aging in place' to older people. Gerontologist **52**(3), 357–366 (2012)
4. Toms, G., Verity, F., Orrell, A.: Social care technologies for older people: evidence for instigating a broader and more inclusive dialogue. Technol. Soc. **58**, 101111 (2019)
5. Australian Institute of Health and Welfare: Patterns in Use of Aged Care 2002–03 to 2010–11. Data Linka, Canberra, Australia: Australian Institute of Health and Welfare (2014)
6. Bottery, S., Varrow, M., Thorlby, R., Wellings, D.: A Fork in the Road: Next Steps for Social Care Funding Reform. The Health Foundation, London, UK (2018)
7. Local Government Association: Transforming social care through the use of information and technology. Local Government Association, London, UK (2016)
8. Maguire, D., Evans, H., Honeyman, M., Omojomolo, D.: Digital Change in Health and Social Care. The King's Fund, London (2018)
9. Ribeiro, V.S., Martins, A.I., Queirós, A., Silva, A.G., Rocha, N.P.: Usability evaluation of a health care application based on IPTV. Procedia Comput. Sci. **64**, 635–642 (2015)
10. Queirós, A., Pereira, L., Dias, A., Rocha, N.P.: Technologies for ageing in place to support home monitoring of patients with chronic diseases. In: HEALTHINF, pp. 66–76. INSTICC, Setúbal (2017)
11. Bunn, S., Crane, J.: Electronic Health Records. https://researchbriefings.parliament.uk/ResearchBriefing/Summary/POST-PN-0519. Accessed 02 Jul 2020
12. Llewellyn, S., Procter, R., Harvey, G., Maniatopoulos, G., Boyd, A.: Facilitating technology adoption in the NHS: negotiating the organisational and policy context – a qualitative study. Heal. Serv. Deliv. Res. **2**(23), 1–132 (2014)
13. Daly Lynn, J., Rondón-Sulbarán, J., Quinn, E., Ryan, A., McCormack, B., Martin, S.: A systematic review of electronic assistive technology within supporting living environments for people with dementia. Dementia **18**(7–8), 2371–2435 (2019)
14. Novitzky, P., et al.: A review of contemporary work on the ethics of ambient assisted living technologies for people with dementia. Sci. Eng. Ethics **21**(3), 707–765 (2015)
15. Ienca, M., Wangmo, T., Jotterand, F., Kressig, R.W., Elger, B.: Ethical design of intelligent assistive technologies for dementia: a descriptive review. Sci. Eng. Ethics **24**(4), 1035–1055 (2018)
16. Cavallo, F., Aquilano, M., Arvati, M.: An ambient assisted living approach in designing domiciliary services combined with innovative technologies for patients with Alzheimer's disease: a case study. Am. J. Alzheimer's Dis. Other Dementiasr **30**(1), 69–77 (2015)
17. Black, A.D., et al.: The impact of ehealth on the quality and safety of health care: a systematic overview. PLoS Med. **8**(1), e1000387 (2011)
18. Gill, R., Borycki, E.M.: The use of case studies in systems implementations within health care settings a scoping review, vol. 234. IOS Press, Amsterdam (2017)
19. Meiland, F., et al.: Technologies to support community-dwelling persons with dementia: a position paper on issues regarding development, usability, effectiveness and cost-effectiveness, deployment and ethics. JMIR Rehabil. Assist. Technol. **4**(1), e1 (2017)

Digitization of Medicine in Russia: Mainstream Development and Potential

Irina Mirskikh[1] , Zhanna Mingaleva[2(✉)] , Vladimir Kuranov[1] ,
and Svetlana Matseeva[1]

[1] E.A. Vagner Perm State Medical University, Perm 614990, Russian Federation
[2] Perm National Research Polytechnic University,
Perm 614990, Russian Federation
mingall@pstu.ru

Abstract. The Digital technologies play an important role in various spheres of activities. Information and communication technologies, digital technologies as well as telecommunication services are getting more widespread in the field of medical services. Digitization in medicine can provide implementing the system of remote health monitoring, visual interaction in-between patient and doctor or between doctors from different hospitals and regions.

At the same time the development of digitization of medicine gives rise to a number of problems. One of the most serious problems is the problem of misuse of personal data and information from medical documents of patients. Unfair access to such personal confidential information can be potentially dangerous and cause serious mistakes in medical treatment.

The aim of the research is to identify the main directions and tendencies in digitization of medical field, reveal the potential of telecommunication medicine in Russia.

Research methods. The study has a research character. It examines the key problems in the process of development of digital technologies use in medicine in order to characterize the main tendencies.

Keywords: Digitization of medicine · Protection of confidential information

1 Introduction

Digital technologies are spreading in various areas of market and society nowadays. Medical services field is not an exception since information and communication technologies, digital technologies as well as telecommunication services are getting more widespread.

According to the assessment of Global Market Insights the volume of the world market for digital medicine in 2016 was 51,3 billion US dollars and it is expected to be 116 billion US dollars by 2024. That is why the expected growth is 200% [1]. According to the forecast of Deloitte the healthcare expenditure growth rates in the whole world will be 5, 4% annually, and from 2017 to 2022 the worldwide healthcare expenditures will increase from 7,724 up to 10,059 trillion US dollars [2]. It must be noted, that earlier in 2012 Deloitte pointed out in its research there is an independent

© Springer Nature Switzerland AG 2021
T. Antipova (Ed.): ICIS 2020, LNNS 136, pp. 337–345, 2021.
https://doi.org/10.1007/978-3-030-49264-9_30

on-line segment of consumers in the area of healthcare, that is measured as 17% [3]. It includes active users of electronic applications and Internet resources in the field of medicine.

The research made by another famous consulting company Arthur D. Little proves that world market of digital health care will reach 233,3 billion US dollars in 2020 [4]. The company analysts forecast the growth of mobile healthcare to be more than 130% in the period of 2017–2020. The conclusion about active development of telecommunication services in medicine is also confirmed by research of the company Zion Market Researches [5].

As for digitalization of medical services in Russia this process is facing some difficulties caused by many factors one of which is legitimacy of delivering medical telecommunication services.

It is necessary to mention that the lawful use of digital technologies in Russia as well as delivering medical telecommunication services have become possible since 2018 when the Federal Law № 242 "On amendments to the current legal acts of the Russian Federation on the issues of using information technologies and delivering medical telecommunication services in the field of health care" was enacted. It determines the significance of analytical researches on finding out possible and the most promising directions in developing digital technologies and medical telecommunication services in Russia and defining the potential of their development.

2   Theory and Methodology of Research

The theoretical basis for research is a modern conceptual approach according to which in the process of forming directions of socio-political development of the society the priority is given to demands of people. In other countries the concept of enhancement of the rights and opportunities for patients, their empowerment is spreading all over. A patient is also encouraged to make an individual choice in the sphere of treatment. These priorities are based on works of J. Khuntia, D. Yimb, M. Tanniru, S. Lim [6]. In these works, the importance of defining the prospects of digital technologies for providing patients with a free choice is specially emphasized. In 2013 A. Jutel and D. Lupton held content-analysis of medical applications, available on Google and AppStore [7]. The key objective of that research was to identify the applications for medical diagnostic that are freely accessible on Internet.

On the other hand, some researches are devoted to analysis of both opportunities and advantages of using mobiles, mobile applications and wireless integrated medical assistance system by medical personnel. Among them there are works of C.L. Ventola [8], E. Lee and S. Han [9], N. Menachemiand and T. Collum [10], T. Antipova and I. Shikina [11], Y. Voskanyan with colleagues [12], and others.

One more very important area of research and the concept that is getting more widely-spread in digitalization of medicine is theoretical and practical reasoning for importance of attracting people, society as a whole, public organizations to create a mechanism for providing high quality medical help [13]. These issues are also discussed in the general context of healthcare as a social institution [14], and in the framework of issues of corporate social responsibility of employers for the health of their staff [15]. Modern means of remote monitoring of the main indicators of the health of workers, digital data transmission to medical institutions (including specialized hospitals and laboratories) and obtaining results make it possible to organize control over the health of personnel at any enterprises in close cooperation with medical institutions. At the same time, many researchers pay special attention to the need for the widespread introduction of digital educational technologies in the training of medical students, as well as in the training and professional development of practicing doctors - "the examination of the participants' educational process of the usage of media innovations, that is, digital technologies, would show a step further to the systematization of the media and information literacy, which would have a great effect on the educational system and creation of competences for active and responsible citizens" [16, p. 1098]. This is naturally determined by the general trends of globalization and the digitalization of the education system as a whole [17–20].

Active studies of opportunities and potential of digital medicine in Russia are held by Russian scientists also. These are the works by G.A. Polinskaya and M.G. Mesropian [21], A.S. Kozlova and D.S. Taraskin [22] and other authors. There are first works to assess the prospects for the development of certain areas of medical care in Russia based on the Blockchain - for example, the dental industry [23].

The attempts to estimate the prospects of digitalization of Russian medicine are made by means of Blockchain technology [24]. The mechanism of percolation suggested by M.A. Kantemirova and Z.R. Alikova makes it possible to improve the focus of digital medicine development in the regions [25]. There is also detailed study of processes of digital technology adaptation in regional medical complexes, including percolation that ensures medical digital technologies spread in regions taking into account objectives regarding decreasing the financial burden on the budget [26]. These studies are closely related to the issues of structural modernization of the regional economies and are carried out as part of the study of the general problem of territorial security through digitization, globalization, innovation development and structural changes, the problem of legal regulation of relations in the field of healthcare with the help of the norms of professional medical organizations [27–29].

This study has a research character. It has to detect main directions and tendencies in digitalization of medical field. It is also aimed to reveal the prospects of developing telecommunication medicine in Russia. These objectives define the method of the study.

The main method of the research is qualitative content-analysis using such search systems as Google and Yandex, Internet sites connected to offering digital medical services and medical goods as well as content-analysis of various medical web-applications.

The following regulatory acts of the Russian Federation in the area of digitalization of economy and some of its sectors are used as the legal sources of the research: the Federal Law №242 "On amendments to the current legal acts of the Russian Federation on the issues of using information technologies and delivering medical telecommunication services in the field of health care" of 01.01.2018 [30], the National Program "Digital economy of Russian Federation" (RF Government Directive of 28.07.2017 №1632) [31], the Convention of Human Rights and Dignity Protection in Sphere of Biomedicine [32].

3   Results of the Study

Based on the analysis the main factors that influence the development of the market of electronic applications and Internet resources in medicine in Russia there were revealed (see Fig. 1).

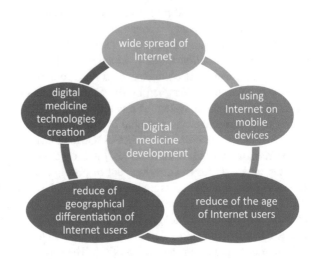

Source: compiled by the authors

Fig. 1. Factors of positive influence on the development of digitalization of medical services in Russia

Combination of these factors has a positive impact on spreading of digital technologies in medicine in Russia.

Statistics shows significant growth of investments in this field. Thus in 2018 the total investment amount into digital medicine sector was 14,6 billion US dollars, that is 10 times higher than in 2010 [33]. Almost 23% of that are investments into development of digital and mobile applications for patients, it also includes telemedicine (about 3,3 billion US dollars). Considerable investments are also directed to create technologies of telecommunication diagnostic and screening (2,3 billion US dollars – 16%) [33].

The analysis of main directions of digitalization of medical services in Russia shows the following.

The most widely spread directions of digitalization of medical services in Russia at the present moment are delivering telemedicine consultations. It refers to organization of collecting, processing and transmission of medical information by means of information-communication technologies and tools of data transmission. Nowadays there are two ways applied for organizing such consultations (see Fig. 2).

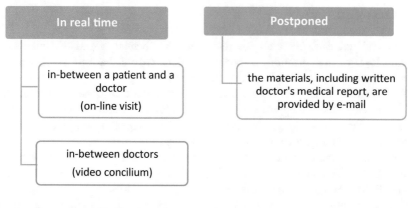

Source: compiled by the authors

Fig. 2. Ways of organizing telemedicine consultations delivery

The main fact defining, how often this or that way is used, is the equipment level of medical institutions and patients themselves with various means of information-communication technologies, such as computers, video cameras, skype and other computer programs providing visual interaction in-between patient and doctor or between doctors from different hospitals and regions.

It must be mentioned that this direction turns to be the most widely spread in the whole world. Thus, in 2016 1,25 million consultations like that were held in the USA [34].

The second major direction of digitalization of medical services development in Russia is implementing the system of remote monitoring. The mail contingent of consumers interested in such kind of medical services is patients who need some of their medical parameters to be controlled constantly. This group of digital technologies users is formed by people suffering from chronical diseases, for whom constant control of definite health conditions (blood pressure, blood structure, etc.) is vital.

The second big group of digital technologies users consists of people working in potentially dangerous industries. These may be various occupations connected to radiation, working in workshops with high level of air pollution with particles of gas, raw materials (many of chemical, oil-chemical industries, pulp and paper industry, electro technical industry, machinery etc.). These industries require constant control not only on air condition at working places but also control and estimation of respiratory tracts conditions of their workers. Applying special monitoring devices at working places will make it possible to monitor in real time how the people affected by hazardous factors feel and prevent diseases at an early stage.

Finally, the third direction of digitalization of medical services in Russia that is developing drastically in recent years is delivering educational medical programs. It stands, however, not only for education as it is but for wider interaction, re-education, skills development and advanced training for medical personnel and doctors. Video conferences with experienced doctors in real time, on-line consulting during surgeries refers to this direction also.

4 Discussion

Foreign and domestic research studies of major advantages and disadvantages of growing level of digitalization of medical services proved that this process has both positive and negative consequences of its development.

The main advantage of mobile health care is significant expanding of access to medical services and the opportunity to apply individual solutions in health care [28], that ensures faster medical service delivery, minimize expenses and gives wide access to services in health care with higher level of quality.

For instance, the calculation results held by specialists of Child medical system "Nemours" (the USA) proves that applying of tele-medical consultations have makes it possible for the Americans to save about 1 h on each visit to a doctor due to no need for queuing, commuting to clinics. It also saved 50 US dollars per month on commuting (paying for taking taxes or public transport etc.) [35].

As for disadvantages they are mostly connected to possible negative consequences of misuse or loss of information from medical documents of patients. The problem of adhering to the ethical principles of counseling as digital technology develops also moves from the general field of consulting to medicine [34, 35].

It is very important to create a progressive legal regulation in the sphere of digital medicine taking into account the necessity to protect personal data and confidential information from medical documents of patients. These types of confidential information can be easily stolen and sold. Some researchers consider that the cost of medical confidential information concerning patient's health can be much higher than the cost of trade secrets and other types of confidential information. Misuse of personal data and information from medical documents of patients becomes one of the most serious problems in the sphere of digital medicine. Unfair access to such personal confidential information can be potentially dangerous and cause serious mistakes in medical treatment.

5 Conclusions

Digitization of medicine in Russia refers to organization of collection, processing and transmission of medical information by means of information-communication technologies and tools of data transmission.

One of the most important directions of digitalization of medical services in Russia is implementing the system of remote monitoring. It is suitable for the patients who need permanent control of some of their medical parameters, i.e. people suffering from chronical diseases, for whom constant control of definite health conditions (blood pressure, blood structure, etc.) is vital.

The equipment level of medical institutions and patients is very important. Various means of information-communication technologies, such as computers, video cameras, skype and other computer programs can provide visual interaction in-between patient and doctor or doctors from different hospitals and regions.

Mobile health care provides the access to medical services, opportunity to apply individual solutions in health care, that ensure faster medical service delivery, minimize expenses.

The use of digital technologies in Russia and delivering medical telecommunication services is based on such legal acts of the Russian Federation as the Federal Law №242 "On amendments to the current legal acts of the Russian Federation on the issues of using information technologies and delivering medical telecommunication services in the field of health care" and the State program "Digital economy of Russian Federation".

Acknowledgment. The work is carried out based on the task on fulfilment of government contractual work in the field of scientific activities as a part of base portion of the state task of the Ministry of Education and Science of the Russian Federation to Perm National Research Polytechnic University (topic No 0751-2020-0026).

References

1. The digital revolution in healthcare: achievements and challenges, TASS, SPIEF-2017 website, 29 May. https://tass.ru/pmef-2017/articles/4278264. Accessed 01 Dec 2019
2. Forecast of the development of the global healthcare industry in 2019. "The future is shaping today", Deloitte. https://www2.deloitte.com/en/ru/pages/life-sciencesand-healthcare/articles/global-health-care-sector-outlook.html. Accessed 11 Dec 2019
3. Greenspun, H., Coughlin, S.: mHealth in a mWorld. How mobile technology is transforming health care, Deloitte Center for Health Solutions. https://www2.deloitte.com/content/dam/Deloitte/us/Documents/life-sciences-health-care/us-lhscmhealth-in-an-mworld-103014.pdf. Accessed 23 Oct 2019
4. Succeeding with digital health. Winning offerings and digital transformation, Arthur D. Little. http://www.adlittle.com//sites/default/files/viewpoints/ADL_2016_Succeeding_With_Digital_Health.pdf. Accessed 04 Nov 2019
5. Zion Market Research. https://www.zionmarketresearch.com. Accessed 01 Nov 2019
6. Khuntia, J., Yimb, D., Tanniru, M., Lim, S.: Patient empowerment and engagement with a health infomediary. Health Policy Technol. **6**(1), 40–50 (2017)
7. Jutel, A., Lupton, D.: Digitizing diagnosis: a review of mobile applications in the diagnostic process. Diagnosis **2**(2), 89–96 (2015)
8. Ventola, C.L.: Mobile devices and apps for health care professionals: uses and benefits. Pharm. Ther. **39**(5), 356–364 (2014)
9. Lee, E., Han, S.: Determinants of adoption of mobile health services. Online Inf. Rev. **39**(4), 556–573 (2015)
10. Menachemi, N., Collum, T.: Benefits and drawbacks of electronic health record systems. Risk Manag. Healthc. Policy **4**, 47–55 (2011)
11. Antipova, T., Shikina, I.: Informatic indicators of efficacy cancer treatment. In: 12th Iberian Conference on Information Systems and Technologies (CISTI), pp. 1–5 (2017)
12. Voskanyan, Y., et al.: Medical care safety - problems and perspectives. In: Antipova, T. (ed.) ICIS 2019. LNNS, vol. 78, pp. 291–304 (2020)
13. Carman, K.L., Workman, T.A.: Engaging patients and consumers in research evidence: applying the conceptual model of patient and family engagement. Patient Educ. Couns. **100**(1), 25–29 (2017)

14. Nasibullin, R.T.: Health as a social institution. Bull. PNIPU: Socio-Econ. Sci. 4, 161–173 (2019)
15. Kozlova, O., Makarova, M., Mingaleva, Z.: Corporative social responsibility as a factor of reducing the occupational health risk of personnel. Int. J. Appl. Bus. Econ. Res. (IJABER) 14(14), 683–693 (2016)
16. Cvetković, B.N., Stošić, L., Belousova, A.: Media and information literacy the basis for applying digital technologies in teaching from the discourse of educational needs of teachers. Croatian J. Educ. 20(4), 1089–1114 (2018)
17. Mingaleva, Z., Mirskikh, I.: Globalization in education in Russia. Procedia-Soc. Behav. Sci. 47, 1702–1706 (2012)
18. Akhmetova, S.G., Nevskaya, L.V.: Experience in introducing new technologies in higher professional education. Bull. PNIPU. Socio-Econ. Sci. 2, 62–69 (2018)
19. Ivanova, A.D., Murugova, O.V.: The release of professional standards and the development of professional competencies of social workers. Bull. PNIPU. Socio-Econ. Sci. 3, 184–196 (2018)
20. Mingaleva, Z., Mirskikh, I.: On innovation and knowledge economy in Russia. World Acad. Sci. Eng. Technol. 66, 1032–1041 (2010)
21. Polynskaya, G.A., Mesropyan, M.G.: Identification of patterns and trends in patient behavior when using electronic applications and Internet resources for self-diagnosis. Bus. Inform. 1 (43), 28–38 (2018)
22. Kozlova, A.S., Taraskin, D.S.: Telemedicine development trends and its impact on the Russian insurance market. Bull. Saratov State Socio-Econ. Univ. 2(71), 144–148 (2018)
23. Dimitrakiev, D., Molodchik, A.V.: The business model of the dental industry based on blockchain and self-organizing community. Bull. PNIPU. Socio-Econ. Sci. 2, 107–115 (2018)
24. Kadirov, A.O., Smykalo, N.V.: Digitization of Russian medicine using blockchain technology: retrospective analysis and development prospects. Innov. Invest. 12, 246–250 (2019)
25. Kantemirova, M.A., Dzakoev, Z.L., Alikova, Z.R., Chedgemov, S.R., Soskieva, Z.V.: Percolation approach to simulation of a sustainable network economy structure. Entrepreneurship Sustain. Issues 5(3), 502–513 (2018)
26. Kantemirova, M.A., Alikova, Z.R.: Digital economy: the development of digitization of medicine in the region. Bull. North Ossetian State Univ. Named After K. L. Khetagurov 1, 92–95 (2019)
27. Morozova, Y.: Digitization as a global, country and industry process in improving the effectiveness and efficiency of healthcare and medicine. Intellect. Innov. Invest. 4, 44–53 (2019)
28. Mingaleva, Z., Gataullina, A.: Structural modernization of economy and aspects of economic security of territory. Middle East J. Sci. Res. 12(11), 1535–1540 (2012)
29. Semeshko, A.I., Kuranov, V.G.: Regulation of the relationships in healthcare by the norms of professional medical organizations on the international regional level. Perm Univ. Herald. Ser.: Yuridical Sci. 3(25), 196–206 (2014)
30. Federal Law №242 "On amendments to the current legal acts of the Russian Federation on the issues of using information technologies and delivering medical telecommunication services in the field of health care" of 01.01.2018. http://www.consultant.ru/document/cons_doc_LAW_221184/. Accessed 23 Oct 2019
31. National Program "Digital economy of Russian Federation" (RF Government Directive of 28.07.2017 №1632). http://government.ru/docs/28653. Accessed 23 Oct 2019
32. Convention of Human Rights and Dignity Protection in Sphere of Biomedicine. http://conventions.coe.int/Treaty/RUS/Treaties/Html/195.htm. Accessed 23 Oct 2019

33. Shevchenko, R.: Expert: the digital medicine market in Russia will grow to 90 billion rubles by 2023. https://medvestnik.ru/content/news/Ekspert-rynok-cifrovoi-mediciny-v-Rossii-vyra stet-do-90-mlrd-rublei-k-2023-godu.html. Accessed 21 Nov 2019
34. Telemedicine. https://telemedicina.ru. Accessed 13 Oct 2019
35. American Telemedicine association. http://www.americantelemed.org. Accessed 28 Oct 2019
36. Kuranov, V.G.: The concept of the quality of medical services: civil-legal aspect. Perm Univ. Herald. Ser.: Yuridical Sci. 4(26), 128–135 (2014)
37. Declaration of doctors of Russia. http://www.rmass.ru/publ/info/deklar.vrachey. Accessed 21 Nov 2019
38. Kuranov, V.G., Semeshko, A.I.: The legal significance of medical ethics. Med. Law: Theory Pract. 1(1), 83–88 (2015)

Impact of Macro Factors on Effectiveness of Implementation of Medical Care Safety Management System

Yuriy Voskanyan[1] ⓘ, Irina Shikina[2,3(✉)] ⓘ, Fedor Kidalov[4] ⓘ,
Olga Andreeva[2,5] ⓘ, and Tatiana Makhovskaya[3] ⓘ

[1] Russian Medical Academy of Continuing Professional Education
of the Ministry of Health of Russia, Bld. 1, 2/1 Barrikadnaya Street,
Moscow 125993, Russia
[2] Federal Research Institute for Health Organization and Informatics of Ministry
of Health of the Russian Federation, Bld. 11, Dobrolyubova Street,
Moscow 127254, Russia
[3] Central State Medical Academy Office of the President of the Russian
Federation, Bld. 1A, 19, Marshal Tymoshenko Street, Moscow 121359, Russia
shikina@mednet.ru
[4] Moscow State Budgetary Institution "Information and Analytical Center
of Healthcare", Bld. 10, Basmannaya Novaya Street, Moscow 107078, Russia
[5] National Research University Higher School of Economics,
Bld. 20 Myasnitskaya Ulitsa, Moscow 101000, Russia

Abstract. The work analysed the impact of macro factors on effectiveness of medical care safety management system implementation in a medical organization. The authors show that existing macro factors in Russia hinder objective assess the amount of damage associated with the provision of medical care. In addition, they do not allow for setting up an effective system for adverse events risk management in medicine. Foremost, it is an imperfect economic model of the state health care system in Russia, the crisis of medical education system, government regulation of the industry, which does not take into account the main scope of medical workers activities. Formation of a general negative image of health care system in the media and the growing practice of criminal prosecution of doctors is important. It is shown that the solution of the described problems is impossible without effective interaction between the state, health care authorities, society and medical organizations. The economic model of state health care should be changed on a state level. The legislative and executive bodies should adopt a series of normative legal acts. Those acts should recognize the inevitability of adverse events, define their true causes, and classify health care as a high-risk service sector. They should also ensure rights and freedoms for health care providers, as well as formalize rights and obligations of patients. At the level of the health care authorities, national security programs should be developed, state regulation of the industry should be optimized and profound reform of the medical education system should be carried out. The latter implies the creation of university clusters, a change of the paradigm of value formation in health care and a change of models of competencies for graduate and specialist doctors. At the societal level, an open discussion regarding medical care safety should be organized and every effort should be

© Springer Nature Switzerland AG 2021
T. Antipova (Ed.): ICIS 2020, LNNS 136, pp. 346–355, 2021.
https://doi.org/10.1007/978-3-030-49264-9_31

made to create a unified team in safety management consisting of health care providers, patients, and their families.

Keywords: Medical care safety · Adverse events · Incident · Medical care safety management system · Patient-doctor communication

1 Introduction

In recent years, domestic and foreign publications have increasingly focused on medical care safety as an integral part of medical care quality. The main focus has been on the introduction of an effective management system in medical organizations. The aim of the article was to show the impact of political, legal, economic and social macro factors on viability and effectiveness of such system implementation using data from scientific publications and official sources of information.

2 Results

In 2020 at a board meeting of rectors of medical universities, Mikhail Murashko declared that annually more than 70,000 patients in Russia die or receive serious complications due to problems associated with medical care provision. The Minister also noted that doctors have poor doctor-patient communication skills [1]. In the same year, the head of the Investigative Committee of Russia, Alexander Bastrykin reported an increase in the number of medical crimes reported: 4497 in 2016; 6050 in 2017; 6623 in 2018 and more than 6,500 in 9 months of 2019 [2]. In fact, both those and other figures are far from the real picture. According to the analytical data provided by the all-Russian Public Organization "League of Patients' Advocates" among the main reasons for citizens' appeals are complaints about communication between doctors and patients (violation of ethical norms, unwillingness to explain risks, comments about health status) and quality of medical care, including "medical errors", harm to health or life [3]. In our previous publications [4], we drew attention to the high incidence of adverse events and resulting deaths according to publications from countries where the registration of medical errors is a national health policy priority and does not entail criminal prosecution. We define adverse events as cases of unintentional mental or physical injury resulting in temporary or permanent disability, death, extended inpatient treatment, and more likely associated (in the opinion of experts) with the provision of medical care than with the course of the underlying disease or associated conditions [5]. Based on the data of later publications [6–10], the cumulative frequency of inpatient adverse events is currently 12.7%, and over the last 30 years the trend line of adverse events probability has remained horizontal (Fig. 1).

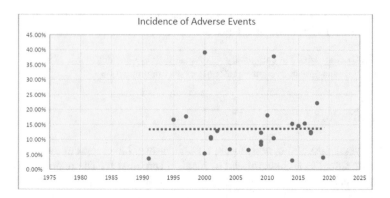

Fig. 1. Incidence of adverse events in multidisciplinary medical hospitals over the past 30 years

The cumulative hospital mortality rate, calculated by scientists at Johns Hopkins University (USA) is 0.71%, representing in absolute terms 251,454 deaths per year in United States' clinics [10]. Taking into account the described statistics and extrapolating it to the 29,578,000 cases of annual hospitalizations in Russia [11], with certain assumptions we will obtain figures far different from the official statements of health authorities and the investigation committee: more than 3,756,000 patients with complications and more than 210,000 deaths caused by them. Taking into account much higher depreciation of infrastructure assets, deficit of industry funding, and significant differences in competence, we can assume that the real figures will be even higher. However, the existing system of criminal prosecution, administrative, disciplinary and punitive measures does not allow for the official recording of threats, incidents and adverse events because of entrenched practices of disguise and concealment by the medical community. Thus, all attempts to implement an effective security management system for medical care are doomed to failure in advance, as it is impossible to manage what cannot be measured.

In our earlier publication [4] we described the basic principles and stages of building a safety management system in a medical organization. Medical care safety as a relationship between the chances of benefit and harm has been considered in four aspects: patient safety, personnel safety, working environment safety and environmental safety. It was shown that implementation of medical care safety management systems in a medical organization should be based on an in-depth understanding of the causes of adverse events, an analysis of a safety model and selection of an adequate safety management model, and use of design management principles in the building of a medical care safety management system. Medical care safety management systems should become an integral part of a quality management system in a medical organization. It should be based on a new safety culture, a system for accounting of threats, incidents and adverse events, and risk management. The use of pilot conversion in the introduction of safety management models, principles of continuous learning and continuous improvement in the development of medical organizations are prerequisites for success.

In the existing set of macro factors only a few clinics had the opportunity to build such a security management system, and even then with very strong limitations. From our point of view, this is due to the following reasons. First, the existing economic model of public health care is not viable. This is due to the fact that the practice of defining normative indicators of medical care needs, resources and workforce, calculation of medical services costs, operational and financial models have not changed since the times of the USSR. This has nothing to do with sources of financing (budget, compulsory medical insurance and citizens' personal funds). All calculations are based on the determination of Russian average indicators (with minor adjustments for individual subjects), which are not connected to the real need and, moreover, to the real capacity of the health care system in each federal subject and, especially, in each municipal entity. Tariffs for medical services are formed on the basis of normative prices and the residual principle; they are not related to the cost of labour and resources on the market. Depreciation of fixed assets (buildings, structures, equipment with the cost of over 100,000 rubles) is recorded only in the accounting system and is not included in the tariffs' costs. The result is obvious as only in children's health care out of 118,865 buildings where medical care is provided, 14% are in critical condition, 30.5% have no water supply, 52.1% have no hot water supply, 41.1% have no central heating, 35% - canalization [12]. The depreciation of fixed assets in health care (excluding moral depreciation) at the end of the year amounted to 52.7%, while in high income countries it did not exceed 35% [13]. Profit is prohibited in the economic activity of state medical organizations, which means there is a lack of funds for development, personnel training, infrastructure renovation, quality and safety management. Planning of budget expenditure of medical organizations is linked to historical costs without taking into account inflation of gross domestic product (GDP), foreign exchange rates, risks of uncertainty, and trends in increasing the volume of medical care. As a result government expenditure on health care in Russia in dollar purchasing power parity ($ PPP) is 11 times lower than in the United States, 4.5 times lower than in the old EU countries, and 1.9 times lower than in the new EU countries [14]. This leads to the share of the wage fund in tariff structure for medical services being 70% or more, which practically excludes the possibility of purchasing quality medicines and medical devices, staff training and development of the medical organizations. At the same time, sanctions policy of medical insurance organizations and compulsory insurance funds are impressive: fines are imposed for any violation of documentation nature unrelated to financial reporting, as well as for verification of any adverse event. A paradoxical situation emerges: a medical organization needs additional resources for the treatment of complications but instead it loses part of its finances, which in the conditions of initial underfunding already infringes on the rights of other patients to receive quality and safe medical care.

Secondly, over the past 30 years, the system of medical education in Russia has not been brought into line with the best international standards. With rare exceptions, most medical higher education institutions (HEIs) do not have their own clinics. The existing regulatory legal framework excludes the possibility of medical interventions by department staff if they are not employees of medical organizations at the same time. All clinical departments working within the framework of a formal contract concluded between the university and a medical organization, in fact are "guests" in the territory of these organizations and have severely limited rights. The inclusion of medical students in the patient treatment is prohibited by the existing normative legal base. The result of the above is a loss of the role of clinical departments in training medical personnel and clinical research management, and the progressive outflow of the best specialists at the medical universities into practical health care. Thus, the main principle of ensuring the stability of quality assurance and safety in health care, which is associated with the need to maintain continuity in turnover of generations of medical workers is violated. Today we are forced to acknowledge the fact that after graduation from university, a doctor comes to the healthcare system and is deprived of any independence in the decision-making process. It should be noted that in the system of medical education in the USSR there were three levels of preparation of a doctor for independent practice: subordination (last year at the institute), internship and residency. In Russia, first the subordination was eliminated, then the internship. Today, district therapists and pediatricians in general can start practicing immediately after graduation. It should be noted that postgraduate education of a doctor in high income countries takes from three to eight years. This training is conducted by the best specialist teachers at university clinics, which are referenced medical organizations in terms of quality and safety, which manage the risks associated with the admission of students to the medical care process. These factors ensure the target level of patient loyalty to the additional burden associated with the need for the participation of students and residents in the medical care provision. The catastrophic lack of competence associated with the progressive depletion of the medical education system in Russia is being compensated for by the strict legal regulation of medical activity. But it is clear to any educated doctor that no clinical recommendation and, moreover, no standard or internal regulations can compensate for the lack of basic medical knowledge, inability to think clinically, critically evaluate medical information and elementary collection of patient data. In addition, it is necessary to introduce a mandatory communication skills training program for medical professionals into the curriculum of medical education in Russia in order to develop practical skills needed to provide high-quality medical care, improve communication, safety and quality [15]. Along with such aspects as leadership, policies and procedures, staffing and reporting, effective communication highlights the necessary elements of a medical organization's culture to ensure patient safety [16].

Thirdly, in recent years state regulation of medical activity in Russia has reached an incredible scale. Every element of infrastructure and medical supplies (such as suture material), every doctor's action is subject to a regulatory legal act. The initial goal of such acts was to improve quality and safety. But have they achieved their goal? After all, any requirements must be supported by funding, competence and time. But in most cases, this does not happen. As a result, we have an underfunded infrastructure where medical doctors are obliged to provide medical care in conditions which do not meet neither basic sanitary nor special medical requirements. Furthermore, medical personnel and doctors in particular do not have a part to play in the decision-making process. All decisions for it are made by procedures, standards and clinical recommendations. The latter are now being elevated to the rank of regulatory legal documents, although, these are in fact recommendations based on expert opinion, and not on the results coming from randomized clinical trials or meta-analysis. One of the most common misconceptions is that methods and forms of management of one organizational macrostructure are to be made the key principles of coordination in another organization quite different in principles and forms. Medical organizations are a typical representative of adhocracy, where the key employees are people with heuristic characteristics, working in conditions of high uncertainty, the basis of whose work is a constant analysis and synthesis of information with the collective, including historical, experience and with the subsequent decision, the most appropriate to each situation, because every patient is unique. The leading coordination mechanisms in such organizations are standardization of qualifications (diploma and certificates of accreditation), which gives the right to make decisions independently, and mutual agreement (consultations, Concilium, clinical studies, other horizontal informal communications), which allows for an exponential increase in the accuracy of collective decisions [17]. The attempt to replace these coordination tools with those used in production line (standardization of processes, etc.) has led to the fact that on the one hand we get a non-independent doctor, while on the other hand we have a standard that cannot take into account the uncertainty and unique features of a particular patient. Of course, standardization in medicine has allowed us to significantly increase the efficient use of resources to optimize planning and logistics. The role of clinical guidelines in supporting clinical decision making is difficult to underestimate. At the same time, neither one can replace the main functions of the doctor, the making of an independent clinical decision and the utilization of medical technology chosen. This should remain the specialists' right and be granted by a standard of qualification, the value of which is entirely determined by the quality of existing medical education system and a level of competence of accreditation commissions.

Fourth, another macro factor that has a strong impact on medical care safety management is public attitude and as a result, a law enforcement system towards a health care worker. The health care provider is "obliged to fulfill his duty under any conditions (even not suitable for this purpose), working 8, 12 and even 24 h a day", and "has no right to make a mistake, and any mistake made is a crime for which he/she must be severely punished because it resulted or may have resulted in additional physical or mental injury to the patient". No one cares that the system is underfunded, that health care providers work in conditions that are often deprived of not only special medical but even basic sanitary infrastructure. The result is described above: medical errors and complications associated with medical care are carefully hidden. The medical community resorts to defending itself, since it transitions from being a subject to an object of mental and physical trauma and criminal prosecution. Let us compare health care with other sectors that are at high risk of human activity, such as civil aviation. Between 2014 and 2016, the largest number of major air disasters were recorded: 17 in which 1,762 people died (587 per year). Taking into account the average number of civil flights (36.4 million) and the average number of passengers carried per year (3.6 billion), we get the probability of one air crash - 1 in 5 million flights - and the probability of death for 1 in 5 million passengers carried. Let us recall the above-mentioned probability of death due to complications related to medical care - 0.71% or 1 out of 140 hospitalized patients. The comparison is clearly not in favor of the health care system. These figures come from the U.S. health care system where the infrastructure, quality and safety of medical care is most likely not worse than in Russia. The emerging sharply negative public attitude towards the health care worker leads to the development of antagonism between a doctor, a patient, and their relatives during treatment. But the degree and quality of the patient's and his family members' participation in the treatment and diagnostic process accounts for a third of all complications associated with the provision of medical care [18].

From our point of view, effective implementation of safety management systems in a medical organization and its scaling within the industry is possible only with the effective cooperation of four subjects: the state, health care authorities, society and the medical organization itself. At the state level, first of all, the economic model of healthcare should be changed. The medical service tariff should include all types of direct and indirect costs at market prices, depreciation of fixed assets regardless of their cost, additional costs associated with quality and safety management, and costs associated with medical risks (development of complications, inefficient treatment). The tariff should also include profit, the intended use of which should be strategic planning and strategy implementation, personnel training and development, project management, research and development. Budgeting for medical organizations should include the real needed volume of medical care, GDP inflation, exchange rate differences, potential fines and penalties, as well as the need for major repairs, major renovation of existing infrastructure, or construction of new facilities. The standards for medical care volume should be linked to the real need of each federal subject or better each municipality. As an example, we would like to underline that even in relatively compact Europe the number of visits to clinics and general practitioners' offices per 1 resident varies from 2.9 (Sweden) to 9.9 (Germany). In Russia, taking into account geographical peculiarities this gap will be even higher, so it is impossible to correct the average standard set in the state guarantees program by any correction factors. Actual standards should determine infrastructure, human resources and financing, taking into account the technical capacity of equipment, physical capacity of personnel and the market value of direct and indirect costs. The quality criteria used must be linked to the target function of medical activity. From this perspective, it is clear that coverage indicators cannot be used as a quality criterion for a process such as adult dispanserization (as an example). Firstly, because a medical organization cannot manage a healthy adult's willingness to be examined within a specified period of time (at best, this is the prerogative of the employer or the government, which can be realized through the amount of insurance contributions if there is an agreement with the Federal Fund for Mandatory Medical Insurance). Secondly, because the main purpose of screening is early detection of socially significant diseases and, above all, malignant neoplasms (MN) [19]. Therefore, the percentage of people with newly diagnosed MN should be taken as a quality criterion of the clinical examination of the adult population, which should not be lower than 0.6% of MN incidence in the population [20]. The next very important step is adoption of federal regulatory legal acts that recognize the possibility and inevitability of medical errors and adverse events in health care and take into account their true causes (latent threats). Health care itself should be classified as a high-risk service sector. It is necessary to formalize responsibilities of a patient and guarantee rights and freedoms to a medical staff in case of medical errors and adverse events registration. An example of such an approach to governance at the state level is the Danish Patient Safety Act [21], analogous of which have been adopted at the federal level in almost all developed countries. Let us pay attention to a short extract from the act: "Front-line personnel who report an adverse event cannot as a result be subjected to investigation or disciplinary action by the employer, the Board of Patient Safety or a court of justice". It is impossible to construct an effective accounting system of threats, incidents and adverse events which is a key element of a control system of medical care safety without normative legal acts at a Federal laws level.

There are three important things that should be done at a health authorities level. First, optimizing the state regulation of medical activity taking into account the leading mechanisms for its coordination (qualification standard and mutual coordination) and the possibilities of infrastructure provision. Secondly, developing a government program to introduce a national medical care safety management system, including the development of national safety guidelines, which are now available in almost all countries with a high level of health care development. Thirdly, immediately reforming the medical education system, with the construction of university clusters, changing the competencies model of a graduate doctor, changing the institute of medical training and the institute of postgraduate education. The reform should be based on the paradigm that the formation of basic value in the provision of medical care (human health) does not begin from the moment when a patient crosses the threshold of a medical organization, but from the moment when a student crosses the threshold of a medical university or institute. A modern university cluster should include a modern educational complex (institute and medical college) with all technologies of online and offline education; a multidisciplinary hospital and clinical and diagnostic center of reference class in terms of quality and safety; a simulation center; a research institute; a scientific and production complex; a library; a dormitory; a hotel; a sport complex, a catering and recreation network. The university cluster should become a network that meets all the needs of trainees and students. The competency model of a graduate doctor should be based on the fact that they are independent in decision-making and in their actions with good communication skills. A doctor with a diploma and a certificate of accreditation should be assigned to a patient. For this purpose, it is necessary to move to a training program that has long proven its effectiveness in the U.S. and Europe. This program includes fundamental education (3 years), clinical education (3–4 years) and step residency (from 3 to 8 years depending on the final specialty). Step residency means, for example, that it is impossible to pass the residency in cardio-thoracic surgery without having completed the residency in ambulance and emergency care, anesthesiology and intensive care, general surgery, etc. Training in residency is provided by the mentor institute with a competitive system of selection of teachers and subsequent attachment to them no more than 1–3 residents (depending on the type of specialty). Training in quality and safety management of medical care should be integrated at all stages of a training system. The decision to award a diploma, certificate of accreditation should be made by a team of experts. At the same time, the university clinic should become a key in creating professional associations and expert groups. The share of state funding for both pre- and post-graduate education should not be less than 80%, otherwise we will either not get doctors with the necessary level of competence or we will lead to a catastrophic shortage of not only primary health care doctors, but even focused specialists. The system of postgraduate education should be maximally optimized and focused on the doctor's own desire to acquire new knowledge and skills. The main focus should be on short cycles, team training, training and obtaining new information online. The Accreditation Panel should not be interested in the number of points the applicant has scored or the number of courses he has attended, but in the real competencies he has at the time of accreditation and the real results of his scientific and practical work [22, 23].

An effective open dialogue between doctors, nurses, patients, charitable and other public organizations should be organized at the society level. In this dialogue, health care should be defined as a high-risk service sector, where a favorable outcome is a combined result of efforts of health care workers, safe infrastructure and the right actions of a patient and family members. Health care providers, patients and their family members should become members of a unified team to manage the risks of adverse events, and their interactions should be based on absolute trust, empathy and effective communication [24].

3   Conclusion

An analysis of existing macro factors in the Russian Federation has shown that introduction of a modern medical care safety management system in medical organization is experiencing significant difficulties due to the impossibility of open recording of medical errors and related adverse events, the imperfect economic model of healthcare and medical education system, and accumulated funding shortage in the industry. The solution of the described problems requires an active participation of the state, health care authorities, society and medical organizations themselves, based on open cooperation and recognition of the potential harms related to medical care provision and understanding of its systematic causes. The described macro factors are conditions unique to each country, thus the situation analysis typical for Russian health care cannot be directly extrapolated to the health care systems of other countries.

References

1. The head of the Ministry of Health announced the statistics: medical failures lead to complications in 70 thousand patients a year. Interfax, 08 February 2020. https://www.interfax.ru/russia/694577
2. Medvestnik: The number of complaints to the Investigative Committee against doctors doubled in 2019 (2019). https://medvestnik.ru/content/news/Chislo-jalob-grajdan-v-Sledstvennyi-komitet-na-deistviya-vrachei-v-2019-godu-vyroslo-vdvoe.html
3. League of Patients' Advocates. Reasons for citizens' appeals to the Patients' League: http://ligap.ru/articles/analitika/uroven/prichina/
4. Voskanyan, Y., Shikina, I., Kidalov, F., Davidov, D.: Medical care safety - problems and perspectives. In: Lecture Notes in Networks and Systems, vol. 78, pp. 291–304 (2020). https://doi.org/10.1007/978-3-030-22493-6_26
5. Zegers, M., Bruijne, M.C., Wagner, C., et al.: Adverse events and potentially preventable deaths in Dutch hospitals: results of a retrospective patient record review study. Qual. Saf. Health Care. **18**, 297–302 (2009)

6. Nilson, L., Risberg, M.B., Montgomery, A., Sjodahl, R., Schldmeijer, K., Rutberg, A.: Preventable adverse events in surgical care in Sweden. A nationwide review of patient notes medicine. Med. (Baltim.) **95**(11), (2016). https://doi.org/10.1097/md.0000000000003047
7. Halfon, P., Staines, A., Burnand, B.: Adverse events related to hospital care: a retrospective medical records review in a Swiss hospital. Int. J. Qual. Health Care **29**(4), 527–533 (2017). https://doi.org/10.1093/intqhc/mzx061
8. Rafter, N., Hickey, A., Conroy, R.M., Condell, S., O'Connor, P., Vaughan, D., Walsh, G., Williams, D.J.: The Irish National Adverse Events Study (INAES): the frequency and nature of adverse events in Irish hospitals - a retrospective record review study. BMJ Qual Saf. **26** (2), 111–119 (2017)
9. Atkinson, M.K., Schuster, M.A., Feng, J.Y., Akinola, T., Clark, K.L., Sommers, B.D.: Adverse events and patient outcomes among hospitalized children cared for by general pediatricians vs hospitalists. JAMA Netw. Open. **1**(8), e185658 (2018). https://doi.org/10.1001/jamanetworkopen.2018.5658
10. Forster, A.J., Huang, A., Lee, T.C., Jennings, A., Choudhri, O., Backman, C.: Study of a multisite prospective adverse event surveillance system. BMJ Qual. Saf. 1–9 (2019). https://doi.org/10.1136/bmjqs-2018-008664
11. Makary, M.A., Daniel, M.: Medical error—the third leading cause of death in the US. BMJ **353**(3), 1–5 (2016)
12. Report on the results of the expert and analytical event "Evaluation of spending efficiency in 2018–2019, federal budget funds allocated to develop the material and technical base of children's clinics and children's clinics departments of medical organizations providing primary health care". Bulletin of the Accounts Chamber of the Russian Federation, no. 2 (2020)
13. Healthcare in Russia: Statistical collection (2019)
14. Ulumbekova, G.E.: Healthcare of Russia. What to do?, 3rd edn. Moscow (2019)
15. Sonkina, A.A.: Skills of professional communication in doctor's work. ORGZDRAV: News. Opinions. Teaching. Herald Higher School, no. 1(1) (2015)
16. Colla, J.B., Bracken, A.C., Kinney, L.M., Weeks, W.B.: Measuring patient safety climate: a review of surveys. BMJ Qual. Saf. **14**(5), 364–366 (2005)
17. Mintzberg, G.: Management: nature and structure of organizations. Moscow (2018). (in Russian)
18. O'Hagan, J., MacKinnon, N.J., Persaud, D., Etchegary, H.: Self-reported medical errors in seven countries: implications for Canada healthcare quarterly **12**(Sp), 55–61 (2009). https://doi.org/10.12927/hcq.2009.2096
19. Antipova, T., Shikina, I.: Informatic indicators of efficacy cancer treatment. In: 12th Iberian Conference on Information Systems and Technologies (CISTI) Lisbon, Portugal, 21–24 June 2017, pp. 1–5. https://doi.org/10.23919/cisti.2017.7976049. http://ieeexplore.ieee.org/document/7976049/
20. Malignant neoplasms in Russia in 2018 (morbidity and mortality)/Under edition of A.D. Kaprin, V.V. Starinskiy, G.V. Petrova. Herzen Research Institute, Branch of FSBI "National Medical Research Radiological Center" of the Ministry of Health of the Russia, Moscow (2019)
21. Act on Patient Safety in the Danish Health Care System ACT No. 429 of 10/06/20037
22. Macchi, L., Pietikäinen, E., Reiman, T., Heikkilä, J., Ruuhilehto, K.: Patient safety management. Available model and system. VTT Technical Research Centre of Finland (2011)
23. Nordin, A.: Patient safety culture in hospital setting. Measurements, health care staff perceptions and suggestions for improvement. Karlstad University, Faculty of Health, Science and Technology Department of Health Sciences, Sweden (2015)
24. WHO patient safety curriculum guide: multi-professional edition. World Health Organization (2011)

Coronavirus Pandemic as Black Swan Event

Tatiana Antipova$^{(\boxtimes)}$ (iD)

Institute of Certified Specialists, Perm, Russia
fakademia@mail.ru

Abstract. Nowadays coronavirus is the hottest break news around the world. This paper aims to study why coronavirus became so meaningful for worldwide life. On one side many countries closed their boarders to prevent coronavirus spreading. On another side some people said that coronavirus is no scared than simple influence. Where is true? The health effects and mortality of coronavirus and influenza were compared. Situation Reports of World Health Organization have been analyzed and signs of Black Swan event considered. Total confirmed cases, total, deaths, and Rate of coronavirus distribution was calculated based on Situation Reports of the World Health Organization. To conclude this paper Coronavirus Pandemic recognized as Black Swan event according to considered figures and facts from recent references.

Keywords: Coronavirus · COVID-19 · 2019-nCov · Black Swan event · Economy impact · Social impact · Housing Boom · Coronavirus impact · Financial consequences · Collapse · Coronavirus Pandemic

1 Introduction

There is a worldwide concern about the new coronavirus 2019-nCoV as a global public health threat [1]. The 2019-nCoV infection is spreading and its incidence is increasing worldwide. Although the current mortality is lower than that of the SARS-CoV and the MERS-CoV, it seems that the 2019-nCoV is very contagious. SarsCoV (severe acute respiratory syndrome coronavirus) was recognized in November 2002, MERS-CoV (Middle East respiratory syndrome coronavirus) in June 2012, and 2019-nCoV in December 2019 [2]. The first deaths occurred mostly in elderly people, among whom the disease might progress faster [3]. But now it established that the young people are also prone to severe coronavirus disease so the society should still be cautious in dealing with the virus and pay more attention to protecting all of people from the virus.

Coronavirus (COVID-19) spread worldwide with the pandemic according to data of World Health Organization (WHO). Since Coronavirus is spreading very fast, 11.03.2020 WHO Director-General's characterized Coronavirus as a pandemic and this pandemic has the social and economic impacts [5, report 51, 71].

There are signs that governments' authorities are still trying to conceal the true scale of the problem with Coronavirus, but at this point the virus appears to be more contagious than the pathogens behind diseases such as Ebola or SARS—though some experts say SARS and coronavirus are about equally contagious. Events like the coronavirus epidemic, and its predecessors—such as SARS, Ebola and MERS—test

© Springer Nature Switzerland AG 2021
T. Antipova (Ed.): ICIS 2020, LNNS 136, pp. 356–366, 2021.
https://doi.org/10.1007/978-3-030-49264-9_32

Health Care systems and force us to think about the unthinkable. What national and international systems need to be in place to minimize the chance of catastrophe on this scale? So far, the 21st century has been an age of **black swans**. That age isn't over, and of the black swans still to arrive, the coronavirus epidemic is unlikely to be the last to materialize around the world [6].

Taleb, in his book [21], coined the term "black swan" to describe random events that form part of our lives. These events have the following three key attributes/signs: (1) outlier, being outside the realm of regular expectations; (2) carries an extreme impact; and (3) explanations for the occurrence are concocted after the fact, making it explainable and predictable.

In identifying the attributes, the classification of a black swan event appears to depend on individual interpretation. Such events are large-scale shocks which can severely challenge economic activity, social cohesion and even political stability. These different patterns of risk can cascade and spread across global systems, where they arise in health, climate, social or financial systems [20]. For example, infectious virus such Bird Flu, Asia (2008) and SARS, Hong Kong (2002) were recognized as Black Swan events by D. Higgins [20].

This paper considers some signs of coronavirus impact in following part.

2   Nature of Coronavirus

On 31 December 2019, the WHO Country Office in China was informed of cases of pneumonia with unknown etiology (unknown cause) found in Wuhan City, Hubei Province, China. No causal link was established from 31 December 2019 to 3 January 2020. One week later the World Health Organisation (WHO) received further details from the Chinese National Public Health Commission that the outbreak was linked to radiation at a seafood market in Wuhan City. And then Chinese authorities identified a new type of coronavirus that had been isolated on 7 January 2020. The name "coronavirus" was given because this virus visually looks like it is surrounded by a crown. It has been officially named COVID-19. The virus affecting people now is what's called a novel coronavirus (nCoV) because it's the first time this particular strain is being seen in humans. So, we have two officially medical abbreviations for Coronavirus: COVID-19 and 2019-nCoV (for laboratory tests).

As coronavirus cases continue to be reported around the world, WHO says countries still have a chance of containing the outbreak. Officials have also sought to differentiate Covid-19 from other viruses, as part of efforts to quell public panic. People want to know the answer on two simple questions: "What is coronavirus?" and "What makes COVID-19 different to influenza?"

What is coronavirus?

For COVID-19, data to date suggest that 80% of infections are mild or asymptomatic, 15% are severe infection, requiring oxygen and 5% are critical infections, requiring ventilation. These fractions of severe and critical infection would be higher than what is observed for influenza infection. Those most at risk for severe influenza infection are children, pregnant women, elderly, those with underlying chronic medical conditions and those who are immunosuppressed [5].

According to "Contacts investigation protocol for coronavirus disease 2019 (COVID-19)" (https://www.who.int/publications-detail/the-first-few-x-(ffx)-cases-and-contact-investigation-protocol-for-2019-novel-coronavirus-(2019-ncov)-infection) WHO listed following symptoms: Fever (>38 °C); Sore throat; Runny nose; Shortness of breath, Chills; Nausea; Diarrhoea; Headache; Rash; Conjunctivitis; Muscle aches; Joint ache; Loss of appetite; Nose bleed; Fatigue; Seizures; Altered consciousness. Some of those symptoms are illustrated on Fig. 1.

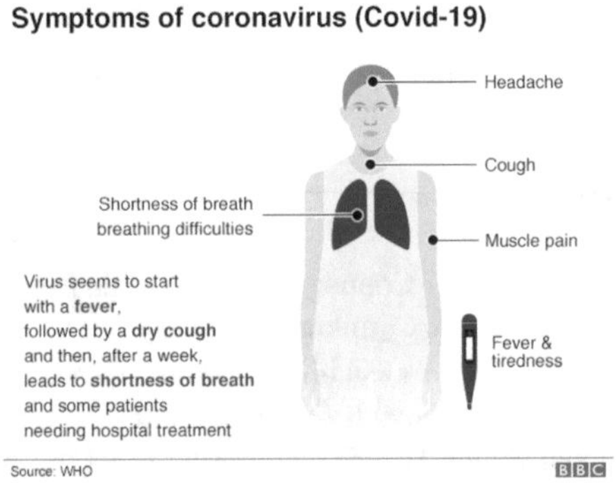

Symptoms of coronavirus (Covid-19)

Headache

Cough

Shortness of breath breathing difficulties

Muscle pain

Virus seems to start with a **fever**, followed by a **dry cough** and then, after a week, leads to **shortness of breath** and some patients needing hospital treatment

Fever & tiredness

Source: WHO

BBC

Fig. 1. Symptoms of coronavirus.

What makes COVID-19 different to influenza?

As the COVID-19 outbreak continues to evolve, comparisons have been drawn to influenza. Both cause respiratory disease, yet there are important differences between the two viruses and how they spread. This has important implications for the public health measures that can be implemented to respond to each virus [5].

Let's consider similarities and differences between COVID-19 and Influenza. The result of this comparison is shown in Table 1.

Table 1. Similarities and Differences between COVID-19 and Influenza.

Similarities	Differences
They both cause respiratory disease, which presents as a wide range of illness from asymptomatic or mild through to severe disease and death	Influenza has a shorter median incubation period and a shorter serial interval than COVID-19 virus. The serial interval for COVID-19 virus is estimated to be 5-6 days, while for influenza virus, the serial interval is 3 days. This means that influenza can spread faster than COVID19
Both viruses are transmitted by contact, droplets and fomites	The reproductive number for COVID-19 is higher than for Influenza– is understood to be between 2 and 2.5 for COVID-19 virus
	Children are important drivers of influenza virus transmission in the community. For COVID-19 virus, children are infected from adults, rather than vice versa
	Mortality of COVID-19 is between 3–5%, but for seasonal influenza, mortality is usually well below 0.1%

Source: Author's elaboration based on 46th WHO Situation Report [5].

So, we can see that COVID-19 and Influenza have similar symptoms, but transmissions but COVID-19 is more scared with higher Mortality. Strangely, perhaps, the most effective general procedure for stopping an outbreak from becoming a full-fledged pandemic is simple common sense; in other words, education about elementary procedures for health care and sanitation. For example, wash your hands when handling food, keep your home and outdoor areas clean, properly take medications. Such procedures go a very long way toward stopping infectious diseases in their tracks [22].

3   Coronavirus Spreading

WHO daily reports incidence of confirmed 2019-nCoV cases on https://www.who.int/emergencies/diseases/novel-coronavirus-2019/situation-reports/. The first Situation Report was dated as 20.01.2020. The author has analyzed some selection WHO situation reports to define trend of Coronavirus spreading. As a result of those reports' study, total numbers of registered cases of Coronavirus infection and total deaths are shown on Fig. 2.

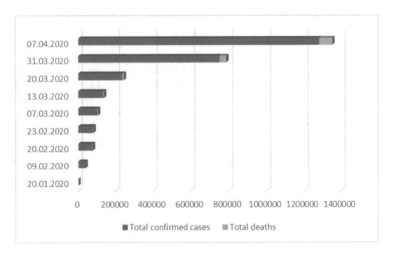

Fig. 2. Total confirmed cases of coronavirus infection causing total deaths. *Source: Author's analysis based on WHO Situation Reports* [5]: *1–77.*

To date (Fig. 2), the total number of confirmed coronavirus cases globally topped over 1 200,000 (1 279 722 as of 07.04.2020), causing 72 614 deaths as of 07.04.2020. During the study period (20.01.2020–31.03.2020), the coronavirus covered 205 Countries/Territories/Areas with average rate of about 10441 people per day for 78 days (20.01.2020–07.04.2020). But the good news is that more than 56,000 people are now free of the virus in China. So, around 70% of coronavirus patients in China have recovered, as per the recent WHO report.

As shown above, Coronavirus became very widespread, so we can see economic and social impacts. If people become very worried that they can catch the disease if they go out in public, this will mean many fewer people will go to restaurants, sports events, movie theaters and concerts, or anyone else where they are likely to be in close proximity to large numbers of people. Many of these businesses are likely to shut down, at least until the major threat of the virus has passed [9].

Mortality of Coronavirus is shown on Fig. 3 and we can see that maximum Mortality of COVID-19 is 5,7% (Share of Total deaths in Total confirmed cases).

Graph on Fig. 3 shows that Mortality of COVID-19 increased from 1,1 to 5,7% just for 78 days and nobody knows when and at what magnitude will stop growing yet.

Covid-19 is spreading from China to other regions causing human suffering, economic disruption, and Globalization. Some people said that if their country did not get foreigners from abroad, coronavirus would never come inside their country. It is raising health concerns and the risk of wider restrictions on the movement of people, goods and services, falls in business and consumer confidence and slowing production. The Interim Outlook presents both a best-case scenario in which the extent of the

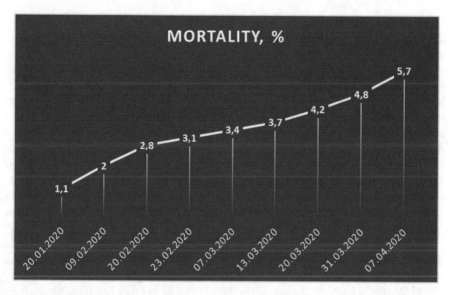

Fig. 3. Mortality of Coronavirus. *Source: Author's analysis based on WHO Situation Reports* [5].

coronavirus is broadly contained and a "domino" prospect of contagion that is more widespread. In both cases, it is reasonable to call governments to act immediately to limit the spread of the coronavirus, protect people and businesses from its effects and shore up demand in the economy. One possible way to protect people from unreasonable decision is using patented method of identification/authentication of citizens [11].

4   Coronavirus Social and Economic Impacts

The novel coronavirus (Covid-19) is a major global public health emergency that has brought tragedy to many lives. Its impact is still unfolding globally. There is already a major slowdown in the wider economy around the world [25]. Many now fear the coronavirus will become a global pandemic. The consequences of a Chinese economic meltdown would travel with the same sweeping inexorability. Commodity prices around the world would slump, supply chains would break down, and few financial institutions anywhere could escape the knock-on consequences. Recovery in China and elsewhere could be slow, and the social and political effects could be dramatic [8].

Businesses are dealing with lost revenue and disrupted supply chains due to China's factory shutdowns, tens of millions of people remaining in lockdown in dozens of cities and other countries extending travel restrictions. The shortage of products and parts from China is affecting companies around the world, as factories delayed opening after the Lunar New Year and workers stayed home to help reduce the spread of the virus. Apple's manufacturing partner in China, Foxconn, is facing a production delay. Some carmakers including Nissan and Hyundai temporarily closed factories outside

China because they couldn't get parts. The pharmaceutical industry is also bracing for disruption to global production. Many trade shows and sporting events in China, Asia and across the world have been cancelled or postponed [26].

Today, it is central to global supply chains and there has been an enormous increase in travel to and from the country, thus heightening the risk of the virus spreading [25]. It means that the tourism and travel industry have suffered from the coronavirus outbreak. Plane travel has been dramatically reduced, as few people would want to be in a crowded plane that could house several people with the virus. This is a huge blow to the tourism industry as people are putting off their holidays until the epidemic subsides [9]. Global airline revenues are expected to fall by $4–5 billion in the first quarter of 2020 as a result of flight cancellations, according to a report from the UN's International Civil Aviation Organization (ICAO). ICAO also forecasts that Japan could lose $1.29 billion of tourism revenue in the first quarter due to the drop in Chinese travelers while Thailand could lose $1.15 billion [26].

The consequences of Covid-19 for global oil demand will be significant. Demand is now expected to contract by 435 kb/d in Q20, the first quarterly decrease in more than a decade. For 2020 as a whole, we have reduced our global growth forecast by 365 kb/d to 825 kb/d, the lowest since 2011. Growth in 2019 has been trimmed by 80 kb/d to 885 kb/d on lower-than-expected consumption in the OECD [25].

The impact of Covid-19 for oil prices have been sharp (see Table 2). Before Covid-19 came along, crisis the market was expected to move towards balance in the second half of 2020 due to a combination of the production cuts implemented at the start of the year, stronger demand and a tailing off of non-OPEC supply growth. Now, the risk posed by the Covid-19 crisis has prompted the OPEC+ countries to consider an additional cut to oil production of 0.6 mb/d as an emergency measure on top of the 1.7 mb/d already pledged. The effect of the Covid-19 crisis on the wider economy means that it will be difficult for consumers to feel the benefit of lower oil prices [25].

The Table 2 represents some signs of social and economic impacts of Coronavirus Pandemic.

Table 2 demonstrates that Chinese and Indian GDP tend to decrease on 1.5 and 2.1% respectively, oil consumption and price fell to lower-than-expected amounts. So, worldwide outbreak plus the fall of important economic indicators such: Global Trade, Economy, Housing Boom, Fuels and energy, Finance, GDP, Global Economic shows the significant consequences of Coronavirus spreading, which proves the possibility of recognizing it as Black Swan event. This event has severely challenge economic activity and social cohesion.

Prospects for a big hit to the economy and the spread of Covid-19 to hundreds of countries have chilled financial markets, government planners and corporate executives. In March 2020 a plunge in Chinese factory output is expected to destroy the disease disrupted supply chains – with damaging consequences for companies around the world. China's Commerce Ministry this week said 90% of the 7,000 exporters it surveyed reported difficulty shipping goods, as counterparts cancel contracts or don't pay. The country's largest steelworks, China Baowu Steel Group Corp., predicted a first-quarter loss of $428 million. [19] At the same time China's president Xi Jinping warned that the coronavirus would have a "relatively big impact on the economy and society". Adding that it would be short-term and controllable, Xi said the government

Table 2. Signs of belonging Coronavirus Pandemic to Black Swan event

Signs	Coronavirus impact/consequences
Global economic growth	Dun & Bradstreet estimates in a new report that if a vaccine for the virus is not invented by the second quarter of 2020, and containment of the disease is therefore delayed until the fourth quarter or beyond, the global economic cost could be as much as a one-percentage-point slowdown in global economic growth [12]
Global trade indicators	Canada's deficit in the global trade of merchandise goods widened in January, to 1.47 billion Canadian dollars ($1.10 billion) from C$732 million in the previous month, on a broad-based drop in exports to both the U.S. and China. Exports of goods fell 2% in January, to C$48.14 billion, or the lowest level in 11 months. Imports declined 0.5%, and on a year-over-year basis decreased 4.9%. Exports to China, Canada's second-largest trading partner, fell 7.8% in January, and 8.6% on a one-year basis. Imports from China fell by a steeper 12.1% in the month. [18] Indian trade proved to be a nightmare for, local markets with frontline gauges losing over three and a half percent in a single day after industry body PHDCCI said that the coronavirus outbreak may negatively impact global growth by 30 basis points or $250 billion [16]
Economy indicators	Commodities-focused companies fell in Tokyo where the Topix shed 0.6%. Singapore's FTSE Straits Times dropped 1.9%. Seoul's benchmark Kospi index closed down 3.1%, with stocks that are exposed to a probable fall in spending by Chinese tourists as a result of the virus among the hardest hit. AmorePacific, South Korea's largest cosmetics company, dropped 8.5% while duty-free chain Hotel Shilla tumbled almost 10%. Smartphone and chip manufacturer Samsung fell more than 3%. [7] Five industry sectors—business and personal services, wholesale trade, manufacturing, retail, and financial services—account for more than 80% of the businesses within the affected provinces [12]
Housing Boom	New home sales data for January comes out at 10 a.m. ET and economists are expecting 711,000 single-family homes sold last month, up from 694,000 in December, according to Dow Jones. This would be a 2.4% jump in home sales month-over-month [14]
Fuels and energy prices	Brent crude was down $2.42, or 4.1%, to $56.09 a barrel. U.S. crude futures fell by $2.12, or 4%, to $51.26. [15] In Sydney the S&P/ASX 200 index slipped 1.8% as energy and mining stocks sold off on concerns over the pathogen's impact on Chinese demand. [7] Brent values fell by about $10/bbl, or 20%, to below $55/bbl [25]
Finance indicators	Bank of America analysts, for their part, have already downgraded their eurozone growth prospects for this year, from their previous 1% to 0.6%. [6] Gold was down 0.9%. [7] Several major companies including Apple (ticker: AAPL), Mastercard (MA), and United Airlines Holdings (UAL) have already warned that they'll be unable to meet prior financial guidance due to Covid-19. Corporate earnings will be hit, but the broader economy can still survive a quarter or two of that. [10] The Dow Jones Industrial Average fell 12.4% last week, its worst showing since the financial crisis, as fear built that the globalized epidemic will damage trade and pull the world economy toward recession [19]

(continued)

Table 2. (*continued*)

Signs	Coronavirus impact/consequences
GDP	Torsten Sløk, expects the outbreak to shave 1.5% points off of **Chinese gross domestic product** this year (to 4.6% from 6.1%) and said he thinks the virus will take a 0.5% point off of global growth this year. [17] **India's Gross Domestic Product** (GDP) growth at 4.5% for the third quarter (Q3) of current fiscal year (FY20), which is lower than 6.6% GDP growth recorded in the corresponding period a year ago [16]
Social impacts	The most of country in lockdown for three weeks and effectively could be many more, brings several concerns in its wake. First, living in isolation is difficult, because humans are social creatures. Secondly, fear and anxiety can take a toll on both young and the old. The young fear losing a living. The old fear emptiness and a lack of purpose. A research published in the American Journal of Epidemiology, reports that there is robust evidence that social isolation and loneliness significantly increase risk for premature mortality, and the magnitude of the risk exceeds that of many leading health indicators. This in turn can lead to difficulties with decision-making and memory storage and recall. They are also susceptible to illness. Researchers found that a lonely person's immune system responds differently to fighting vi-ruses, making them more likely to develop new illness. 1.3 billion people confined to homes for a month and more can create a new dimension of a yet unfolding problem for both the health workers and the government [27]

would step up efforts to cushion the blow. The country has taken a number of measures in recent weeks to prop up its economy [24].

The International Monetary Fund (IMF) forecasts the epidemic will reduce China's economic growth this year by 0.4% point to 5.6% [19]. IMF head, Kristalina Georgieva, said that the global lender of last resort was ready to provide additional support, particularly to poorer countries by way of grants and debt relief. Speaking at a G20 meeting of finance leaders and central bank chiefs, she said the IMF assumed the impact would be relatively minor and short-lived, although she warned that the continued spread of the virus could have dire consequences. She added: "Global cooperation is essential to the containment of the Covid-19 and its economic impact, particularly if the outbreak turns out to be more persistent and widespread" [24].

Currently, there is no registered treatment or vaccine for Coronavirus disease unfortunately. In the absence of a specific treatment for this novel virus, there is an urgent need to find an alternative solution to prevent and control the replication and spread of the virus [4]. Since there is no cure for coronavirus yet, the country that will receive this medicine at first may influence world politics. And this is the most important Coronavirus Pandemic impact on world political life.

The social impact of coronavirus can also be attributed to the fact that many countries have closed their borders and introduced quarantine/self-isolation for their citizens for one month at least. And people cannot move not only from country to country but leave their own home during quarantine in these countries (most often in March–April 2020). Cultural institutions such as museums, theatres and libraries, etc. have also been closed. Major sporting events have been cancelled or postponed. Thus, the Tokyo Olympics has been postponed by one year.

5   Conclusions

As it shown above, Coronavirus Pandemic recognized as Black Swan event according to considered signs from recent references and we can see severely challenges economic activity, social cohesion and even political stability.

To reduce Coronavirus consequences official recommendations from both the Centers for Disease Control and the World Health Organization encourage the use of telemedicine apps. Public companies that offer telemedicine, including Teladoc, have seen their stock surge in the past week, and many private start-ups, from AmericanWell to Plushcare, tell CNBC they are bracing for increased usage [14]. Guidance on how to manage patients with COVID-19 must be delivered urgently to healthcare workers in the form of workshops, online teaching, smart phone engagement, and peer-to-peer education. Equipment such as personal protective equipment, ventilators, oxygen, and testing kits must be made available and supply chains strengthened [23].

As we can see from study result, Coronavirus Pandemic has significant impacts on social and economic life. In forecasting future socio-economic indicators, we will have to make adjustments to the black swan effect, which is Coronavirus Pandemic.

The crisis caused by Coronavirus Pandemic calls on our self-confident technological civilization to assess the limits of its capabilities and to recognize the fragility of this civilization. We have amazing scientific research and its results translated into reality, but we remain very fragile and truly weak. And this crisis shows us today the limitations of our capabilities in order not to give up, not to fall into confusion, but to restrain the pandemic and remain sapiens human.

References

1. Benvenuto, D., Giovanetti, M., Ciccozzi, A., Spoto, S., Angeletti, S., Ciccozzi, M.: The 2019-new coronavirus epidemic: evidence for virus evolution. J. Med. Virol. **92**, 455–459 (2020). https://doi.org/10.1002/jmv.25688
2. Stein, R.A.: The 2019 coronavirus: learning curves, lessons, and the weakest link. Int. J. Clin. Pract. (2020). https://doi.org/10.1111/ijcp.13488
3. Wang, W., Tang, J., Wei, F.: Updated understanding of the outbreak of 2019 novel coronavirus (2019-nCoV) in Wuhan. China J. Med. Virol. **92**, 441–447 (2020). https://doi.org/10.1002/jmv.25689
4. Zhang, L., Liu, Y.: Potential interventions for novel coronavirus in China: a systematic review. J. Med. Virol. 1–12 (2020). https://doi.org/10.1002/jmv.25707
5. Coronavirus disease 2019 (COVID-19). Situation Report, pp. 1–77. https://www.who.int/docs/default-source/coronaviruse/situation-reports/

6. Briançon, P.: Are markets putting too much hope in ECB's capacity to help Europe deal with coronavirus crisis? Barron's (2020)
7. Georgiadis, P., Lockett, H., Rocco, M.: US stocks rebound day after coronavirus fears hit markets. FT.Com (2020)
8. Walter, R.M.: China is the real sick man of Asia; its financial markets may be even more dangerous than its wildlife markets. Wall Street J. (2020)
9. Baker, D.: Coronavirus, the Stock Market, and the Economy. Beat the Press [BLOG] (2020)
10. Jasinski, N.: What Comes After the Coronavirus? It Could Be a 'V-Shaped Market'. Barron's (2020)
11. Konyavsky, V., Ross, G.: New method for digital economy user's protection. In: ICIS 2019. LNNS, vol. 78, pp. 221–230 (2020). https://doi.org/10.1007/978-3-030-22493-6_20
12. McCann, D.: Coronavirus Crimps Supply Chains, May Harm World Economy. CFO.Com (2020)
13. Coronavirus watch, earnings, new home sales: 3 things to watch for in the markets on Wednesday. https://www.cnbc.com/2020/02/25/outlook-for-wednesday-coronavirus-watch-earnings-new-home-sales.html
14. Anonymous Stock Review: Two-bagger teladoc health jumps 7.8% on strong volume. News Bites Pty Ltd., Melbourne (2020)
15. Anonymous Stock Review: Baker hughes climbs 4.5%, issued 9 new patents. News Bites Pty Ltd. Melbourne (2020)
16. Anonymous "Markets witness mayhem as coronavirus pandemic fear intensifies" Accord Fintech (2020)
17. Beilfuss, L.: China's coronavirus data look too good to be true. Barrons **100**(7), 35 (2020)
18. Vieira, P.: Canada data offer warning signs on coronavirus; job creation, wage gains continued in February, but trade report for January highlights economic risk posed by epidemic. Wall Street J. (2020)
19. Areddy, J.T.: China's economy reels as coronavirus hits manufacturing worse than financial crisis; the fate of China's economy is of crucial importance to a world with few solid drivers of growth. Wall Street J. (2020)
20. Higgins, D.M.: The black swan effect and the impact on Australian property forecasting. J. Financ. Manag. Property Constr. **18**(1), 76–89 (2013)
21. Taleb, N.: The Black Swan: The Impact of the Highly Improbable, 2nd edn. Penguin Book, London (2009)
22. Casti, J.: Four Faces of Tomorrow. OECD International Future Project on Future Global Shocks (2011). https://www.oecd.org/futures/globalprospects/46890038.pdf
23. COVID-19: Too little, too late? EDITORIAL. Lancet **395**(10226), 755. https://doi.org/10.1016/S0140-6736(20)30522-5
24. https://www.theguardian.com/business/2020/feb/23/economic-impact-of-coronavirus-outbreak-deepens
25. https://www.iea.org/reports/oil-market-report-february-2020
26. https://www.weforum.org/agenda/2020/02/coronavirus-economic-effects-global-economy-trade-travel/
27. Mantha, S.S.: The cost of corona. Lokmat Times **29**(03), 6 (2020)

Information Management

Use of Information Management Technologies in Housing and Communal Utilities

Larisa Akifieva$^{(\boxtimes)}$ ⓘ, Mikhail Polyakov ⓘ, Natalia Sutyagina ⓘ,
Irina Zvereva ⓘ, and Irina Zhdankina ⓘ

Nizhny Novgorod State Engineering and Economic University,
Knyaginino 606340, Russia
laraakif@mail.ru

Abstract. The article considers the importance of studying the introduction of information technologies in the management of housing and communal utilities. Studying of the state of housing and communal utilities in the Russian Federation indicates the need to search new technologies to activate this area. As a result, we proposed the algorithm for using the crowdsourcing method to improve the quality of housing and communal services, which is a universal tool for solving tasks and becomes possible at low costs for ongoing activities. Based on compliance with this algorithm, it is possible to implement crowdsourcing project that allows to achieve set goals.

Keywords: Housing and communal utilities informatization · Information and communication tools · Housing and communal utilities · Quality of housing and communal utilities · Innovations · Management of innovative activity · Crowd technology · Crowdsourcing · Crowdsourcing project · Crowdsourcing platform

1 Introduction

Housing and communal utilities provided to the population of Russia occupy the largest share of the volume of all services rendered. That is why one of the main tasks is to improve the quality of this type of services [1, p. 254]. Since housing and communal utilities are social and economic sphere, it constantly experiences problems with population, and population, in turn, considers this sphere as a system that is regulated and financed by the state.

The reform in housing and communal sphere has been going for quite a long time (more than 20 years), significant changes haven't been happened, as evidenced by the ruin and liquidation of management companies providing housing and communal utilities, wear and tear of infrastructure at public utilities facilities, and others.

The elements of the structure should not be used until complete wear and tear. In the course of functional use of the object, works aimed at compensating for normative wear and tear are performed [2, p. 2512].

In the sphere of management of housing and communal enterprises, innovative ways of management, including quality management, should be introduced. In order to improve the quality of housing and communal utilities, it is necessary to introduce the

© Springer Nature Switzerland AG 2021
T. Antipova (Ed.): ICIS 2020, LNNS 136, pp. 369–374, 2021.
https://doi.org/10.1007/978-3-030-49264-9_33

latest equipment, use materials and technologies related to improving the productivity of housing management through innovation, and attract investors [3, p. 24].

That is why it is relevant to study informatization technologies as a way of managing the housing and communal sphere, aimed at achieving improvement of the quality of provision of housing and communal utilities.

2   Use of Information Technologies in Housing and Communal Utilities

In order to improve the quality of housing and communal utilities based on information technologies related to improving management efficiency, it is necessary to solve a set of management tasks.

In order to ensure the competitiveness of management companies, it is necessary to introduce continuous improvements in all spheres of housing and communal utilities. That is why each stage of the innovative process of management of housing and communal sphere has its own peculiarities and corresponding difficulties. It, in turn, involves coordinating the actions of the creative team at all stages of the innovation process: from the formation of an idea to its introduction to the market.

The integrated introduction of not only innovative information technologies, but also modern management in the management companies of the housing and communal sphere will make it possible to balance social and economic aspects of housing and communal utilities, which will necessarily lead to the renewal of the housing stock of housing and communal utilities and to the removal of tensions in society, accordingly.

Management companies, which are competitive in the housing and communal sphere, solve the issue of formation of creative innovation idea with the help of instruments of open innovation. Open innovation is a modern tool in formation, developing and enhancing innovation. The process of formulating an idea or solving a complex problem involves the mental abilities of people outside of management organizations. Crowdsourcing is just one of the tools of open innovation. Crowdsourcing will help to be an effective and creative tool for finding management solutions that can formulate an innovative idea taking into account the current needs of consumers.

With the help of crowdsourcing, it is possible to solve a huge number of tasks at various levels – from personal to organizational and even state. The concept of crowdsourcing is a tool that can be used to attract the public who do not have specialized knowledge in any area through info communication networks to solve a variety of tasks and problems.

Crowdsourcing suggests that the main catalyst for people is not material stimulation, but moral one – the ability to epitomize their creative thought in actual activity. In society, as a rule, there are always creative people ready for initiative or for a symbolic fee to formulate ideas, solve problems and conduct scientific research on a voluntary basis.

Crowd technologies are quite effective tools that are actively used by many commercial organizations, which make it possible to consider their application in the management of housing and communal sphere. Crowd technologies are technologies of interaction with the public, allowing to solve various tasks related to increase of social activity of citizens and involvement of them in processes of self-government.

Nowadays the housing and communal complex of the Russian Federation more than ever needs for substantial investments in order to improve the level and quality of services provided, ensure stable and effective activity of communal companies, form urban infrastructure, and also improve the environmental situation and health of people [4, p. 23].

At the final stage of crowdsourcing, the project accepted for implementation goes to crowdfunding, where it determines the method of financing.

The crowdfunding process is usually implemented through a crowdfunding platform on the Internet and gives each user the right to get acquainted with existing projects, assist those who like them, activate the desire to get original content for its financing, and also come up with and implement their own idea and form on the basis of the project, in order to subsequently purchase funding from users for its implementation. Thus, it can act as a specially created platform and as a site for the project, in which various crowdfunding projects are organized [5, p. 12].

The project "State Information System of Housing and Communal Services" (SIS HCU), which is a website where every consumer of housing and communal utilities can leave his offers, requests and creative ideas on any issue in any sphere of activity of housing and communal utilities, is considered an initial example of the application of crowdsourcing in Russia. The purpose of SIS HCU is to collect the best offers from consumers, which contribute to solving topical problems of housing and communal sphere [6, p. 46].

Also at the moment in Moscow and Moscow region there is an internet portal "Dobrodel," which is aimed at the general solution of problems in the sphere of housing and communal utilities by state authorities and consumers. The Ministry of Economic Development of the Russian Federation collects opinions of citizens, conducts a survey on the presented Strategy of Development of the Russian Federation for the period up to 2035.

However, the use of crowdsourcing as practice technology in municipal management is not quite popular and developed.

The authorities of housing and communal services have primarily to think not about implementation of information technologies in this sphere, but about increasing the efficiency of their activities, profitability, quality of housing and communal services, etc. Almost in all municipalities, housing and communal problems are not solved due to the presence of low incomes and corruption in this sphere.

Often inadequate provision of housing and communal utilities to the population is due to the lack of understanding of the actions taken by municipalities (managing companies in the process of maintaining and managing the housing stock). To solve this problem, a number of actions are proposed that will help make the activities of municipalities as transparent and open as possible: in particular, the availability of relevant information at housing and communal facilities will increase loyalty and confidence on the part of homeowners by coordinating the work, plans and activities for the construction of housing.

From the point of view of public administration, crowdsourcing is a method by which you can distribute the labor of a relatively large number of people (active citizens, volunteers) who participate in the implementation of a project or are interested in this project. The purpose of this distribution is to increase the efficiency of decisions taken, perform complex and particularly complex tasks, and implement projects together [7, p. 70].

It follows that crowdsourcing is a form of such activity, which is based on the open participation of citizens by the level of interests and by the level of knowledge of citizens, non-profit or commercial organizations, the implementation of which brings mutually beneficial results to performers and initiators from the solution of the problem.

3   Results

It is possible to solve at least a small part of the problems related to the quality of housing and communal services by using the crowdsourcing method. The algorithm for using the crowdsourcing method is presented further.

At the first stage – the consumer's application for a problem – not satisfactory quality of provision of housing and communal services is reduced to the fact that when the consumer has this problem, he describes it on a crowdsourcing platform (for example, on the official website of housing and communal services enterprises). They, in turn, identify socially significant problems that can only be solved with the participation of other consumers.

To implement a crowdsourcing project, it is necessary to have a manager who must set the goals and objectives of the project; determine the terms of implementation, responsible persons; control the stages of development and implementation of the project.

Then there is selection of performers, including setting requirements for performers of the project, selection of performers according to the criteria, and direct creation of a group of performers.

At the next stage, the project performers generate ideas using the following methods: brainstorming, Edward De Bono's thought caps, mental maps, and others.

At the "expert assessment" stage, the authorized persons analyze the received ideas and offers from the performers; calculate the costs of time and resources, expected social and economic effect of the project implementation, choice of the project financing method.

The final decision on the choice of the method of solving the stated problem is made by voting of citizens on the crowdsourcing platform or by making the final decision of experts. This stage ends with the approval of the final decision by manager of the crowdsourcing project, after which the execution of the decision begins, which is controlled by manager of the crowdsourcing project and experts.

The introduction of innovations in the procedure for creating housing and communal utilities should be carried out only if it can give a significant economic result, but it must be beneficial to all members of this process.

It is important to use characteristics that allow assessing the current changes and clearly demonstrating the benefits of innovation.

The economic efficiency of managing the quality of housing and communal utilities on the basis of innovation can be expressed by the corresponding indicators: the ratio of the reduction of costs of housing and communal companies from the introduction of innovation in the procedure of production of housing and communal services to the costs of their purchase; dynamics of income level of housing and communal utilities company from services sold to consumers.

In general, crowdsourcing is a tool, not a universal mean for solving all problems, and along with other management technologies, it has both advantages and disadvantages.

The advantages of crowdsourcing include:

- saving time;
- easy implementation and use;
- saving money;
- broad audience coverage;
- development and adoption of non-standard solutions;
- involving only stakeholders and others.

The disadvantages of crowdsourcing include:

- inability to monitor the process of operation;
- poor quality of decisions;
- need for constant motivation;
- lack of development of crowdsourcing culture in housing and communal services;
- possibility of information leaks and others.

4   Conclusion

Crowdsourcing has a chance to become a new tool for the development of information and communication technologies of management in the field of housing and communal utilities and making effective management decisions, to become an effective tool in the selection and implementation of innovative projects of housing and communal utilities.

Crowdsourcing is an effective management tool in housing and communal utilities, which is provided for the use of innovation bodies and the identification of sources of its financing.

Thus, with the help of crowd technologies, which are a powerful source of financing used for the implementation of innovative projects, it is possible to build effective management of innovation activities of the housing and communal sphere. Crowdsourcing is an information and communication tool for moving to the modern level of housing and communal complex formation, which will be the initial stage of moving away from traditional management methods while providing quality housing and communal services.

References

1. Sutyagina, N.I.: Models of public-private partnership in housing and communal sphere. Vestnik of Lobachevsky University of Nizhniy Novgorod (1-2), 254–261 (2012)
2. Smirnova, Z.V.: Maintenance and repair of engineering systems of an apartment building. In: Smirnova, Z.V., Rudenko, A.A., Cherney, O.T., Provalenova, N.V., Bystrova, N.V., Permovsky, A.A., Semakhin, E.A. (eds.) International Journal of Innovative Technology and Exploring Engineering (IJITEE), vol. 9, no. 3, pp. 2511–2515. Exploring Innovation (2020). https://doi.org/10.35940/ijitee.l3264.019320
3. Kolesnikova, K.S., Kalinina, N.M.: Innovation in the sphere of housing and utilities as a way to improve the quality of services. In: 4th Inter-University Scientific and Practical Conference Service Economy: Problems and Prospects, pp. 24–25 (2018)
4. Gabaydullina, R.A.: Modern crowd technology in management. In: International Youth Symposium on Management, Economics and Finance, Kazan, pp. 23–24 (2016)
5. Efimova, T.B.: Housing and communal utilities as one of the directions of crowdsourcing in municipal administration. In: Scientific Forum of Technical and Mathematical Sciences: Materials of XVI International Scientific and Practical Conference, vol. 6, no. 16, pp. 11–19 (2018)
6. Pakhomova, O.A.: Crowdsourcing as a way to solve social problems in small cities. Discussion **3**(66), 46–49 (2016)
7. Ivanov, P.A.: About the need to introduce innovations in housing and communal utilities. Almanac World Sci. **11–3**(14), 70–71 (2016)

A Comparative Study of Color Spaces for Cloud Images Recognition Based on LBP and LTP Features

Ha Duong Thi Hong and Vinh Truong Hoang$^{(\boxtimes)}$

Ho Chi Minh City Open University, Ho Chi Minh City, Vietnam
{hadth.178i,vinh.th}@ou.edu.vn

Abstract. The texture classification problem is widely applied for ground-based cloud images recognition due to its efficiency. Local Binary Pattern and its variants are usually investigated to represent cloud images. The appropriate choice of color space might enhance performance of the recognition system for many applications. In this paper, we propose a comparative study to select the best candidate's space for coding cloud images by fusing multiple texture description methods. The proposed approach is evaluated on the SWIMCAT database.

Keywords: Ground-based cloud images · LTP-LBP · Cloud images recognition · Data fusion · Pattern recognition · Cloud classification

1 Introduction

According to the World Meteorological Organization, cloud observation and identification are very important to meteorology and climate research. In the beginning, the shortage of radio waves and observation satellites have been used to observe the whole sky on the ground [1]. These available images have low cost and high resolution, and they provide accurate information for local cloud [2, 3]. Recently, ground-based cameras were opened new opportunities to track the earth's atmosphere. These cameras are used as an addition for image of satellite which can provide more data to geologists efficiently [4].

One of the most major topic in computer vision is the ground-based cloud image recognition which have various application such as weather prediction, solar energy generation [5], local weather forecast, tracking condensation streaks [6] and communication signal reduction [7]. Ground-based cloud classification is extreme difficult due to the variations in cloud appearance under various atmospheric conditions. In order to address this issue, many works in literature have been proposed. Wang et al. [8] proposed an effective feature extraction method by ranking all frequencies of occurrence of rotation-invariant Local Binary Pattern (LBP) for cloud images recognition. Cheng and Yu [9] proposed a block division and block-based classification based on LBP features. Zhen et al. [10] extract spectral and texture feature based on analyzing statistical tone and gray level cooccurrence matrix (GLCM). Zhang et al. [11] proposed a feature extraction technique called transfer deep local binary patterns to solve the

© Springer Nature Switzerland AG 2021
T. Antipova (Ed.): ICIS 2020, LNNS 136, pp. 375–382, 2021.
https://doi.org/10.1007/978-3-030-49264-9_34

change of views Convolutional Neural Network and extract local features based on feature maps.

One important inheritable feature is color information which is an addition to shape and texture, usually leads to better performance [12]. Thus, combining multiple texture feature sets is required to achieve the robust recognition accuracy. They can be features extracted from several color channels or color spaces and different types of local or global features [13]. Recently, various works are proposed for investigating color information in pattern recognition. For example, Hoang and Ali [14] presents a comparative study of the impact of color spaces for fabric defect recognition. The fabric image is characterized by LBP descriptor based on eight color spaces. Bianconi et al. [15] compare the performance of twelve color spaces for stones images classification. Van and Hoang [16] investigate fourteen color spaces to extract features in order to verify the kinship. Duong and Hoang [17] apply to extract rice seed images based on features coded in multiple color spaces using HOG descriptor. Sandid et al. reveal that the choice of the best color space depends on the used image database, and hence the considered application [18].

In this work, we apply to extract LBP and its variant, namely Local Ternary Pattern (LTP) feature to represent color cloud images. The feature fusion strategy is then applied and encoded in different color spaces. The rest of this paper is organized as follows. Section 2 briefly introduces the feature extracting methods based on LBP and LTP descriptors. Section 3 reviews the common color spaces used in pattern recognition. Section 4 introduces the experimental results on SWIMCAT dataset. Finally, the conclusion is discussed in Sect. 5.

2   Local Binary Pattern and Its Variant

Local Binary Pattern (LBP) proposed by Ojala et al. has been known as one of the most successful statistical approaches due to its efficacy, robustness against illumination changes and relative fast calculation [19]. This operator compute LBP code via the local neighborhood structure by characterizing the texture around each pixel of the image from a square neighborhood of 3×3 pixels. The $LBP_{P,R}(x_c, y_c)$ code of each pixel (x_c, y_c) is computed by comparing the gray value g_c of the central pixel with the gray values $\{g_i\}_{i=0}^{p-1}$ of its p neighbors, as follows:

$$LBP_{P,R}(x_c, y_c) = \sum_{i=0}^{p-1} \Phi(g_i - g_c) \times 2^i \qquad (1)$$

where Φ is the threshold function which is defined by:

$$\Phi(g_i - g_c) = \begin{cases} 1 & \text{if } (g_i - g_c) \geq 0, \\ 0 & \text{otherwise} \end{cases} \qquad (2)$$

LBP is resistant to lighting effects in the sense that they are invariant to monotonic gray-level transformations, and they have been shown to have high discriminative power for texture classification. However, because they threshold at exactly the value of the central pixel they tend to be sensitive to noise, especially in near-uniform image regions. Many authors have proposed several LBP variants by changing the thresholding scheme or the number of quantization level to gain noise robustness and discrimination power. One of the common extension for characterizing cloud images is the Local Ternary Patterns (LTP) which proposed by Tan et al. [20]. They propose to encode the central pixel by three values $\{-1, 0, 1\}$ as follows:

$$\Phi'(g_i - g_c) = \begin{cases} 1 & \text{if } (g_i - g_c) \geq t, \\ 0 & \text{if } |g_i - g_c| < t \\ -1 & \text{if } (g_i - g_c) \leq t \end{cases} \tag{3}$$

In this approach, an additional parameter t of user is used to define a tolerance for similarity between different gray intensities to be robust to noise and reduce the dimensionality. Each ternary is then split into positive and negative parts, which are subsequently treated as two separate LBP components for which histograms are computed and finally concatenated.

3 Color Spaces

Color in the machine vision system is defined by a combination of 3 color channels specified by a color space. For visual recognition tasks like object detection, retrieval and recognition, different color spaces possess significantly different physical, physiologic and psycho-visual properties in terms of discriminating power. Several effective color spaces including have been proposed to achieve better recognition performance by computing from the RGB space via means of either linear or nonlinear transformations [21]. They can be divided into four groups:

- the primary color spaces: RGB, XYZ
- the independent axis color spaces: I1I2I3
- the luminance-chrominance color spaces: Luv, Lab, YUV, YIQ
- the perceptual color spaces: HSI, HLS, HSV, bwrgby, HLS.

4 Experimental Results

Although the ground-based cloud images classification problem has been studied for many years, few publicly available clouds image datasets exist. SWIMCAT [22] stands for Singapore Whole sky IMaging Categories Database, containing 784 images categorized in 5 distinct classes: clear sky, patterned clouds, thick dark clouds, thick white clouds, and veil clouds (see Fig. 1).

| Clear sky | Patterned | Thick dark | Thick white | Veil |

Fig. 1. Example images of SWIMCAT dataset.

We use fourteen color spaces (RGB, bwrgby, HLS, HSV, I1I2I3, IHLS, Lab, Luv, rgb, XYZ, YcbCr, YIQ, YUV) to represent images of SWIMCAT dataset. Three type of features are used (LBP, LTP and LBP+LTP) are used to characterized cloud images in order to compare the performance. The 5-fold cross validation technique is applied to create the initial dataset into two training and testing set. As we mentioned in the Sect. 1 that color information is very useful to extract features. Wang et al. [8] consider grayscale image to extract apply LBP and LTP features. Here, we extract features from 3 color channel of each color space and finally concatenate to build a feature vector of each cloud image. This makes the dimension space is three times longer than in [8] for a single descriptor LBP or LTP in use. By changing the R and P value, we can compute LBP features for dealing with the texture at different scales. For example, $LBP_{16,2}$ refers to 16 neighbors in a circular neighborhood of radius 2. The LBP feature produces 2^P different output values and gives rise to a 2^P-dimensional histogram. In this work, we vary the value of $R \in \{1, 2, 3, ..., 5\}$ and $P \in \{4, 8, 12\}$ for LBP, LTP and LBP+LTP features. Here, we set the value of $t = 1$ for LTP descriptor. Table 1 presents the classification results on the SWIMCAT dataset. We only report the best accuracy given by specific color space for each type of features. The value in bold indicates the best accuracy of each row. The value in italic indicates the best accuracy of each column. By observing this table, we can see the absence of RGB color space while the HLS, HSV and IHLS achieve a good performance. So, the luminance-chrominance and perceptual color spaces are suitable for characterizing cloud texture. The best accuracy is 99.2 ± 0.8 given by LBP(2,12). This result confirms again that an appropriate of LBP and LTP need to be adjusted for a considered application as in [14].

Table 1. The classification results of three types of features (LBP, LTP and LBP+LTP) on 5-fold cross validation. The accuracy is reported based on the best results performing by specific color space or the selected value of (R, P). The value in bold indicates the best accuracy of each row. The value in italic indicates the best accuracy of each column. The value in red color indicates the best value of each type of features.

(R,P)	LBP HLS	HSV	IHLS	Lab	rgb	YIQ	YUV	LTP HLS	HSV	IHLS	ISH	LBP_LTP HLS	HSV	IHLS	ISH	YIQ
(1,4)	97.6 ±0.6									97.7 ±1.6				96.3 ±2.1		
(2,4)		*98.3 ±0.9*									98.2 ±1.4	98.1 ±0.5				
(3,4)		97.7 ±1.3	97.7 ±1.0							98.1 ±0.9	98.1 ±1.0			98.1 ±0.9		
(4,4)	98.1 ±0.8							97.7 ±1.0				97.4 ±1.1				
(5,4)	98.2 ±0.9							98.2 ±0.8				97.7 ±1.1				
(1,8)						98.7 ±0.4		98.2 ±0.5								98.3 ±0.7
(2,8)		98.7 ±1.2									98.7 ±1.0			98.6 ±1.4		
(3,8)			98.5 ±1.2						98.6 ±1.3						98.6 ±0.9	
(4,8)				98.3 ±1.1				99.0 ±0.7							98.1 ±0.9	
(5,8)		98.0 ±0.5						98.8 ±0.8					98.1 ±0.8			
(1,12)			99.0 ±0.5								98.3 ±0.7			98.6 ±0.9		
(2,12)							99.2 ±0.8		98.7 ±1.1					99.0 ±0.9		
(3,12)				98.9 ±0.5				99.0 ±0.9							98.3 ±0.8	
(4,12)	98.9 ±0.8							98.7 ±1.0							98.6 ±0.8	
(5,12)		98.2 ±0.3						98.7 ±0.8				98.2 ±0.8				

The detailed comparison of theses color spaces is then given in Fig. 2. We select the best accuracy obtained for each couple (R, P) for comparing the performance of different color spaces. The primary color spaces including RGB and XYZ does not give a good performance comparing with other spaces. The performance is stable at 98.0% when we concatenate LBP and LTP features.

We summarize our results and compare with previous works in the state-of-the-art in Table 2. By incorporating color information and investigated different color space, we improve more than 2% comparing with other works. It is worth to repeat that we apply the same decomposition strategy in [8] to split data. The statistical meaning of our results has more confident since the standard deviation is fewer.

Table 2. The comparison with previous works

Features	Color space	Accuracy
LBP(1,8) [23]	Gray	79.3 ± 3.9
DLBP [24]	Gray	81.8 ± 4.2
CLBP [25]	Gray	84.2 ± 1.0
SaLBP [24]	Gray	89.1 ± 1.9
SLBP [8]	Gray	97.1 ± 1.7
LBP [19]	YUV	99.2 ± 0.8
LTP [20]	HLS	99.0 ± 0.9
LBP+LTP (this paper)	IHLS	99.0 ± 0.9

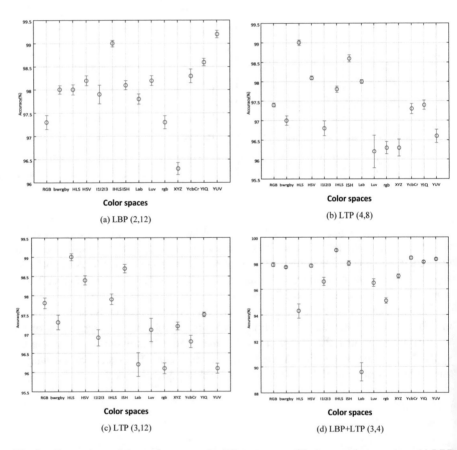

Fig. 2. Comparison of the performance via different types of features and parameters: (a) LBP (2,12), (b) LTP(4,8), (c) LTP(3,12), (d) LBP+LTP(3,4).

5 Conclusion

We proposed to incorporate color information to extract feature from LBP and LTP descriptor. Various color spaces are also investigated to compare the performance for cloud image classification. The experiment is evaluated on the SWIMCAT dataset which have shown the efficiency of proposed approach. We improve more than 2% comparing with previous works in literature. However, the dimension space is increased by adding color information and concatenating features. This issue should be solved in the future to remove irrelevant features.

References

1. Feister, U., Moller, H., Sattler, T., Shields, J., Gorsdorf, U., Guldner, J.: Comparison of macroscopic cloud data from ground-based measurements using VIS/NIR and IR instruments at Lindenberg, Germany. Atmos. Res. **96**, 395–407 (2010)
2. Dev, S., Savoy, F.M., Lee, Y.H., Winkler, S.: WAHRSIS: a low-cost high-resolution whole sky imager with near-infrared capabilities, Baltimore, Maryland, USA, p. 90711L, May 2014
3. Dev, S., Savoy, F.M., Lee, Y.H., Winkler, S.: Design of low-cost, compact and weather-proof whole sky imagers for high-dynamic-range captures. In: 2015 IEEE International Geoscience and Remote Sensing Symposium (IGARSS), Milan, Italy, pp. 5359–5362. IEEE, July 2015
4. Dev, S., Wen, B., Lee, Y.H., Winkler, S.: Machine learning techniques and applications for ground-based image analysis, June 2016. arXiv:1606.02811 [cs]
5. Fu, C.-L., Cheng, H.-Y.: Predicting solar irradiance with all-sky image features via regression. Sol. Energy **97**, 537–550 (2013)
6. Schumann, U., Hempel, R., Flentje, H., Garhammer, M., Graf, K., Kox, S., Losslein, H., Mayer, B.: Contrail study with ground-based cameras. Atmos. Meas. Tech. **6**, 3597–3612 (2013)
7. Yuan, F., Lee, Y.H., Meng, Y.S.: Comparison of radio-sounding profiles for cloud attenuation analysis in the tropical region. In: 2014 IEEE Antennas and Propagation Society International Symposium (APSURSI), Memphis, TN, USA, pp. 259–260. IEEE, July 2014
8. Wang, Y., Shi, C., Wang, C., Xiao, B.: Ground-based cloud classification by learning stable local binary patterns. Atmos. Res. **207**, 74–89 (2018)
9. Cheng, H.-Y., Yu, C.-C.: Block-based cloud classification with statistical features and distribution of local texture features. Atmos. Meas. Tech. **8**, 1173–1182 (2015)
10. Zhen, Z., Wang, F., Sun, Y., Mi, Z., Liu, C., Wang, B., Lu, J.: SVM based cloud classification model using total sky images for PV power forecasting. In: 2015 IEEE Power & Energy Society Innovative Smart Grid Technologies Conference (ISGT), Washington, DC, USA, pp. 1–5. IEEE, February 2015
11. Zhang, Z., Li, D., Liu, S., Xiao, B., Cao, X.: Multi-view ground-based cloud recognition by transferring deep visual information. Appl. Sci. **8**, 748 (2018)
12. Khan, F.S., Anwer, R.M., Weijer, J.D., Felsberg, M., Laaksonen, J.: Compact color-texture description for texture classification. Pattern Recogn. Lett. **51**, 16–22 (2015)
13. Nhat, H.T.M., Hoang, V.T.: Feature fusion by using LBP, HOG, GIST descriptors and Canonical Correlation Analysis for face recognition. In: 2019 26th International Conference on Telecommunications (ICT), pp. 371–375, April 2019

14. Hoang, V.T., Rebhi, A.: On comparing color spaces for fabric defect classification based on local binary patterns. In: 2018 IEEE 3rd International Conference on Signal and Image Processing (ICSIP), pp. 297–300, July 2018
15. Bianconi, F., Bello, R., Fernandez, A., Gonzalez, E.: On comparing colour spaces from a performance perspective: application to automated classification of polished natural stones. In: Proceedings of International Workshops New Trends in Image Analysis and Processing, pp. 71–78. Springer International Publishing (2015)
16. Van, T.N., Hoang, V.T.: Kinship verification based on local binary pattern features coding in different color space. In: 2019 26th International Conference on Telecommunications (ICT), pp. 376–380, April 2019
17. Duong, H., Hoang, V.T.: Dimensionality reduction based on feature selection for rice varieties recognition. In: 2019 4th International Conference on Information Technology (InCIT), pp. 199–202, October 2019
18. Sandid, F., Douik, A.: Robust color texture descriptor for material recognition. Pattern Recogn. Lett. **80**, 15–23 (2016)
19. Ojala, T., Pietikäinen, M., Mäenpää, T.: Multiresolution gray-scale and rotation invariant texture classification with local binary patterns. IEEE Trans. Pattern Anal. Mach. Intell. **24**, 971–987 (2002)
20. Tan, X., Triggs, B.: Enhanced local texture feature sets for face recognition under difficult lighting conditions, p. 15 (2007)
21. Hoang, V.T.: Multi color space LBP-based feature selection for texture classification. Ph.D. thesis (2018)
22. Dev, S., Lee, Y.H., Winkler, S.: Categorization of cloud image patches using an improved text on based approach. In: 2015 IEEE International Conference on Image Processing (ICIP), Quebec City, QC, Canada, pp. 422–426. IEEE, September 2015
23. Ojala, T., Pietikainen, M.: A comparative study of texture measures with class based on feature distribution, p. 9 (1996)
24. Liao, S., Law, M.W.K., Chung, A.C.S.: Dominant local binary patterns for texture classification. IEEE Trans. Image Process. **18**, 1107–1118 (2009)
25. Guo, Z., Zhang, D., Zhang, D.: A completed modeling of local binary pattern operator for texture classification. IEEE Trans. Image Process. **19**, 1657–1663 (2010)

Knowledge Representation Models Application and Fuzziness in Innovation Projects Assessment

Denis A. Istomin$^{(\boxtimes)}$ ⓘ and Valerii Yu. Stolbov

Perm National Research Polytechnic University, Perm, Russian Federation
dai@pstu.ru, valeriy.stolbov@gmail.com

Abstract. Application of three knowledge representation models are presented: fuzzy production rules, frames, semantic networks. Assessment of innovative project is described as a two-stage process. The first step is processing the project attributes using knowledge models in order to build a set of optimality criteria. The second step is the construction of a comprehensive criterion of optimality and the use of a specific metric for comparing and ranking innovative projects.

The model of production rules using fuzzy inference is considered first. The use of production rules for the logical inference of more complex criteria is proposed.

The frame model of knowledge representation is considered for calculating the parameters of an innovative project, since this model supports calculations in the process of inference. An application of semantic networks is considered with a combination of node-based and arc-based inference styles. Complex rules and relationships between entities in a complex process of evaluating an innovative project are presented using a semantic network. As a last step, the combination of all the evaluated characteristics of an innovative project into one complex criterion of optimality using fuzzy sets is described.

Keywords: Knowledge representation · Production rules · Fuzzy set · Frames · Semantic network · Innovation project · Innovation

1 Introduction

Every company must be innovative to meet the rapidly changing market needs. BCG's survey of senior executives [1] shows that total percentage of respondents who say that innovation is one of the top-three main priorities of their companies is 72%. R&D projects implementation is also a key component in the development of new competitive advantages. When a company's portfolio contains a large number of innovative projects, it is necessary to solve two main tasks: assessment and selection of projects for implementation [2].

Project portfolio selection methods have evolved over time [3]. In the beginning, methods were strongly focused on mathematical techniques [4]. Over time, as the business and technology environment became more complex, more complex methods were created.

© Springer Nature Switzerland AG 2021
T. Antipova (Ed.): ICIS 2020, LNNS 136, pp. 383–391, 2021.
https://doi.org/10.1007/978-3-030-49264-9_35

The most popular methods are financial methods. In these methods, profitability is the most important selection criterion, and the goal is to create a portfolio that maximizes expected return [5]. These methods use basic financial indicators, such as: net present value (NPV), profitability index (PI), internal rate of return (IRR), discounted payback period (DPP) [6].

But there are more complex methods:

1. Strategic methods.

   These methods are mainly determined by strategic objectives. Since choosing the right projects is extremely important not only to maintain competitiveness, but also to achieve global financial goals.

2. Scoring methods.

   These methods are based on subjective manager ratings. These methods are also popular in practice because of their ease of execution.

3. Behavioral methods.

   These methods are aimed at reaching a consensus based on the opinions of experts and managers in order to decide which projects should be implemented [7].

Expert systems based on knowledge representation models can be used in these models. The use of expert systems can ensure the transparency of decisions made during this process. Assessment of an innovation project can be based on expert knowledge, and the entire project can be represented as a set of characteristics – a set of optimality criteria. The process of assessing and ranking innovation projects may include the following steps (see Fig. 1):

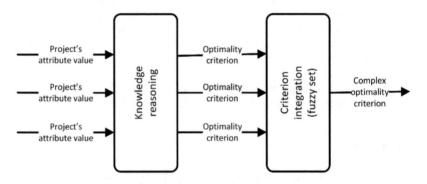

Fig. 1. Innovation project assessment steps

1. Gathering facts about innovation project: duration, costs, resources, etc.
2. Converting these values to optimality criteria using knowledge in an expert system.
3. Building a comprehensive optimality criterion that fully describes an innovation project.
4. Application of the operator for ranking projects based on a comprehensive criterion of optimality.

2 Knowledge Representation Models

The application of three models of knowledge representation will be considered: production rules with fuzzy logic, frames and semantic networks.

2.1 Fuzzy Production Rules

Fuzzy logic is often used in innovation management [8]. Fuzzy logic could be used in risks assessment [9]. But not only risks can be assessed using production rules. Consider the case of defining new criteria for an innovative project, which are derived on the basis of existing ones. For example, criteria "Strategy fit" could be defined by rule:

IF "Duration" = "<= 10 months" AND "Type of project" = "Product improvement" AND "New technology used" = "True" then "Strategy fit" = "3"

This rule can be rewritten using linguistic variables, which better corresponds to how experts make their decision:

IF "Duration" = "Short" AND "Type of project" = "Product improvement" AND "New technology used" = "True" then "Strategy fit" = "Normal" CF = 0.9

Each company can create its own set of fuzzy rules in the knowledge base The logical inference process according to the generated rules will contain several steps (see Fig. 2).

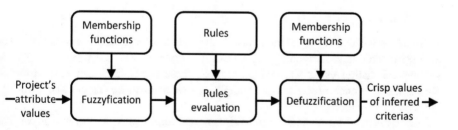

Fig. 2. Fuzzy inference process

For this example, membership functions for fuzzification (see Fig. 3) and defuzzification (see Fig. 4) can be defined. The certainty factor (CF) can be used in accordance with the standard approach in fuzzy logic.

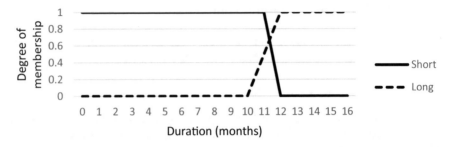

Fig. 3. Membership function for duration

Fig. 4. Membership function for strategy fit

Using this approach, each enterprise can define its own set of optimization criteria and a set of fuzzy rules, which will be used in innovative projects assessment.

2.2   Frames

A frame knowledge representation model [10] can also be used for assessment. An enterprise can define its typology of innovative projects. [11]. Since frames natively support calculation using attached procedures, these calculations can be performed during the inference process. For example, a risk assessment can be made during inference. (see Fig. 5).

Assessment of an innovative project using a frame representation model can be performed in two stages:

1. Creating a new frame and assigning values to available slots.
2. Logical inference through the frame hierarchy with assignment of values to the remaining slots.

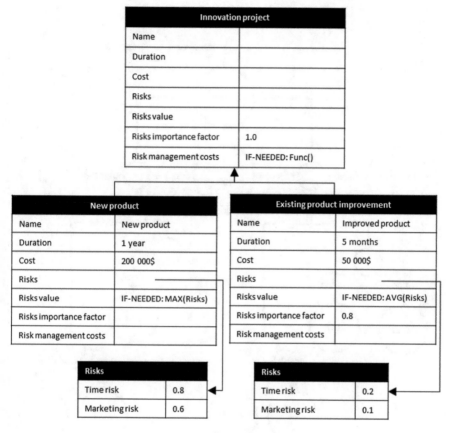

Fig. 5. Frames example

2.3 Semantic Networks

Semantic networks are often used in definitions of binary relations between entities (see Fig. 6).

Fig. 6. Simple semantic network

But semantic networks can display more complex relationships between entities. To achieve this, the semantic network must be built in a certain way, and also a way of logical inference. There are two inference styles: path-based and node-based [12]. For example (see Fig. 7), semantic network could contain generic rules and specific rules together.

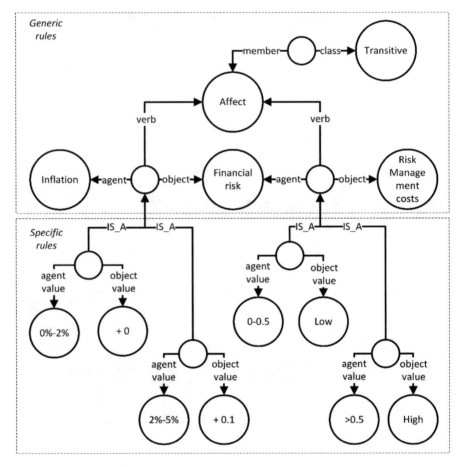

Fig. 7. Semantic network

Each unnamed node represents a rule. There is a generic rule: "Inflation" affects "financial risks", "financial risks" affect "risk management costs", so "inflation" affects "risk management costs". Specific rules could be defined using inheritance.

3   Ranking

Each innovation project has a corresponding set of optimality criteria (OC), which are evaluated using knowledge reasoning. The combination of these criteria can be presented (1) as a comprehensive optimality criterion (COC).

$$COC = (OC_1, OC_2, \ldots, OC_N) \tag{1}$$

Each optimality criterion may have each own priority/weight in comparison. Even the same optimality criterion for different projects can have different weights. Comprehensive optimality criterion can be presented as a fuzzy set (2).

$$\widehat{A} = \bigcup_{u \in U} \mu_A^i / u, i = \overline{1, n} \tag{2}$$

where n-number of innovation projects, and μ_A^i – membership function for i-th project. Comprehensive optimality criterion for projects A and B can be defined using Eqs. (3–4).

$$COC_A = A = \left(\mu_1^A(a_1)/a_1, \mu_2^A(a_2)/a_2, \ldots, \mu_n^A(a_n)/a_n \right) \tag{3}$$

$$COC_B = B = \left(\mu_1^B(b_1)/b_1, \mu_2^B(b_2)/b_2, \ldots, \mu_n^B(b_n)/b_n \right) \tag{4}$$

To be able to perform comparison ranking metric should be introduced [13]. It must be noted that different optimality criterions have different nature (costs, risks, etc.) and they should not be mixed together. Custom ranking metric can be introduced (6–7).

$$H(A, B) = sign \max |C_i|, \text{ where } C_i = \frac{\mu_i^A a_i - \mu_i^B b_i}{\max(a_i, b_i)} \tag{5}$$

$$\begin{cases} H(A, B) > 0 \Rightarrow A > B \\ H(A, B) < 0 \Rightarrow A < B \end{cases} \tag{6}$$

If there are multiple l equal maximum values, Eqs. (7–8) can be used.

$$\lambda = \sum_{i=1}^{l} sign \max |C_i| \tag{7}$$

$$\begin{cases} \lambda = 0 \Rightarrow A = B \\ \lambda > 0 \Rightarrow A > B \\ \lambda < 0 \Rightarrow A < B \end{cases} \tag{8}$$

3.1 Example of Ranking

There are two projects (A and B):

1. Project A has duration (τ) 28 days with costs (ε) \$2M
2. Project B has duration (τ) 40 days with costs (ε) \$0.5M

$$\mu_1^A(\tau) = \begin{cases} 0, 8, \tau < 30 \ days \\ 0, 1, \tau \geq 30 \ days \end{cases}, \mu_2^A(\varepsilon) = \begin{cases} 0, 5, \varepsilon < \$1M \\ 0, 4, \varepsilon \geq \$1M \end{cases}$$

$$\mu_1^B(\tau) = \begin{cases} 0, 9, \tau < 41 \ days \\ 0, 1, \tau \geq 41 \ days \end{cases}, \mu_2^B(\varepsilon) = 0, 1$$

Using Eq. (5):

$$C_1 = \frac{0,8 \cdot 28 - 0,9 \cdot 40}{\max(28,40)} = -0,34; C_2 = \frac{0,5 \cdot 0,5 - 0,1 \cdot 2}{\max(0,5,2)} = -0,025$$

And $|-0,34| > |-0,025| \Rightarrow H(A,B) < 0 \Rightarrow A < B$

4 Conclusion

Application of three models of knowledge representation in the assessment of innovative projects is presented. An example of the application of production rules with a fuzzy logic inference is presented. The application of the frame knowledge representation to represent the hierarchy of innovative projects is proposed. For more complex rules and relations between entities, the use of semantic networks is considered. These models of knowledge representation are considered only for the evaluation of innovative projects, but it may be possible to find application in other aspects of innovative project management. In future studies, the applicability of frames for different stereotypical situations can be investigated. Also, in future studies, it is also necessary to describe the specification of an information system that would allow the use of the described models.

Acknowledgements. The reported study was funded by the Government of Perm Krai of the Russian Federation (the project «Computer biomechanics and digital technologies in biomedicine»).

References

1. Andrew, J.P., et al.: A return to prominance-and the emergence of a new world order. BCG Most Innovative Companies (2010)
2. Meredith, J.R., Mantel Jr., S.J.: Project Management: A managerial Approach. Wiley, Hoboken (2009)
3. Chaparro, X.A.F., de Vasconcelos Gomes, L.A., de Souza Nascimento, P.T.: The evolution of project portfolio selection methods: from incremental to radical innovation. Revista de Gestão (2019). https://doi.org/10.1108/rege-10-2018-0096
4. Baker, N.R.: R & D project selection models: an assessment. IEEE Trans. Eng. Manage. **4**, 165–171 (1974)
5. Bard, J.F., Balachandra, R., Kaufmann, P.E.: An interactive approach to R&D project selection and termination. IEEE Trans. Eng. Manage. **35**(3), 139–146 (1988)
6. Lal, D.: Methods of project analysis: a review. International Bank for Reconstruction and Development, New York (1974)
7. Cooper, R.G., Edgett, S.J., Kleinschmidt, E.J.: New product portfolio management: practices and performance. J. Prod. Innov. Manage. **16**(4), 333–351 (1999)
8. Alfaro-García, V., Gil-Lafuente, A., Calderon, G.: A Fuzzy logic approach towards innovation measurement. Glob. J. Bus. Res. **9**, 53 (2015)
9. Shapiro, A.F., Marie-Claire, K.: Risk assessment applications of fuzzy logic. Casualty Actuarial Society, Canadian Institute of Actuaries, Society of Actuaries (2015)

10. Roberts, R.B., Goldstein, I.P.: The FRL Primer. Massachusetts Institute of Technology, Cambridge (1977). №. AI-M-408
11. Istomin, D.A., Gitman, M.B., Trefilov, V.A.: Frames knowledge representation model of innovative projects assessment methodologies. Neirokompyutery: razrabotka, primenenie [NeuroComputers: Development, Use], no. 2, pp. 12–22 (2018). (in Russ.) ISSN 1999-8554
12. Shapiro, S.C.: Path-based and node-based inference in semantic networks. In: Proceedings of the 1978 workshop on Theoretical Issues in Natural Language Processing. Association for Computational Linguistics (1978)
13. Istomin, D.A., Gitman, M.B.: Simulation model of strategic innovation management at manufacturing enterprises. Bull. Kalashnikov ISTU. **20**(2), 150–153 (2017). https://doi.org/10.22213/2413-1172-2017-2-150-153. (in Russ.)

From Data to Insight: A Case Study on Data Analytics in the Furniture Manufacturing Industry

Sunet Eybers$^{(\boxtimes)}$ and Rameez Mayet

University of Pretoria, Private Bag X20, Hatfield 0028, South Africa
sunet.eybers@up.ac.za

Abstract. The manufacturing sector in South Africa has been dramatically affected by easily accessible and low-cost goods from China and other manufacturing giants in the East. Organizations which have not kept abreast with the technological advances in manufacturing systems and processes have been left behind and often face liquidation. The purpose of this study is to identify solutions to overcome issues faced by manufacturing organizations, through the possible implementation of data analytics. The study furthermore initiates the search for solutions which could revolutionize the manufacturing industry in South Africa. A review of existing literature regarding data analytics capabilities in the manufacturing industry found limited academic literature on the topic. A case study research strategy was adopted to explore the problem at hand through a holistic, in-depth investigation focusing on all the managerial levels in a manufacturing organization. The results indicated that executive and top management were more knowledgeable in the area of data analytics than lower level management and stated that the decisions they make were reactive instead of pro-active. This was attributed to outdated IT infrastructure and information systems. A single system with a central repository of information equipped with real-time analytical dashboards were identified as a good start to improve business processes, reduce time wastage and provide for data-driven decision making.

Keywords: Data analytics · Manufacturing industry · Decision-making · Business process improvement

1 Introduction

Data Analytics is a continuously growing field due to the exponential increase in data being produced by people, processes and organizations. According to previous studies, the digital universe and the data being created has a projected growth of 40% each year and the total amount of data ever created doubles every two years [1, 2].

The manufacturing industry in South Africa is yet to recover from the deindustrialization cycle. According to Andre de Ruyter, Chairperson of the Manufacturing focus group, the manufacturing sector contribution to the Gross Domestic Product (GDP) of South Africa, a developing country, has fallen from 24% in 1980 to below 13% in 2015 [3]. A major contributing factor is the large number of imports from China

© Springer Nature Switzerland AG 2021
T. Antipova (Ed.): ICIS 2020, LNNS 136, pp. 392–405, 2021.
https://doi.org/10.1007/978-3-030-49264-9_36

and other Eastern nations who have boosted support from the state, cheap labor costs and large workforces [4]. This slump has led to a job loss of over 500 000, coupled with a significant drop in consumer demand and lower investor confidence [3].

Advanced technology, low costs and large-scale manufacturing capabilities of China make it difficult for South African manufacturers to compete with. There remains a lack of initiatives to educate and transform the South African manufacturing industry through the use of advanced analytics, data-driven smart manufacturing, predictive learning architectures and big data centric solutions to manufacturing [5]. Minimal research has been carried out to explore the potential impact that implementing analytics into manufacturing organizations could have on the performance of the industry in South Africa. This study will focus on the possible application of analytics to enhance business processes and to provide real-time, relevant information to assist in executive decision making. This will be achieved through a case study approach by analyzing all the information available concerning a single organization in the furniture manufacturing industry in South Africa. The research question is *"how can data analytics be used in the manufacturing industry to improve business processes in order to provide data-driven insights for decision making?"*. The paper starts with a discussion on key concepts such as data analytics and big data, how big data can be used to contribute to a competitive advantage, in particular manufacturing that are data-driven to understand customer behavior and subsequent knowledge creation. The concept of Business Process Management (BPM) is discussed where after the case study is presented and the study concluded with a summary of findings and a proposal for further research.

2 Data Analytics and Big Data

In the digital age we are currently living in, computerization allows for access to data to be readily available. The analysis and processing of data has been carried out by means of statistical and scientific methods [6]. The term data analytics can be defined as the operation of analytical techniques on large data sets [7]. The data sets can be structured, semi-structured and unstructured [8]. Structured data refers to the data having a defined structure and format, for example a database. Semi-structured data refers to textual data files that have an apparent pattern which enables analysis, such as spreadsheets. Unstructured data has no inherent structure and originates in various formats and such as images, documents and videos to name a few [8]. Due to the exponential increase in growth of data available online, companies are turning to analytics to provide them with a competitive advantage [6].

The current demand for data analytics is driven by the need for businesses to use the available data in order to identify potential threats to their businesses and to drive business decisions which will improve revenue and lower costs for the organization [9]. Data analytics involves the use of advanced statistical and mathematical models to analyse large, complex, unstructured datasets.

Having access to large quantities of data is useless if no action is taken. The value hidden within the data can only be unlocked by means of using the data to drive decision making. To enable data to become the driving force, businesses require efficient processes to transform large volumes of various, real time data into meaningful information for decision making [10]. The process of deriving knowledge from big data involve two main stages; data management, which includes using technologies to acquire, store, and prepare the data for analysis; and analytics, which refers to techniques used to derive intelligence from the data [10].

While there are a multitude of benefits of data analytics, there are several challenges as well. According to a survey conducted by the IDC, commissioned by the Irish Department of Jobs, Enterprise and Innovation [11], the major challenges faced by data analytics projects are, in order of most challenging; a lack of skilled data analysts, insufficient amount of relevant data, cost, lack of corporate sponsorship and data protection issues [6]. This is similar to barriers of Big Data Analytics identified in a study by Moktadir et al. [12] highlighting data related issues (complexity of integration, privacy, analytical tools) combined with technological challenges including infrastructure and financial challenges.

Big Data is a term used to describe the ever-increasing large amounts of data available from various sources. Whilst there are multiple definitions for big data, a fitting one resides from researchers at McKinsey who offer a definition of big data as "datasets whose size is beyond the ability of the typical database software tools to capture, store manage and analyze" [1]. The data available due to the internet and computerization of industries and societies originates from vastly different sources and as such is of different formats and includes structured and unstructured data.

Big data is often suggested to have five characteristics; variety, volume, velocity, value and veracity [13]. In simple terms, the five V's can be explained (in order of listing) as, the data produced is of various formats, the volume of data produced increases exponentially, traditional analytics systems do not have the capability to cope with the speed at which data flows, the value of big data is limited to the information which can be extracted from the data and the quality of the data determines the accuracy of the information extracted [13].

3   Leveraging Data for a Competitive Advantage

In organizations, analytics continues to have an increasing impact on decision making and company performance. The digital age of today has brought with it challenges and data-centric opportunities. Companies which have embraced the digital era have access to large amounts of data and their ability to transform that data into an asset of knowledge for making data-driven decisions creates a platform from which to disrupt and differentiate in its market [14].

Organizations which are still primarily using analytics for their baseline analytic procedures such as forecasting, budgeting and supply chain management are falling behind [15]. Making analytics a primary objective in an organizations' business processes does not solely allow the organization to have a refined view of future trends and business opportunities, but to know the why, what and how of the outcome of every business decision as well [6]. According to a study performed by MIT Sloan Management Review together with the IBM Institute for Business Value which involved surveying more than 3000 business executives, managers and analysts around the globe, top performing organizations incorporated analytics in every business department possible. Additionally, the top performing organizations corroborated with MIT Sloan and IBM Institute for Business Value in that the use of analytics and business intelligence differentiated them within their industry. Senior executives and organizational leaders want their businesses to perform on data-driven decision [15].

It is suggested that in addition to having the right personnel to analyze the data, organizational factors such as top management support, organizational culture and sponsors being open to change and new ideas have an effect on whether the organization will gain a competitive advantage from using analytics [14]. Organizational culture can be defined as a set of shared beliefs and aims around which expected practices, behaviors and norms are adhered to [15]. In an organization with a data-driven culture, employees who are data-oriented and those who are not utilize analytical insights in their decisions [15].

In order for analytics to provide an organization with a competitive advantage, a data-oriented culture at an enterprise level must be adopted. According to a study performed by Kiron and Shockley [15] three characteristics were identified in a data-oriented culture; analytics is seen and used as a strategic asset, top management supports the enterprise-wide use of analytics and insights are available to those who need them. Companies which adopt this culture whilst incorporating data analytics will likely achieve the competitive advantage they desire. Once an organization has achieved competitive advantage using insights produced through data analytics, the next step is to maintain that advantage by continuously revisiting those insights. In some instances, once an organization in a particular market finds a unique method to gain an advantage using data, competitors swiftly follow suit and eradicate that advantage [15].

An important driving force behind competitive advantage is innovation [15]. Leading organizations understand the potential for technological capabilities to assist them in predicting customer needs and they are investing substantially in innovation that leverages opportunities to collect new data and combines internal and external data upon which data analytics is applied [16]. In the study performed by Marshall et al. [16] the three strategies which leading organizations in innovation employ revolve around data, skills and culture are; promote excellent data quality and accessibility, include innovation and analytics in every aspect of the organization and build a quantitative innovation culture.

3.1 Data-Driven Manufacturing

In the recent past, manufacturing companies have had the ability to reduce raw material waste and inconsistencies in their production processes, together with an increase in product quality by means of 'Lean Six Sigma' methodology [17]. The 'Six Sigma' part refers to the tools and methods used to improve an organizations' manufacturing process and the 'Lean' part refers to any method or technique that assists in the process of identification and elimination of waste [17]. This approach has proved valuable for organizations in the manufacturing industry thus far. However due to continuously evolving and complex manufacturing processes, a more detailed diagnosis of business processes is required to improve and correct process shortcomings [17].

In this era of big data which we currently live, technology has evolved the way in which organizations are able to transform raw data into meaningful information by means of data analytics, machine learning, business intelligence and data science to name a few. The Internet of things (IoT) has brought about advanced information technologies such as smart sensors and product embedded information devices (PEIDs) which are used in manufacturing organizations to monitor, control and manage the product lifecycle (PLM) which produces large amounts of data [18]. According to Tao [19], big data generated by manufacturing processes can include (1) management data collected by management information systems such as Enterprise Resource Planning (ERP), Supply Chain Management (SCP) and Customer Relationship Management (CRM) systems, (2) data collected by IoT technologies relating to real-time performance and operating conditions, (3) user data collected from internet sources or social networking sites, (4) product data collected from smart products and product-service systems, and (5) Publicly accessible data through open databases. Manufacturing businesses can leverage this vast variety of data to expand their understanding of customers, suppliers, products and processes [19].

The Product Lifecycle Management (PLM) approach generates large quantities of data which creates challenges for organizations to develop efficient data mining processes to improve business forecasts [18]. Advanced analytics will also need to be implemented to assist management in making more informed decisions.

3.2   Data Analytics and Understanding the Customer

In the last 20 years, marketing personnel have been provided with numerous, innovative marketing research tools and techniques thanks to technological advances. Marketing analytics is the newest addition to the marketing research process and has gained a large market adoption due to technology improving the feasibility for marketers to mine and analyse large data sets [20].

The world of social media has brought with it millions upon millions of pieces of data arising from Facebook posts, Instagram likes, tweets, pins and so on. The ability to analyse, correlate and extract value from this data is enhancing the relationship between company and customer [21]. Technology such as smart phones and laptops use data plans to connect to the internet which provides companies with a way of capturing user data in real time 24/7. Social network sites continuously map user data to identify trends in user behavior and are constantly correlating the analysed data to provide real time targeted advertising [22].

An important technique used by businesses is Text Analytics. Textual Analytics refers to techniques, often in the form of statistical analysis and machine learning, which extract information from textual data such as social network feeds, blogs, online forums and corporate documents [10]. Text analytics allows business to transform large quantities of data generated by people, processes and organizations into summaries to be used for data-driven decisions. An example of text analytics would be to predict the next commercial product based on information extracted from social networking feeds. In terms of the relationship between customer and company, being able to analyse customer feedback, which are often textual, helps to identify the problems facing the customer and the business strategist will be able to improve customer experience by improving problem areas of operation [6].

3.3   Creating Knowledge Through Analytics

According to Liberatore et al. [23], analytics is the process of transforming data into actionable information in the context of the organizations decision making culture. From a manufacturing perspective, Tao et al. [19] identified a framework for deriving knowledge from the data produced in the product management lifecycle beginning with the source of the data. This refers to the varying source data created throughout the manufacturing value-chain. Followed by the collection of data from all IoT devices in the organization which enable real-time data to be captured. Third is the storage of the structured and unstructured data produced which can be expensive for an organization. Cloud storage has created an objective, cost-effective means for organizations to store the created data. Thereafter the stored data must be cleaned to remove redundant and duplicate data in order to provide data of an excellent quality to allow for manufacturers to well-informed strategic business decisions. Finally, data visualization through means of charts and graphs enable users to comprehend the derived information more easily.

Through analytics organizations can reveal unexpected insights in their data which can lead to improved efficiency and icrease in yield [17]. According to a study performed by Auschitzky et al. [17], manufacturers need to first recognize the amount of data at their disposal and their objectives for that data. It is illogical for an organization to invest heavily in analytics without investing in the required skills needed to optimize their business processes through the existing process information [17]. The effective implementation of analytics and skilled employees has proven to significantly increase organization performance [24].

4   Business Process Management

In the last two decades, technology has diversified the way business is conducted around the globe which forced organizations to adapt and evolve in order to remain relevant. Incorporating new technologies into existing business cultures and finding new means for improving operations remains at the forefront of many organizations. A business process can be defined as set of logically related activities which are performed together in series to achieve an organizational goal [25]. The advancement of technology requires businesses to adapt and align their business processes with their strategic objectives as the organization continues to evolve.

For the purpose of this study, Business Process Management (BPM) will be explained in two ways. Firstly, BPM aims to improve business processes by modelling existing processes and analysing it after which management may suggest possibilities to reduce costs or increase production output [26]. Secondly, BPM may refer to the software which has been developed to manage and support a business's operational processes which initialised a boom for a new, innovative technology known today as a Business Process Management System [26]. The revolutionary aspect of BPM systems is their ability to communicate with various existing legacy systems in an organization as well as emerging technology such as smart devices which the organization may be looking to incorporate into its' operations.

5   Research Approach

A case study approach was adopted for the purpose of this study focusing on a furniture manufacturing organization located in Pretoria, Gauteng Province, South Africa. The Gauteng province is the largest in population of nine provinces in South Africa, classified as a developing country.

The focus of the study was on a family owned, well known furniture manufacturing organisation and involved observations of all employees across all the functions of the organisation. The data from the observations were substantiated by interviews focusing on middle, senior and top management. Two semi-structured interview templates were created to cater for top management (CIO) as well as one for middle and senior management. A total of 50 participants were approached by the researcher to participate in the research. Of the 50 participants only 36 gave their consent. All the participants were knowledgeable on the overall furniture manufacturing value chain and processes.

Organizational documentation, such as current manufacturing process document, information delivery artefacts (reports, dashboards and Excel spreadsheets) were studied to understand various business processes in the organization as well as the decision making process followed by employees on all organizational levels.

6   Discussion of Findings: Case Study

The manufacturing company is a family owned business who started their operations in 1982. It is classified as a Small medium sized organisations (SME). It focuses on the manufacturing of furniture and supplies various large furniture chain stores with customized furniture.

It was also important to investigate the source of data that the various participants used to obtain trustworthy data for decision making. All participants in the study interact with and get their data from either one or both systems. The majority of the participants were from the manufacturing job function with 78% obtaining data from the financial system. In order to understand their current business and the influence of data and decision-making process within the organization, documentation was studied to obtain a high-level overview of the process of manufacturing a product (the main focus of their business). In addition, an interview with the CIO were used to verify the process and obtain more detail about the process. Figure 1 depicts, on a high-level, how the organization processes an order for furniture from a customer and the cross organizational departments involved in the manufacturing of each order.

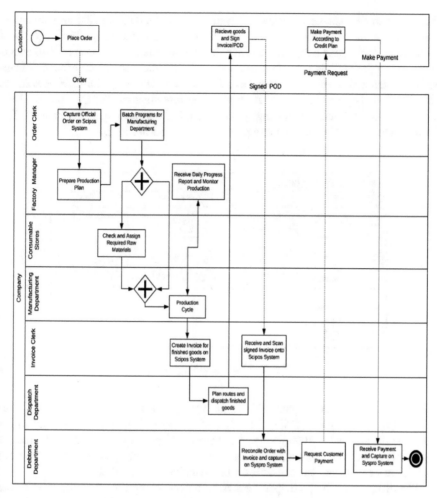

Fig. 1. Manufacturing/production business process – flow diagram

Further observations and interviews carried out by the researcher this time focusing on senior and middle management. It became evident that the organization has certain rules and assumptions by which it abides in order to conduct business. The organization only manufactures goods to fulfil a customer order and does not manufacture goods for stock purposes. Each furniture design has a unique bill of materials which remains standard after it has been programmed into the existing system. The organization purchases raw materials which are required across all products, such as screws and staples, in advance and basis this purchasing on previous working knowledge and manual forecasting. The invoicing and manufacturing processes/functions are performed on a system called Scipos (operational manufacturing system). The debtors' department, creditors department and consumable stores department are required to capture all information on a system called Syspro (an accounting system). Focusing on the two main data sources, participants were asked if the data which was available to them was sufficient to perform their job effectively and efficiently. The majority of participants (61%) stated that the data they have is sufficient while 39% of participants stated the opposite. It is important to note that all middle, senior and executive management stated that the data they have is not sufficient to perform their tasks as effectively as they would like to. As part of the investigation, the CIO provided the researcher access to certain documentation regarding the manufacturing process. The documented organizational business process is depicted in Fig. 2.

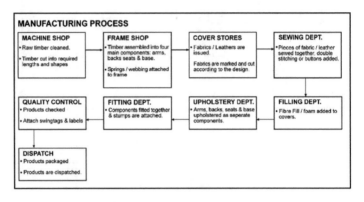

Fig. 2. Manufacturing process

The majority of participants fell in the age group between 40–49 years (36%) followed by 50–59 years (31%), older than 60 years (8%), whilst 25% were younger than 30 years of age. Subsequently, the majority of participants fulfilled the role of supervisor in the organization (39%), followed by middle-management (25%), assistant (14%), senior management and director or executive level (both 11%).

The diversity of the participants provided a good glimpse of the utilization of the current manufacturing process in the organization and the utilization of data in the decision making process. From this participant group, only 22% has heard of the term data analytics whilst the remainder has never heard of the term. Important to note is that no supervisor has heard of the term before.

Based on these two figures (Fig. 1 – business process and Fig. 2 – manufacturing process) possible improvements were investigated to support data collection procedures that can subsequently be used for data driven decision making.

There are differing levels of 'sufficient' data depending on the job or task to be completed. Participants were asked if the data available was sufficient to perform their job effectively and efficiently. The majority of participants (61%) stated that the data they have is sufficient while 39% of participants stated the opposite. It is important to note that all middle, senior and executive management stated that the data they have was not sufficient to perform their tasks as effectively as they would like to. Furthermore participants were asked if they could make adequate decisions based on the data they have. The majority of participants, 56%, stated that their decisions could be better if they had more data available, whilst 44% of participants stated they could not make the correct decisions given the available data (see Fig. 3).

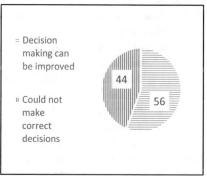
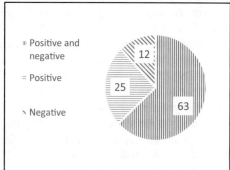

Fig. 3. Quality of decision making   **Fig. 4.** Decision making effect on organisation

A set of questions targeting senior and executive management only focused on themes of decision making, multiple data sources in the decision making process, their perception on the impact and quality of their decisions and the achievement of competitive advantage based on their decisions. All participants indicated that they use documentation, raw data, people as a source of data, management information systems, intuition, experience and their emotions to make decisions. Furthermore 63% of senior and executive management agree that their decisions affected the organization in both a positive and negative manner. The remaining 37% of senior and executive management stated that their decisions affected the organization positively (25%) and negatively (12%) (see Fig. 4). It is important to note that all of the participants agreed that their decisions always had an impact on the organization, whether positive or negative.

Participants were also asked to rate the quality of their decision-making based on the current sources of data at their disposal on scale from 1 to 5. The majority of participants (5) stated that their decision-making is good. None of the participants said their decision-making is perfect nor poor. The other senior and executive managers stated that there decision-making is great (1) and average (2). It is important to note that only one top level manager stated that their decision making was great. 62% of participants said that their decisions in the past have led to some form of competitive advantage for the organization whilst the other 38% stated differently.

Based on the processes specified in Fig. 1 and Fig. 2, all participants were asked their opinion on the sufficiency of the process to supply data for decision making. The majority of participants in the study (61%) stated that they have access to sufficient data to perform their job functions but all the participants voiced shortcomings in their working processes. Using the diagrams (Fig. 1 and Fig. 2) as a reference to depict the general order of business and the various operational units, the issues identified by participants will be briefly discussed:

The **order clerks** are required to manually check (via email or telephonically) with the consumable store manager and covering store manager if the required raw materials are available before a customer order may be processed. This wastes time as feedback provided to the order clerk can take up to two working days.

The **factory manager** oversees all processes relating to the daily production operations of the business. A task of the factory managers is to determine a monthly production plan using the orders received and raw materials available using data from both business systems (Scipos and Syspro) and spreadsheets received from supervisors. According to the relevant participants, the existing systems are not integrated and information from supervisors often include errors and are delayed. The **consumable stores** department is responsible for the procurement and issuing of raw materials. According to the relevant participants, they are required to capture receipts of raw materials on one system and then issue the required materials to the manufacturing supervisors on another system. The **manufacturing department** consists of eight functional areas which the smooth running of are a necessity to fulfil customer orders. Participants apart of the machine shop, frame shop, sewing department, filling department, upholstery department, fitting department and quality control department all stated that they have the required information for production of the goods but do not have data of materials (required quantity and type) on goods which have previously been through production. This requires each supervisor to manually calculate and confirm the required materials with the consumable stores and covering stores. The cover stores department is required to ensure that sufficient stock of leather and fabric is on hand to fulfil customer orders at all three manufacturing sites. According to the relevant participants, they use the order situation obtained from Scipos, bill of materials obtained from Syspro, production plan obtained from the factory manager and shipping status of raw materials obtained from an excel spreadsheet to make decisions. They are always under pressure and are required to assess data from four separate sources which is time consuming and demoralizing. The **invoice clerks** are responsible for capturing invoices and scanning Purchase Order Deliveries using the Scipos system. No issues were revealed by the relevant participants. The **dispatch department** determines when and where to deliver the goods on time based on quantity to be delivered, area of delivery and date of delivery. According to the relevant participants, they are required to determine the above-mentioned tasks manually by using the production plan, order situation and delivery dates which are obtained from the production manager, excel spreadsheet and order department respectively. The **debtors' department** are required to ensure payment by customers is received on time. According to relevant participants, they are required to reconcile invoice information from the Scipos system with the correct order information on the Syspro system. They are then required to re-capture invoicing information from the Scipos system onto the Syspro system as well as capture payment details on the Syspro system. Re-capturing information which already exists on one system onto the other system is unnecessary and time-consuming. Data integration is a challenge acknowledge by existing studies [6, 12]. **Executive Management** are required to make decisions which affect the entire business operation, whether it be for the short or long term. According to the relevant participants, they currently receive information from multiple information systems and people (data challenges). They are then required to make decisions based on an overloaded, error-prone and unstructured pile of data. Executive management of the organization stated that there is a lack of IT infrastructure (also highlighted by [6, 12]) and large amount of information being processed from daily operations (data and process challenges). They are currently unable to make long-term, informed strategic decisions to benefit the organization as they do not have the appropriate reporting tools to provide them with the insight they require. Manually analysing all the data without the assistance of a system/software which automatically identifies trends in the data and shortcomings in processes is time-consuming and negatively impacts the effectiveness of their decision-making. All these challenges are not unique to the manufacturing industry as indicated by [27] who classify these challenges into data (issues with regards to data characteristics), process (process of working with data) and management related challenges (such as ethics and governance).

7 Conclusion

The study explored the current business operations of a single furniture manufacturing organization in South Africa with the aim to understand how data analytics could be used in the manufacturing industry.

Not having access to cross-divisional data at a single source/system was the most commonly expressed concern by executive and senior management and resulted in data integration challenges. This is a common challenge identified by many authors [6, 12]. Participants also indicated a strong need for an organizational level dashboard to display organizational data (across all business processes) to enable management to keep track of current operations. Top management, in particular, indicated a strong need to use the information contained on the dashboard to make decisions that can contribute to their competitive advantage by keeping track of current and future product trends. On an operational level, fabric stores indicated the need to obtain real-time stock level data of both raw materials and current customer orders and requirements. This will allow the procurement division to pro-actively produce the necessary materials therefore contribute to the efficiency and effectiveness of the procurement process. The dispatch, order clerk, debtors clerk all require access to data from the single source system eliminating time wasted on feedback required from other departments. Supervisors in the manufacturing process indicated the need to view the required materials of goods previously run through production having access to historical data.

As described above, data analytics are required throughout the entire furniture manufacturing business process across all organizational levels to reduce wasted time, automate repetitive manual tasks and allow for managers to make data-driven decisions which could lead to a competitive advantage. It would be interesting to conduct subsequent studies in similar organisations to compare the findings.

References

1. Manyika, J., Chui, M.: Big Data: The Next Frontier for Innovation, Competition, and Productivity. McKinsey Global Institute, Lexington (2011)
2. The Digital Universe of Opportunities: Rich Data and the Increasing Value of the Internet of Things Sponsored by EMC. https://www.emc.com/leadership/digital-universe/2014iview/index.htm. Accessed 26 Mar 2020
3. Odendaal, N.: South Africa's manufacturing sector 'woefully underperforming' – De Ruyter. https://www.engineeringnews.co.za/article/south-africas-manufacturing-sector-woefully-und erperforming-de-ruyter-2017-06-27/rep_id:4136. Accessed 26 Mar 2020
4. Place, Y.W. in O.: Deindustrialisation: how we got to this point. https://showme.co.za/lifestyle/deindustrialisation-how-we-got-to-this-point/. Accessed 26 Mar 2020
5. Analytics in Manufacturing: Are South African Manufacturers Ready for "MAnalytics"?. Deloitte, South Africa (2015)
6. Karanth, P., Mahesh, K.: From data to knowledge analytics: capabilities and limitations. Inf. Stud. **21**, 261–274 (2015). https://doi.org/10.5958/0976-1934.2015.00019.1
7. Russom, P.: Big Data Analytics. The Data Warehouse Institute (TDWI) Research (2011)
8. Tanwar, M., Duggal, R., Khatri, S.K.: Unravelling unstructured data: a wealth of information in big data. In: Infocom Technologies and Optimization (ICRITO) (Trends and Future Directions). IEEE, Noida (2015)
9. Big Data Analytics and the Future Ahead (2018)
10. Gandomi, A., Haider, M.: Beyond the hype: big data concepts, methods, and analytics. Int. J. Inf. Manag. **35**, 137–144 (2015)
11. A study on harnessing Big Data for innovation led grwoth: an assessment of Ireland's progress and further policy requirements. IDC, London, UK (2015)
12. Moktadir, A., Ali, S.M., Paul, S., Shukla, N.: Barriers to big data analytics in manufacturing supply chains: a case study from Bangladesh. Comput. Ind. Eng. **128**, 1063–1075 (2019)
13. Prasad, P.S., Rajesh, K.: A novel study on Big Data: issues, challenges. Tools. Int. J. Sci. Eng. Comput. Technol. **7**, 44–49 (2017)
14. Svilar, M.: Adopting right-time analytics for creating a competitive advantage. https://www.dqindia.com/adopting-right-time-analytics-for-creating-a-competitive-advantage/
15. Kiron, D., Shockley, R.: Creating business value with analytics. MIT Sloan Manag. Rev. **53**, 57–63 (2011)
16. Marshall, A., Mueck, S., Shockley, R.: How leading organizations use big data and analytics to innovate. Strat. Leadersh. **43**, 32–39 (2015)
17. Auschitzky, E., Hammer, M., Rajagopaul, A.: How big data can improve manufacturing (2014)
18. Zhang, Y., Ren, S., Liu, Y., Sakao, T., Huisingh, D.: A framework for Big Data driven product lifecycle management. J. Clean. Prod. **159**, 229–240 (2017)
19. Tao, F., Qinglin, Q., Lui, A., Kusiak, A.: Data-driven smart manufacturing. J. Manuf. Syst. **48**, 157–169 (2018)
20. Hauser, W.J.: Marketing analytics: the evolution of marketing research in the twenty-first century. Direct Mark. Int. J. **1**, 38–54 (2007)
21. Ducange, P., Pecori, R., Mezzina, P.: A glimpse on big data analytics in the framework of marketing strategies. Soft Comput. **22**. https://doi.org/10.1007/s00500-017-2536-4
22. Nair, L.R., Shetty, S.D., Shetty, S.D.: Streaming Big Data analysis for real-time sentiment based targeted advertising. Int. J. Electr. Comput. Eng. IJECE **7**, 402–407 (2017). https://doi.org/10.11591/ijece.v7i1pp402-407

23. Liberatore, M.J., Pollack-Johnson, B., Chain, S.H.: Analytics capabilities and the decision to invest in analytics. J. Comput. Inf. Syst. **57**, 364–373 (2017)
24. Ghasemaghaei, M., Ebrahimi, S., Hassanein, K.: Data analytics competency for improving firm decision making performance. J. Strat. Inf. Syst. **27**, 101–113 (2018)
25. What is a Business Process? - Definition from Techopedia. https://www.techopedia.com/definition/1168/business-process. Accessed 26 Mar 2020
26. van der Aalst, W.M., Rosa, M.L., Santoro, F.M.: Business process management. Bus. Inf. Syst. Eng. **58**, 1–6 (2016)
27. Sivarajah, U., Kamal, M.M., Irani, Z., Weerakkody, V.: Critical analysis of Big Data challenges and analytical methods. J. Bus. Res. **70**, 263–286 (2016)

VNPlant-200 – A Public and Large-Scale of Vietnamese Medicinal Plant Images Dataset

Trung Nguyen Quoc$^{(\boxtimes)}$ and Vinh Truong Hoang

Ho Chi Minh City Open University, Ho Chi Minh City, Vietnam
{trungnq.188i,vinh.th}@ou.edu.vn

Abstract. Plant identification is an essential topic in computer vision with various applications such as agronomy, preservation, environmental impact, discovery of natural and pharmaceutical product. However, the standard and available dataset for medicinal plants have not been widely published for research community. This work contributes the first large, public and multi class dataset of medicinal plant images. Our dataset consists of total 20,000 images of 200 different labeled Vietnamese medicinal plant (VNPlant-200). We provide this dataset into two versions of size 256×256 and 512×512 pixels. The training set consists of 12,000 images and the remainder are used for testing set. We apply the Speed-Up Robust Features (SURF) and Scale Invariant Feature Transform (SIFT) for extracting features and the Random Forest (FR) classifier is associated to recognize plant. The experimental results on the VNPlant-200 have been shown the interesting challenge task for pattern recognition.

Keywords: VietNam medicinal plant · SIFT · SURF · Random Forest · Plant identification

1 Introduction

Nowadays, the growth of urbanization and technology makes people has been facing more diseases. The cost of medical examination and treatment is increasingly expensive for people all around the world. In some countries, approximately 80% of the population relies on traditional medicine for their primary health care needs [1]. Vietnam is considered as a country with biodiversity in terms of forestry, agriculture and traditional medicine. There are a lot of research showing that Vietnamese use flora and fauna, of which plants are mainly used as traditional medicine. According to statistics [2] "Dictionary of Vietnamese medicinal plants" there are about 4,700 species of medicinal plants, which are used in traditional medicine in Vietnam. Nonetheless, the usage of these medicinal plants must be based on the experience of specialists. However, to understand correctly the knowledge about medicinal plants and identify them is not an easy task.

Today, the development of science and technology, especially the application of computer vision, has made it easier, faster and more accurate for humans to identify plants. They only need to use a smart device with an integrated camera to access a plant's information from captured images. Currently, there are many studies on medicinal plants in computer vision. Plant identification applications are first based on

© Springer Nature Switzerland AG 2021
T. Antipova (Ed.): ICIS 2020, LNNS 136, pp. 406–411, 2021.
https://doi.org/10.1007/978-3-030-49264-9_37

many elements such as a plant's leaves, roots, branch fruit and veins. Therefore, Le-Viet and Truong Hoang [3] proposed to identify plant via its bark by a new variant of Local Binary Patterns based on image gradient, namely GLBP. Zhang et al. [4] suggested a method to combine flowers and leaves for tree classification. Various works [5–8] focus on recognizing plants via shape and structure of leaves by hand-crafted or deep features. Nevertheless, there are no availability and real-world medicinal plant datasets for community research. Most of these datasets do not reflect the real conditions since the images are captured under the same condition. This article introduces an image dataset of 200 classes of Vietnamese medicinal plants which are captured in different condition such as lighting, geography and scale. All images are labeled manually by botanist and technician. We also apply two well-known local image descriptors such as SIFT [9], SURF [10] for features extraction. Then, the Random forest is used for classifying medicinal plants. This is an extremely challenging dataset for traditional algorithms of computer vision approached-based features extraction.

The following of paper is organized and structured as follows. Section 2 introduces the related works. Sections 3 and 4 present proposed method and experimental results. Finally, the conclusion is discussed in Sect. 5.

2 Related Works

To our knowledge, no such public and large-scale image datasets for medicinal plants have not been published. Naresh et al. [11] released the dataset named UoM Medicinal Plants Dataset which contains leaf images of 33 medicinal plants in 2016. However, this dataset represents only one leaf for each image, all images were captured in the indoor condition under the same light and distance. Hence, the classification performance of this dataset is very high with 93%. Then, Sabu et al. [12] introduced a new medicinal plant dataset at Western Ghats, India in 2017. The authors contribute a dataset with 200 images of 20 medicinal plants, each species has 10 images of leaves of the plant. Therefore, this is also a simple dataset so the experimental results on SURF and HOG algorithms are achieved nearly 100%. Recently, plant recognition based on leaves has become a major topic since plant leaves contain more discriminative features including shape, texture, and color. Turkoglu et al. [13] proposed a new approach based on extraction methods associated with classifiers, handcrafted-features method based on variant of LBP features such as: Region Mean-LBP, Overall Mean-LBP, and ROM-LBP. The experiment is evaluated on the 4 datasets of Flavia, Swedish, ICL, and Foliage and the results were 98.94%, 99.46%, 83.71%, and 92.92% respectively. Similarly, the images of these datasets also captured in a laboratory, so the obtained accuracy is very high. Mostajer and Asghari [14] present a new method of plant identification by using GIST features and Principal Component Analysis (PCA) algorithm. The best result is obtained on the Flavia dataset was 98.7%. Currently, there are a various method used to extract local image features such as SIFT, SURF, HOG, LBP or GIST.

In this article, we are interesting to apply SURF and SIFT descriptors to extract the features via "Bag of Words (BOW)" or "Bag of visual words". Then, we apply Random Forest classifier to identify medicinal plant.

3   Method

3.1   Data Collection

We introduce and release a new Vietnamese medicinal plant dataset, which contains all 20,000 images of 200 different medicinal plants. Each class contains 100 images of each species. The main purpose of this release is to allow computer scientist and botanist can access the large and public dataset for medicinal plant. We try to simulate the real-world environment as mobile user does. Thus, all images of VNPlant-200 dataset are acquired from the National Institute of Medicinal Materials, Ho Chi Minh City and Ngoc Xanh Island Resort, Phu Tho City in the unconstrained condition such as illumination, rotation, scale. They are labeled manually by botanists and technical experts.

This dataset is more challenging and noisier than others because many leaves appear together in a single image and it also contains the background such as soil, tree bark, flower, etc. Fig. 1 illustrates several image examples from four selected class of VNPlant-200 dataset. Two versions of resolutions: 256×256 and 512×512 pixels are provided in Sect. 6. The high-resolution version is also available upon request.

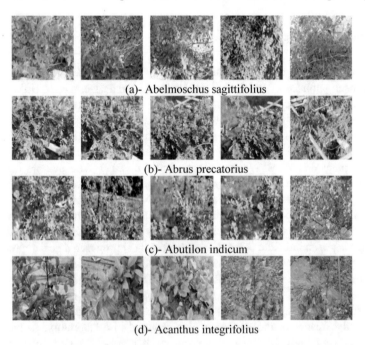

(a)- Abelmoschus sagittifolius

(b)- Abrus precatorius

(c)- Abutilon indicum

(d)- Acanthus integrifolius

Fig. 1. Example images from four selected classes of the VNPlant-200 dataset

3.2   Experimental Setup

In order to provide a benchmark dataset, we divide VNPlant-200 dataset into two training and testing set with the ratio is set at 60:40. We then apply the SURF and SIFT descriptor to extract the features associated with the BOW technique to create feature vectors.

We perform feature extraction on each dataset with Scale-Invariant Feature Transform (SIFT) [9] and the Speed Up Robust Feature (SURF) [10]. The SIFT descriptor detecting points of interest to describe many objects in an image. The local invariant features SIFT of an image are invariant with changing aspect ratios, image rotation, sometimes by changing point of view and adding noise or changing the light intensity of an image. The SURF descriptor is inspired by SIFT descriptor which is faster than SIFT and to be more robust against different image transformations. The feature extraction from SURF or SIFT are combined with BoW using K-means with

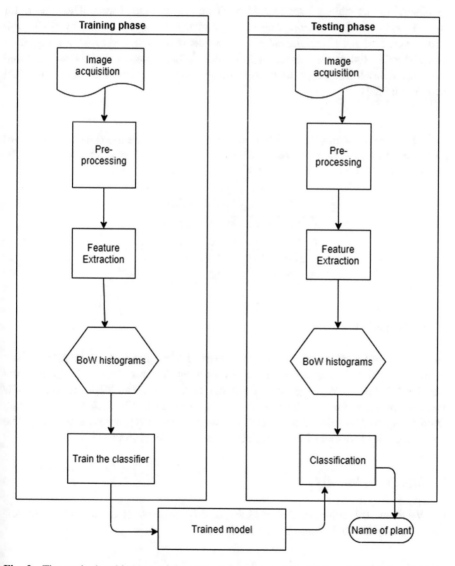

Fig. 2. The method architecture of the proposed approach. The SIFT and SURF descriptors to extract the features via BoW hisograms. The Random Forest is associated for classification.

K = 100. The Random forest classifier [15] with the number of trees n = 1,000. All experiment is simulated on a PC configured by Core-i7 CPU and 16 GB of Ram. The learning and testing stage are summarized in Fig. 2.

4   Experimental Results

Table 1 presents the classification performance of SIFT and SURF features on the two version of VNPlant-200 dataset. We use four evaluation metrics to measure the results. By observing this table, we see that the SIFT descriptor outperforms SURF descriptor on two versions of resolution. The best obtained accuracy is 37.4% by SIFT with images size 512 × 512 pixels. Hence, this result is not impressive for a recognition rate. However, the main purpose of this study is the release a new benchmark and challenge dataset. We might think this contribution can excite and support in plant recognition study.

Table 1. This table shows the classification performance of VNPlant-200 by using SIFT and SURF features. Two versions of image resolution are considered: 256 × 256 and 512 × 512. The bold value indicates the best result achieved.

	256 × 256		512 × 512	
	SURF	SIFT	SURF	SIFT
Precision	19.0	26.0	34.0	**39.0**
Recall	21.0	28.0	35.0	**37.0**
F-score	18.0	25.0	32.0	**36.0**
Accuracy	21.0	28.0	34.7	**37.4**

5   Conclusion

This article introduces a new challenging dataset of Vietnamese medicinal plants which compose of 20,000 images belong to 20 specifies. We also report the preliminary classification results by using two local image descriptors. The future of this work is now continuing to improve the classification performance with background removal, identify objects as medicinal plants before extracting features to improve accuracy. The potential of deep learning framework should be considered in perspective.

6   Data Availability

The VNPlant-200 dataset of this work is publicly available at the repository: https://github.com/kencoca/VietNam-Medicinal-Plant.

References

1. Chen, S.L., Yu, H., Luo, H.M., Wu, Q., Li, C.F., Steinmetz, A.: Conservation and sustainable use of medicinal plants: problems, progress, and prospects. Chin. Med. **11**(1), 37 (2016). https://doi.org/10.1186/s13020-016-0108-7
2. Dictionary of vietnamese medicinal plants. Vietnam (2011)
3. Le-Viet, T., Hoang, V.T.: Local binary pattern based on image gradient for bark image classification. In: Mao, K., Jiang, X. (eds.) Tenth International Conference on Signal Processing Systems, p. 39. SPIE, Singapore, April 2019. https://doi.org/10.1117/12.2522093
4. Zhang, S., Zhang, C., Huang, W.: Integrating leaf and flower by local discriminant CCA for plant species recognition. Comput. Electron. Agric. **155**, 150–156 (2018). https://doi.org/10.1016/j.compag.2018.10.018
5. Hewitt, C., Mahmoud, M.: Shape-only features for plant leaf identification. arXiv:1811.08398 [cs], November 2018
6. Lee, S.H., Chan, C.S., Mayo, S.J., Remagnino, P.: How deep learning extracts and learns leaf features for plant classification. Pattern Recogn. **71**, 1–13 (2017). https://doi.org/10.1016/j.patcog.2017.05.015
7. Lee, S.H., Chan, C.S., Wilkin, P., Remagnino, P.: Deep-plant: plant identification with convolutional neural networks. arXiv:1506.08425 [cs], June 2015
8. Bertrand, S., Ben Ameur, R., Cerutti, G., Coquin, D., Valet, L., Tougne, L.: Bark and leaf fusion systems to improve automatic tree species recognition. Ecol. Inf. **46**, 57–73 (2018). https://doi.org/10.1016/j.ecoinf.2018.05.007
9. Lowe, D.: Object recognition from local scale-invariant features. In: Proceedings of the Seventh IEEE International Conference on Computer Vision, vol. 2, pp. 1150–1157. IEEE, KERKYRA (1999). https://doi.org/10.1109/ICCV.1999.790410
10. Bay, H., Tuytelaars, T., Van Gool, L.: Surf: speeded up robust features. In: Leonardis, A., Bischof, H., Pinz, A. (eds.) Computer Vision - ECCV 2006, pp. 404–417. Springer, Heidelberg (2006)
11. Naresh, Y.G., Nagendraswamy, H.S.: Classification of medicinal plants: an approach using modified LBP with symbolic representation. Neurocomputing **173**, 1789–1797 (2016)
12. Sabu, A., Sreekumar, K., Nair, R.R.: Recognition of ayurvedic medicinal plants from leaves: a computer vision approach. In: 2017 Fourth International Conference on Image Information Processing (ICIIP). pp. 1–5. IEEE, Shimla, December 2017. https://doi.org/10.1109/ICIIP.2017.8313782
13. Turkoglu, M., Hanbay, D.: Leaf-based plant species recognition based on improved local binary pattern and extreme learning machine. Phys. A **527**, 121297 (2019). https://doi.org/10.1016/j.physa.2019.121297
14. Mostajer Kheirkhah, F., Asghari, H.: Plant leaf classification using GIST texture features. IET Comput. Vis. **13**(4), 369–375 (2019). https://doi.org/10.1049/iet-cvi.2018.5028
15. Breiman, L.: Consistency for a simple model of random forests (2004)

Author Index

© Springer Nature Switzerland AG 2021
T. Antipova (Ed.): ICIS 2020, LNNS 136, pp. 413–414, 2021.
https://doi.org/10.1007/978-3-030-49264-9